FUNDAMENTALS OF MATHEMATICS
Skills and Applications

Bryce R. Shaw
Robert B. Kane
Katherine K. Merseth

ADVISERS
Charles E. Allen
Miriam M. Schaefer
Grayson H. Wheatley

CONSULTANTS
Mary E. Lester
George F. Thomson, Jr.

HOUGHTON MIFFLIN COMPANY • BOSTON
Atlanta   Dallas   Geneva, Ill.   Lawrenceville, N.J.   Palo Alto   Toronto

AUTHORS

Bryce R. Shaw, Former Director of Mathematics and Computer Systems, Flint Community Schools, Michigan

Robert B. Kane, Director of Teacher Education and Head of the Department of Education, Purdue University, Indiana

Katherine K. Merseth, Research Assistant, Harvard Graduate School of Education, Harvard University, Massachusetts

ADVISERS

Charles E. Allen, Instructional Specialist in Mathematics and Science, Los Angeles Public Schools, California

Miriam M. Schaefer, Former Mathematics Coordinator, Flint Community Schools, Michigan

Grayson H. Wheatley, Chairman of Mathematics and Science Education and Professor of Mathematics and Education, Purdue University, Indiana

CONSULTANTS

Mary E. Lester, Director of Mathematics, Dallas Independent School District, Texas

George F. Thomson, Jr., Secondary Mathematics Coordinator, DeKalb County School System, Decatur, Georgia

Printed in U.S.A.

ISBN: 0-395-36293-8

DEFGHIJ-D-943210-898

Contents

Addition and Subtraction
Estimation
Problem Solving

Multiplication
Customary System of Measurement
Problem Solving

Division
Metric System
Problem Solving

Basic Fraction Skills
Geometric Definitions
Problem Solving

Add and Subtract Fractions
Pythagorean Theorem, Perimeter
Problem Solving

Multiply and Divide Fractions
Area and Volume
Problem Solving

Rates, Ratios, Proportions Probability and Statistics Problem Solving

Fractions, Decimals, Percents Bar, Line, and Circle Graphs Problem Solving

Percent Skills
Applying Percents
Problem Solving

Positive and Negative Numbers
Using Negative Numbers
Problem Solving

Variable Expressions
Translating Words into Symbols
Problem Solving

Equations
Coordinates and Graphs
Problem Solving

An Introduction to Your Book

FUNDAMENTALS OF MATHEMATICS, SKILLS AND APPLICATIONS is designed to help you refresh or master the basic computational skills, elementary algebra skills, and important applications. Each unit begins with a Unit Preview of the skills and applications that are developed in the unit.

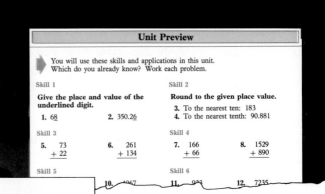

Unit Preview

You will use these skills and applications in this unit. Which do you already know? Work each problem.

Skill 1

Give the place and value of the underlined digit.

1. 6<u>8</u>   2. 350.2<u>6</u>

Skill 2

Round to the given place value.

3. To the nearest ten: 183
4. To the nearest tenth: 90.881

Skill 3

5.   73
   + 22

6.   261
   + 134

Skill 4

7.   166
   + 66

8.   1529
   + 890

Skill 5

10.   1067

Skill 6

11.   9~~   12.   7235

In the following lessons each skill or application is then presented with a completely worked-out model. There are "Get Ready" exercises to get you started in the right direction, followed by practice exercises to help you master the lesson topic.

Skill 7

Adding Decimals

MODEL

Add.

$6.92 + 3.04 + 6.3$

Align.

$$\begin{array}{r} 6.92 \\ 3.04 \\ + 6.3 \\ \hline 6 \end{array}$$

$$\begin{array}{r} 6.92 \\ 3.04 \\ + 6.3 \\ \hline .26 \end{array}$$

$$\begin{array}{r} 6.92 \\ 3.04 \\ + 6.3 \\ \hline 16.26 \end{array}$$

$$\begin{array}{r} 6.92 \\ 3.04 \\ + 6.3 \\ \hline 16.26 \end{array}$$

GET READY.

1.   26.0
   + 43.8

2.   84.3
   + 13.5

3.   4.56
   + 2.13

4.   8.23
   + 1.04

5.   6.08
   + 3.90

6. $34.2 + 5.38$ 7. $5.47 + 82.2$ 8. $91.7 + 4.22$

NOW USE SKILL 7.

9.   2.146
   + 3.915

10.   12.34
   + 38.09

11.   82.49
   + 45.28

12.   671.3
   + 548.9

13.   73.103
   + 9.477

14. $3.79 + 26.34$ 15. $829.4 + 96.2$

16. $67.23 + 8.486$ 17. $103.5 + 6.982$

18. $6.03 + 306.761 + 49.439$ 19. $474.67 + 9.436 + 7.469$

20. $4637.04 + 193.764 + 106.4$ 21. $9.4613 + 27.347 + 4691.9$

22. $\$87.46 + \$29.77$ 23. $\$647.87 + \$278.69$

24. $\$476.43 + \$9.08 + \$47.92$ 25. $\$3789.47 + \$47.03 + \$1.84$

Add the rows and columns to complete the blanks.

26.	9.6	203.15	?
27.	16	0.03	?
28.	?	?	

29.	$5.98	$6.16	?
30.	$4.89	$28.88	?
31.	?	?	

22

For more practice: **Skill Bank**, page 408.

Application A3

PLICATION A3. **Estimate the sum to the near-**

22. $1.68 + 0.29$ 23. $7.46 + 7.29$

25. $31.85 + 10.95$ 26. $56.74 + 56.25$

28. $291.17 + 14.02 + 1.8$ 29. $7.68 + 9.14 + 4.26$

to the nearest whole number.

31. $42.49 + 82.11$ 32. $6.95 + 98.02$

34. $\$159.95 + \$420.59$ 35. $\$.84 + \$9.49$

37. $\$8642.89 + \$1010.39$ 38. $\$49 + \$9.81 + \$11.25$

rence to the nearest tenth.

40. $16.42 - 9.099$ 41. $754.950 - 309.777$

43. $160.07 - 79.13$ 44. $8.0539 - 0.4609$

rence to the nearest whole number.

46. $10.405 - 5.5$ 47. $37.902 - 1.1495$

49. $1089.52 - 750$ 50. $9.95 - 3.95$

52. $\$63.42 - \$19.80$ 53. $\$447.49 - \$32.03$

or difference.

54. You purchase a mop for $3.29, a can of cleanser for $.59, a box of laundry detergent for $2.29, and a can of floor wax for $1.69. To the nearest dime, estimate the amount of your purchases.

55. The median weight of 10-year-old boys is 69.17 pounds, and of 10-year-old girls is 71.61 pounds. To the nearest tenth, estimate the difference of these median weights.

56. In the first six months of the year, the consumer price index rises from 253.5 to 265.1. To the nearest whole number, estimate how much the consumer price index rose.

57. You want to buy four books that cost $12.95, $9.95, $4.39, and $5.25. You have $25.00 to spend. To the nearest whole dollar, estimate the amount of your purchase. Do you have enough money?

For more practice: **Application Bank**, page 449.

25

At the end of each lesson you will find a reference to one of the 61 pages of extra practice in the Skill Bank or the Application Bank at the back of your book.

This textbook is organized so that each unit is broken down into three weekly sections. In addition to the weekly lessons on skills and applications, there are weekly problem solving lessons that take you step-by-step toward the solution of a problem from everyday life.

Color banding makes identifying each lesson easy:

Blue—skills lessons

Green—applications lessons

Red—problem solving lessons

Yellow—testing and review

Orange—special features

A weekly quiz provides a check on your progress.

Special features add interest to the week's work by relating mathematics to the day-to-day world, by providing information about computers, and by sharpening your calculator skills.

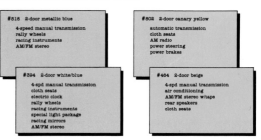

Problem Solving

Clarion Rally Coupe

You have $5600 with which to buy a new car. You decide to purchase a 2-door Clarion Rally Coupe. The base price of the car is $4777.45. A dealer has four 2-door models in stock. You look at these file cards showing the factory-installed options on each.

#518 2-door metallic blue
  4-speed manual transmission
  rally wheels
  racing instruments
  AM/FM stereo

#802 2-door canary yellow
  automatic transmission
  cloth seats
  AM radio
  power steering
  power brakes

#594 2-door white/blue
  4-spd manual transmission
  cloth seats
  electric clock
  rally wheels
  racing instruments
  special light package
  racing mirrors
  AM/FM stereo

#464 2-door beige
  4-spd manual transmission
  air conditioning
  AM/FM stereo w/tape
  rear speakers
  cloth seats

PROBLEM: Select the cars that you are able to buy.

Solution Plan: Estimate the amount that you can spend on options. Estimate the cost of options on the cars in stock. Choose the cars with options that cost less than the amount you can spend.

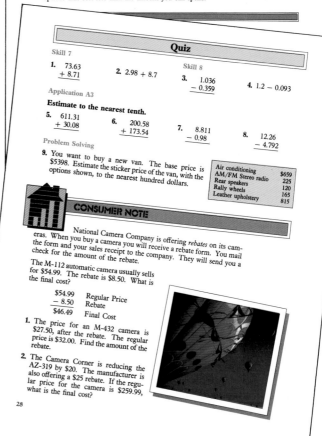

Quiz

Skill 7

1. 73.63
  + 8.71

2. 2.98 + 8.7

Skill 8

3. 1.036
  − 0.359

4. 1.2 − 0.093

Application A3

Estimate to the nearest tenth.

5. 611.31
  + 30.08

6. 200.58
  + 173.54

7. 8.811
  − 0.98

8. 12.26
  − 4.792

Problem Solving

9. You want to buy a new van. The base price is $5398. Estimate the sticker price of the van, with the options shown, to the nearest hundred dollars.

Air conditioning	$659
AM/FM Stereo radio	225
Rear speakers	120
Rally wheels	165
Leather upholstery	815

CONSUMER NOTE

National Camera Company is offering *rebates* on its cameras. When you buy a camera you will receive a rebate form. You mail the form and your sales receipt to the company. They will send you a check for the amount of the rebate.

The M-112 automatic camera usually sells for $54.99. The rebate is $8.50. What is the final cost?

  $54.99   Regular Price
  − 8.50   Rebate
  $46.49   Final Cost

1. The price for an M-432 camera is $27.50, after the rebate. The regular price is $32.00. Find the amount of the rebate.

2. The Camera Corner is reducing the AZ-319 by $20. The manufacturer is also offering a $25 rebate. If the regular price for the camera is $259.99, what is the final cost?

28

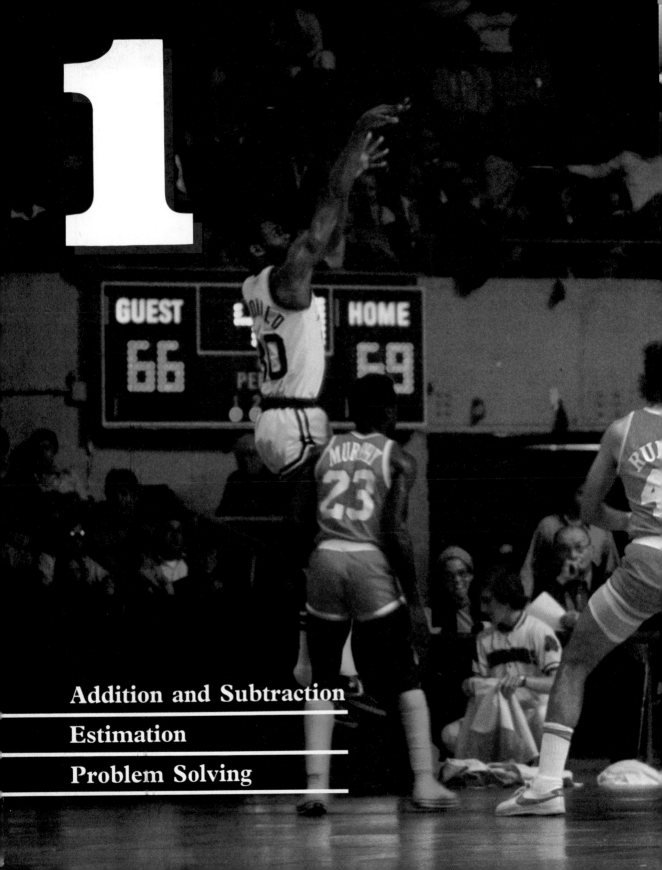

1

Addition and Subtraction

Estimation

Problem Solving

Unit Preview

 You will use these skills and applications in this unit.
Which do you already know? Work each problem.

Skill 1

Give the place and value of the underlined digit.

1. 6<u>8</u> **2.** 350.2<u>6</u>

Skill 2

Round to the given place value.

3. To the nearest ten: 183
4. To the nearest tenth: 90.881

Skill 3

5. $\begin{array}{r} 73 \\ + 22 \\ \hline \end{array}$ **6.** $\begin{array}{r} 261 \\ + 134 \\ \hline \end{array}$

Skill 4

7. $\begin{array}{r} 166 \\ + 66 \\ \hline \end{array}$ **8.** $\begin{array}{r} 1529 \\ + 890 \\ \hline \end{array}$

Skill 5

9. $\begin{array}{r} 3298 \\ - 170 \\ \hline \end{array}$ **10.** $\begin{array}{r} 4067 \\ - 3011 \\ \hline \end{array}$

Skill 6

11. $\begin{array}{r} 923 \\ - 853 \\ \hline \end{array}$ **12.** $\begin{array}{r} 7235 \\ - 653 \\ \hline \end{array}$

Skill 7

13. $22.5 + 1.7$ **14.** $91.32 + 9.784$

Skill 8

15. $71.1 - 6.9$ **16.** $85.2 - 19.78$

Application A1 **Compare the two numbers. Which is larger?**

17. 110 or 99 **18.** 561.27 or 598.01

Application A2

Estimate to the nearest ten.

19. $798 + 91$ **20.** $143 + 88$

Estimate to the nearest hundred.

21. $707 - 378$ **22.** $3562 - 436$

Application A3

Estimate to the nearest tenth.

23. $55.31 + 8.57$ **24.** $83.09 + 10.50$

Estimate to the nearest whole number.

25. $44.29 - 3.80$ **26.** $30.309 - 4.68$

Check your answers. If you had difficulty with any skill or application, be sure to study the corresponding lesson in this unit.

Naming Places and Values of Whole Numbers

MODEL

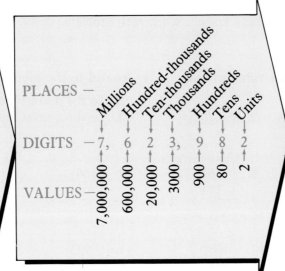

In the number 7,623,982, the 3 is in what place and represents what value?

The 3 is in the thousands' place and represents a value of 3000.

 GET READY. Complete the table for the given number.

893

	Digit	Place	Value
	8	hundreds	800
1.	9	?	?
2.	3	?	?

4672

	Digit	Place	Value
3.	?	thousands	?
4.	?	hundreds	?
5.	7	?	?
6.	?	?	2

 NOW USE SKILL 1. Give the place and value of the under-lined digit.

7. 6<u>7</u>2

8. <u>3</u>198

9. 2<u>3</u>

10. <u>4</u>06

11. 89<u>3</u>0

12. 7,908,<u>8</u>673

13. 300<u>5</u>

14. 3<u>4</u>8,621

15. 68,<u>3</u>04

16. 90<u>3</u>0

17. <u>3</u>,498,002

18. 254,40<u>3</u>

Naming Places and Values of Decimals

MODEL

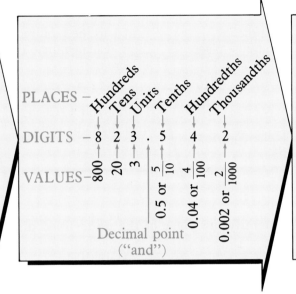

In the number 823.542, the 5 is in what place and represents what value?

PLACES – Hundreds Tens Units · Tenths Hundredths Thousandths

DIGITS – 8 2 3 . 5 4 2

VALUES – 800 20 3 $\frac{5}{10}$ or 0.5 $\frac{4}{100}$ or 0.04 $\frac{2}{1000}$ or 0.002

Decimal point ("and")

The 5 is in the tenths' place and represents a value of 0.5, or $\frac{5}{10}$.

GET READY. Complete the table for the given number.

67.35

	Digit	Place	Value
	6	tens	60
19.	7	?	?
20.	3	?	?
21.	?	hundredths	?

698.035

	Digit	Place	Value
22.	?	hundreds	?
23.	?	?	90
24.	?	units	?
25.	0	?	?
26.	?	hundredths	?
27.	5	?	?

NOW USE SKILL 1. Give the place and value of the under-lined digit.

28. 89.2̲3

29. 624.8̲6

30. 90.71̲

31. 1729.00̲5

32. 34.302̲

33. 6708.89̲

34. 2̲,234,320.3

35. 645,6̲70.027

For more practice: **Skill Bank,** page 405.

3

Rounding Whole Numbers

MODEL

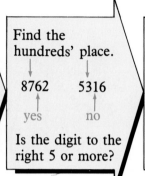

Round to the nearest hundred.	Find the hundreds' place.	Add 1 to the hundreds' digit. Change final digits to zeros.	Rounded to the nearest hundred:
	8762 5316	$8762 \rightarrow 8800$	
8762	yes no	Do not add 1. Change final digits to zeros.	8762 is 8800
5312	Is the digit to the right 5 or more?	$5316 \rightarrow 5300$	5316 is 5300

 GET READY. Identify the place of the underlined digit.

1. 26<u>7</u>3   **2.** 40<u>7</u>6   **3.** 1<u>9</u>,213   **4.** 1<u>2</u>0,632

 NOW USE SKILL 2. Complete the steps to round 18,736 to the nearest thousand.

5. Find the thousands' digit.   **6.** What is the next digit to the right?

7. Is this digit 5 or more?   **8.** Does the thousands' digit change?

9. Do the final digits change?   **10.** What is the rounded number?

Round each number to the nearest ten.

11. 486   **12.** 612   **13.** 2108   **14.** 25

15. 350   **16.** 654   **17.** 96   **18.** 171

Round each number to the nearest hundred.

19. 6123   **20.** 46,209   **21.** 23,055   **22.** 1860

23. 2960   **24.** 764   **25.** 62,817   **26.** 4096

Round each number to the nearest thousand.

27. 1,204,361   **28.** 889,906   **29.** 6,437,293

30. 543,756   **31.** 18,351   **32.** 62,901

Rounding Whole Numbers and Decimals

MODEL

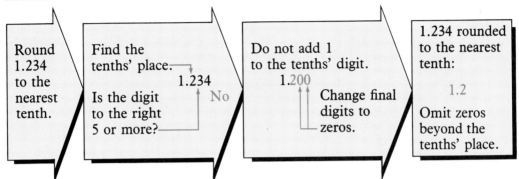

Round 1.234 to the nearest tenth.

Find the tenths' place. → 1.234 Is the digit to the right 5 or more? → No

Do not add 1 to the tenths' digit. 1.200 Change final digits to zeros.

1.234 rounded to the nearest tenth: 1.2 Omit zeros beyond the tenths' place.

 GET READY. Identify the place of the underlined digit.

33. 12.3<u>9</u>   **34.** 23.09<u>4</u>   **35.** 19.<u>2</u>6   **36.** 1<u>4</u>.317

 NOW USE SKILL 2. Complete the steps to round 168.436 to the nearest hundredth.

37. Find the hundredths' digit.   **38.** What is the next digit to the right?

39. Is this digit 5 or more?   **40.** Does the hundredths' digit change?

41. Does the final digit change?   **42.** What is the rounded number?

Round each number to the nearest tenth.

43. 0.38   **44.** 0.81   **45.** 14.27   **46.** 42.93

47. 0.55   **48.** 0.96   **49.** 1.43   **50.** 9.86

Round each number to the nearest hundredth.

51. 6.084   **52.** 49.951   **53.** 0.506   **54.** 10.983

55. 0.899   **56.** 0.009   **57.** 1.997   **58.** 2.068

Round each number to the nearest ten cents.

59. $12.43   **60.** $4200.38   **61.** $1695.89   **62.** $98.99

63. $803.08   **64.** $25.95   **65.** $9.36   **66.** $2.48

For more practice: **Skill Bank,** page 405.

Comparing Numbers

Either one number is larger than another number or the two numbers are equal. To compare two numbers, compare the digits by place value.

MODEL

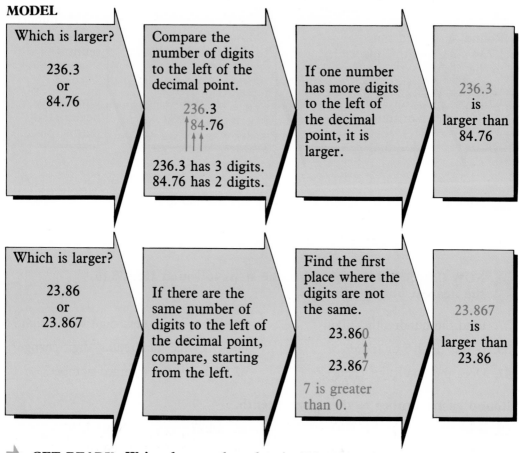

Which is larger?

236.3
or
84.76

Compare the number of digits to the left of the decimal point.

236.3
84.76

236.3 has 3 digits.
84.76 has 2 digits.

If one number has more digits to the left of the decimal point, it is larger.

236.3
is
larger than
84.76

Which is larger?

23.86
or
23.867

If there are the same number of digits to the left of the decimal point, compare, starting from the left.

Find the first place where the digits are not the same.

23.860
23.867

7 is greater than 0.

23.867
is
larger than
23.86

 GET READY. Write the number that is 100 more than the given number.

1. 463   2. 2943   3. 4010   4. 206.54

5. 4238.2   6. 53.19   7. 98.06   8. 690.98

Write the number that is 10 less than the given number.

9. 596   10. 3424   11. 4231.2   12. 21,050

13. 60.091   14. 800.01   15. 102.36   16. 69,703

NOW USE APPLICATION A1. Compare the two numbers. Which is larger?

17. 2067.31 or 2984.20   **18.** $15.98 or $19.88   **19.** 2.943 or 2.940

20. 0.0358 or 0.036   **21.** $40.61 or $39.99   **22.** 7,652,000 or 7,651,900

A library card shows that a book has been checked out. Between which two card numbers should the card for the checked-out book be placed in the card file?

23.

24.

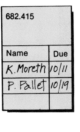

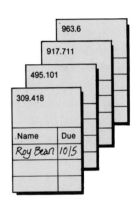

A furniture store has placed this advertisement in the local newspaper. Use the advertisement to answer the questions.

25. Which piece of furniture costs more, the love seat or the deluxe wing chair?

26. Which piece of furniture costs more than $1000?

27. You have budgeted $500 to spend on the purchase of new furniture. You would like to buy a sofa bed. Is the sofa bed within your budget?

28. List the furniture in the advertisement in order from most expensive to least expensive.

sent $18,500 a year to $5.

Best Furniture	
Love Seat	$694.95
Deluxe Wing Chair	695.96
8-pc Modular Sofa	1299.00
Wall System	
Self-assemble, 6' x 3'	263.99
Sofa Bed	499.99
Solid Pine Bunk Bed	259.90
Student Desk	138.95

For more practice: **Application Bank,** page 449.

Sports Banquet

You are the sports writer for the school newspaper. You have been asked to give a speech at the winter sports banquet. You want to present some interesting facts about this year's team. You use the final team statistics below.

Northwood High School Panthers Basketball Team				
Player	**Year**	**Position**	**Total Points Scored**	**Average Points Per Game**
Smith	Sr	Guard	190	7.6
Keen	Sr	Center	280	11.2
Brown	Sr	Guard	355	14.2
Jackson	Jr	Guard	170	6.8
Olvero	Jr	Guard	520	20.8
Mann	Jr	Forward	388	15.5
Segal	Jr	Forward	198	7.9
McCoy	Soph	Guard	293	9.7
Sully	Soph	Center	465	18.6
Samuel	Soph	Guard	308	12.3

PROBLEM: Prepare a speech relating some interesting facts about the basketball team.

Solution Plan: Make your speech interesting by finding who the outstanding players are on this year's team. Compare their records with leading scorers in the school's history.

WHAT DO YOU KNOW? You know the team's statistics. Use the information given to answer these questions.

1. How many players are on this year's team?

2. Name the players who played the following positions.

   a. Guard   **b.** Forward   **c.** Center

3. How many players averaged 10 or more points per game?

4. Who was the leading scorer on this year's team?

WHAT DO YOU WANT TO KNOW? You want to know who the leading scorers are. Complete the table below.

You begin by comparing the point totals for the seniors on this year's team. You find that Brown scored the highest number of points of this year's seniors.

Smith	190 points
Keen	280 points
Brown	355 points

355 is greater than 190 or 280, so Brown is the leading scorer.
Application A1

	5.	**6.**	**7.**	**8.**	**9.**
	Senior	**Junior**	**Sophomore**	**Guard**	**Center**	**Forward**
Leading Scorer	Brown	?	?	?	?	?
Total Points Scored	355	?	?	?	?	?
Per Game Average	14.2	?	?	?	?	?

WHAT MORE DO YOU WANT? You want to compare this year's leading scorers with players of past teams. Use the all-time scoring list at the right to answer these questions.

10. How does this year's leading point scorer rank with the all-time scorers of past teams?

11. Players who average 18 or more points per game are named to the all-time scoring list. Write the names, in order, of the players on the all-time scoring list after this season.

Basketball Hall of Records		
All-time Scoring		
Player	Per Game Average	Season
1. Jones	22.0	78–79
2. Webster	21.2	76–77
3. McKinney	20.1	79–80
4. Cohen	18.8	75–76

NOW SOLVE THE PROBLEM. Use the information you have found to prepare your speech for the banquet.

Welcome to this year's Annual Winter Sports Banquet. The Northwood High Basketball team had another terrific year, winning the division title. Many players consistently scored in double figures. For example, 12. __?__ , the leading junior player, averaged 13. __?__ points per game. Olvero now ranks 14. __?__ among Northwood's all-time scorers. 15. __?__ , the leading sophomore player, averaged 16. __?__ points per game. They will both be back next year. The Central Valley Division will have to make way for the Panthers again next year.

Quiz

Skill 1

1. Give the place and value of the underlined digit in 1<u>2</u>5.

2. Give the place and value of the underlined digit in 54.3<u>2</u>.

Skill 2

3. Round 8327 to the nearest thousand.

4. Round 5201.76 to the nearest tenth.

Application A1

5. Which is larger: 6894 or 6852?

6. Which is larger: 25.84 or 25.72?

Problem Solving

Use the table at the right to answer the questions.

7. Name the player with the highest total yardage for the season.

8. Name the player with the highest yardage per carry.

Player	Yards per Carry	Total Yards for Season
Jones	4.3	814
Lewis	11.6	585
Barnes	7.2	1219

EXTRA!

Many countries use international road signs. Some common road signs you may see when you travel are shown below. What does each sign mean?

1.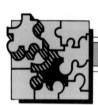

2.

3.

4.

5.

6.

7.

8.

9.

10.

11.

12. STOP

Reading Tables

To read a table, find the row and the column headings that contain the information you *have*. Read down the column and read across the row. You will find the information you *need* at the meeting point.

How much would it cost to ship a package that weighs 5 lb to Zone 3?

The table shows that it costs $2.06.

Find the cost to ship each package to the given zone.

1. 2 lb, Zone 1   **2.** 4 lb, Zone 4

3. 6 lb, Zone 3   **4.** 5 lb, Zone 4

5. 4 lb 3 oz, Zone 2

6. 3 lb 9 oz, Zone 1

📦 PACKAGE SHIPPING SERVICE

Any fraction of a pound over the weight shown takes the next higher rate.

Weight not to exceed	ZONE			
	1	2	3	4
1 lb	1.21	1.24	1.30	1.35
2 lb	1.34	1.41	1.49	1.60
3 lb	1.46	1.56	1.68	1.85
4 lb	1.58	1.72	1.87	2.09
5 lb	1.69	1.87	2.06	2.35
6 lb	1.82	2.02	2.27	2.60

To find the second heading of an entry, first locate all the information you have. Then read the table to find the matching column or row heading.

It costs $6.35 to ship a package to Zone 3. What is the most the package can weigh?

The information $6.35 is located under column heading Zone 3. The matching row heading for this entry is 5 lb. The package weighs, at most, 5 lb.

7. A 4 lb package costs $4.48 to ship. To which zone is it being delivered?

✈ AIR SHIP CORPORATION

Any fraction of a pound over the weight shown takes the next higher rate.

Weight not to exceed	ZONE			
	1	2	3	4
1 lb	—	1.97	2.19	3.44
2 lb	2.75	2.81	3.23	4.42
3 lb	3.54	3.64	4.27	5.39
4 lb	4.33	4.48	5.31	6.37
5 lb	5.13	5.30	6.35	7.38
6 lb	5.93	6.14	7.40	8.33

8. What is the weight of the heaviest package you can ship to Zone 3 for $7.40?

11

Adding Whole Numbers Without Carrying

MODEL

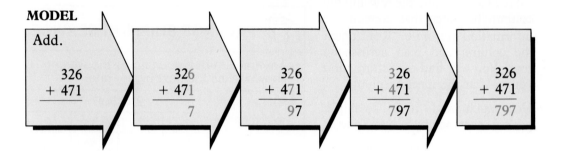

Add.

326	326	326	326	326
+ 471	+ 471	+ 471	+ 471	+ 471
	7	97	797	797

 GET READY.

1.   7
    + 1

2.   4
    + 3

3.   2
    + 6

4.   4
    + 2

5.   4
    + 4

6.   5
    + 1

7. $10 + 20$ **8.** $30 + 30$ **9.** $60 + 10$ **10.** $80 + 10$

 NOW USE SKILL 3.

11.   63
    + 2

12.   25
    + 4

13.   70
    + 2

14.   92
    + 4

15.   53
    + 5

16.   51
    + 23

17.   52
    + 37

18.   26
    + 72

19.   35
    + 62

20.   642
    + 31

21.   400
    + 36

22.   204
    + 30

23.   608
    + 21

24.   412
    + 123

25.   647
    + 251

26. $7 + 2 + 9$ **27.** $3 + 8 + 9$ **28.** $2 + 6 + 5$

29. $414 + 283$ **30.** $573 + 216$ **31.** $214 + 755$

32. $3257 + 5431$ **33.** $9047 + 842$ **34.** $1149 + 2300$

35. $7742 + 2147$ **36.** $7146 + 842$ **37.** $4029 + 5970$

For more practice: **Skill Bank,** page 406.

Adding Whole Numbers With Carrying

MODEL

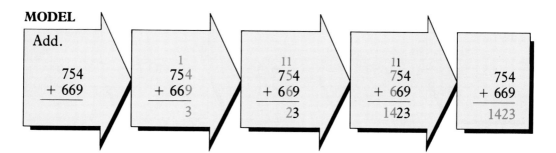

Add.

754	754	754	754	754
+ 669	+ 669	+ 669	+ 669	+ 669
	3	23	1423	1423

(with carried: 1 over 754; 11 over 754; 11 over 754)

 GET READY.

1. 6	**2.** 3	**3.** 9	**4.** 6	**5.** 8	**6.** 5
+ 9	+ 8	+ 9	+ 8	+ 9	+ 9

 NOW USE SKILL 4.

7. 629	**8.** 834	**9.** 196	**10.** 486	**11.** 249
+ 316	+ 185	+ 392	+ 263	+ 538

12. $2576	**13.** $7608	**14.** 6729	**15.** $5198	**16.** $4083
+ 1954	+ 1484	+ 5403	+ 8810	+ 2996

17. 16 + 32 + 64 **18.** 47 + 59 + 213 **19.** 207 + 635 + 79

20. 23,049	**21.** 38,940	**22.** 47,983	**23.** 46,381
+ 48,908	+ 17,893	+ 39,479	+ 19,880

Add the rows and columns to complete the blanks.

24.	329	816	?
25.	240	709	?
26.	?	?	

27.	11	22	33	?
28.	44	55	66	?
29.	77	88	99	?
30.	?	?	?	

For more practice: **Skill Bank,** page 406.

Subtracting Whole Numbers Without Borrowing

MODEL

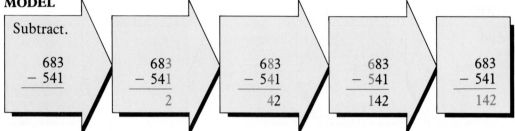

Subtract.

683	683	683	683	683
− 541	− 541	− 541	− 541	− 541
	2	42	142	142

 GET READY.

1. 9
    − 3

2. 7
    − 2

3. 5
    − 2

4. 8
    − 6

5. 5
    − 3

6. 9
    − 9

NOW USE SKILL 5.

7.   45
    − 23

8.   67
    − 52

9.   73
    − 1

10.   84
    − 61

11.   98
    − 34

12.   648
    − 20

13.   872
    − 50

14.   497
    − 40

15.   274
    − 60

16.   549
    − 40

17.   434
    − 213

18.   628
    − 112

19.   486
    − 372

20.   968
    − 253

21.   875
    − 324

22.   3829
    − 2512

23.   6457
    − 2316

24.   8572
    − 2240

25.   6739
    − 2428

26.   9089
    − 5038

27. $7524 − $6013   **28.** 8973 − 5461   **29.** $8404 − $3302

30.   54,271
    − 22,150

31.   63,704
    − 41,201

32.   55,425
    − 32,014

33.   77,398
    − 21,215

34.   63,842
    − 50,131

35.   31,648
    − 20,317

36.   67,684
    − 53,270

37.   94,326
    − 53,014

For more practice: **Skill Bank,** page 407.

Subtracting Whole Numbers With Borrowing

MODEL

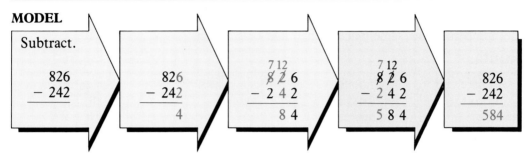

Subtract.

		7 12	7 12	
826	826	8 2 6	8 2 6	826
− 242	− 242	− 2 4 2	− 2 4 2	− 242
	4	8 4	5 8 4	584

 GET READY.

1. 396
− 1

2. 687
− 5

3. 748
− 3

4. 976
− 2

5. 386
− 3

6. 188 − 74   **7.** 947 − 22   **8.** 449 − 36

 NOW USE SKILL 6.

9. 442
− 138

10. 956
− 239

11. 384
− 236

12. 652
− 317

13. 885
− 426

14. 7926
− 873

15. 2452
− 564

16. 3764
− 299

17. 1381
− 694

18. 1463
− 749

19. $539 − $235   **20.** $905 − $188   **21.** $980 − $521

22. 92,786
− 18,149

23. $70,650
− 24,916

24. 50,082
− 36,807

25. $46,002
− 19,998

Subtract the rows and columns to complete the blanks.

26.

84	76	?

27.

68	37	?

28.

?	?

29.

397	148	?

30.

238	129	?

31.

?	?

For more practice: **Skill Bank,** page 407.

Estimating Sums and Differences of Whole Numbers

To estimate the sum or difference of two or more whole numbers, round the numbers to the same place. Then add or subtract.

MODEL

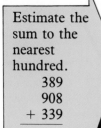

Estimate the sum to the nearest hundred.	Round all numbers to the nearest hundred. Then add.	Skill 2	An estimate of
389 908 + 339		389 → 400 908 → 900 + 339 → + 300 ──── 1600 Skill 3	389 908 + 339 ──── is 1600.

Estimate the difference to the nearest thousand.	Round the numbers to the nearest thousand. Then subtract.	Skill 2	An estimate of
30,385 − 15,590		30,385 → 30,000 − 15,590 → − 16,000 ──── 14,000 Skill 6	30,385 − 15,590 ──── is 14,000.

 GET READY. Round the number to the nearest ten.

1. 29   **2.** 46   **3.** 73   **4.** 15

5. 92   **6.** 307   **7.** 678   **8.** 268

Round the number to the nearest hundred.

9. 230   **10.** 407   **11.** 197   **12.** 298

13. 66   **14.** 1539   **15.** 783   **16.** 21,443

 NOW USE APPLICATION A2. Estimate the sum to the nearest ten.

17. 76 + 34   **18.** 28 + 23   **19.** 186 + 296

20. 17 + 63 + 7   **21.** 42 + 76 + 120   **22.** 636 + 143 + 498

Estimate the sum to the nearest thousand.

23. $12,095 + 8912$ **24.** $5842 + 6980$ **25.** $10,092 + 10,902$

26. $74,399 + 821$ **27.** $6584 + 5581$ **28.** $7821 + 4151$

29. $8462 + 7621 + 1864$ **30.** $1012 + 8176 + 3784 + 4455$

Estimate the difference to the nearest ten.

31. $83 - 49$ **32.** $147 - 49$ **33.** $132 - 76$

34. $784 - 324$ **35.** $1984 - 799$ **36.** $1019 - 605$

Estimate the difference to the nearest hundred.

37. $153 - 120$ **38.** $447 - 190$ **39.** $682 - 327$

40. $905 - 675$ **41.** $1560 - 786$ **42.** $2385 - 980$

43. $4528 - 2119$ **44.** $10,482 - 4670$ **45.** $79,412 - 41,820$

The Magnetics, a rock group, appeared for three performances at the Concert Hall. The attendance for each performance is shown in the table below. Use the table.

46. To the nearest thousand, estimate the total attendance at all three performances.

47. To the nearest hundred, estimate the total paid attendance at all three performances.

48. To the nearest hundred, estimate the number of people who did not pay to attend Wednesday's concert.

49. To the nearest hundred, estimate the number of people who did not pay to attend Friday's concert.

50. The capacity of the Concert Hall is 16,155. To the nearest hundred, estimate the number of empty seats at Wednesday's performance.

Performance	Total Attendance	Paid Attendance
Wednesday	10,863	9,206
Thursday	12,049	10,865
Friday	15,729	14,912

For more practice: **Application Bank,** page 449.

Hi-Tech Electronics

You are the manager of Hi-Tech Electronics. Your store sells electronic games, calculators, and parts. You plan to advertise a sale of your best-selling electronic games and calculators. You use the sales report below to help plan the sale.

Hi-Tech Electronics—Calculator Sales				
Model	1st Quarter	2nd Quarter	3rd Quarter	4th Quarter
C-100	247	260	423	404
C-200	186	172	216	205
C-300	274	289	334	356
C-350	307	251	140	96

PROBLEM: Select the best-selling electronic games and calculators to advertise in the sale.

Solution Plan: To select the best-selling games and calculators, estimate the total annual sales of electronic calculators. Then estimate the total annual sales of electronic games. Choose the products with the highest sales totals.

WHAT DO YOU KNOW? The chart above shows the quarterly sales totals for calculators. Use the chart to answer the questions.

1. How many different models of calculators do you sell?

2. Which model had increasing sales during the entire year?

3. Which model had decreasing sales during the entire year?

4. How many model C-100 calculators were sold during the first quarter? second quarter? third quarter? fourth quarter?

5. How many model C-300 calculators were sold during the first quarter? second quarter? third quarter? fourth quarter?

WHAT DO YOU WANT TO KNOW? You want to know the annual sales of each model of calculator. Use the Calculator Sales chart.

You begin with the Model C-100. You round each quarterly sales figure to the nearest hundred. For the year, you estimate that about 1300 C-100 calculators were sold.

$$
\begin{array}{rcl}
247 & \longrightarrow & 200 \\
260 & \longrightarrow & 300 \\
423 & \longrightarrow & 400 \\
+\ 404 & \longrightarrow & +\ 400 \\
\hline
 & & 1300
\end{array}
$$

Skill 2

Application A2

Estimate annual calculator sales to the nearest hundred.

6. Model C-200   **7.** Model C-300   **8.** Model C-350

WHAT MORE DO YOU WANT? You want to know sales of electronic games. You have been selling electronic games since June.

Estimate total sales to the nearest hundred.

9. Galex Gallop

10. Magic Nets

11. Touchdown

12. Z-man

Sales of Electronic Games		
Game	3rd Quarter	4th Quarter
Galex Gallop	839	1049
Magic Nets	507	631
Touchdown	436	374
Z-man	392	412

NOW SOLVE THE PROBLEM. Use the information you have found to complete your advertisement.

Hi-Tech Electronics — *January Electronics Sale*

Do you need help with your math?
Try our two best-selling calculators!

    Model ___?___ and Model ___?___ , now priced at 30% off!!

13.

Want fun and relaxation? A Challenge?
Play our two best-selling electronic games!

    _____?_____ or _____?_____ , now priced at 25% off!!

14.

Come in today!!! Don't miss this once a year sale!!!!!

19

Skill 3

1. 32 + 15	**2.** 395 + 104

Skill 4

3. 426 + 394	**4.** 53,927 + 70,698

Skill 5

5. 54 − 30	**6.** 1894 − 753

Skill 6

7. 801 − 738	**8.** 12,005 − 1,636

Application A2

Estimate to the nearest hundred.

9. 726 + 398	**10.** 265 + 654

Estimate to the nearest ten.

11. 561 − 226	**12.** 984 − 578

Problem Solving

13. You want to offer rebates to car buyers who purchase your slowest-selling Clarions. To the nearest hundred, estimate sales for the first three quarters. On which model Clarion will you offer a rebate?

Quarterly Sales—Clarions			
Quarter	Hatchback	4-door	2-door
1	3675	4268	2109
2	5831	5820	4110
3	4927	6345	3979

EXTRA!

Magic Squares

A group of numbers arranged so that the sums of the numbers of every row, column, and diagonal are the same is called a magic square.

1. The numbers in the inner box form a magic square with sum 39. Supply the three missing numbers.

2. Supply the missing numbers to make the larger square a magic square.

3. Add 4 to each number in the magic square in Problem 2. Is the result a magic square?

?	23	25	7	8
4	16	9	14	22
21	11	?	15	5
20	?	17	?	6
?	3	1	19	?

Addition Chains

The Fergusons need some new furniture. They plan to spend up to $500. Home Furnishings has its annual sale this week. The Fergusons have checked off the items they need.

They used a calculator to find the cost of these items at the sale price.

KEY-IN	DISPLAY
25.99	*25.99*
+	*25.99*
136.99	*136.99*
+	*162.98*
165.99	*165.99*
+	*328.97*
156.99	*156.99*
=	*485.96*

HOME FURNISHINGS ANNUAL SALE

	Reg.	Sale
√ Desk Lamp	$35.99	$25.99
Fold-up Table		
34" square	20.39	16.99
40" round	26.39	21.99
Bentwood Rocker	117.59	97.99
√ Roll-top Desk	171.59	136.99
√ Oak Curio Cabinet	223.99	165.99
5-piece Dinette	407.99	339.99
Serving Cart		
√ Brass	170.99	156.99
Maple	143.99	119.99
Recliner	229.19	160.99

Answer the questions.

1. The Corcorans bought:
  1 34″ square fold-up table
  1 5-piece dinette
  1 maple serving cart
How much did they spend?
What is the cost of these items at the regular price?

2. Amy Wang bought:
  1 desk lamp
  1 bentwood rocker
  1 roll-top desk
How much did Amy spend?
What is the cost of these items at the regular price?

3. Within the first hour of the sale, Home Furnishings had sold the following:

ITEM	QTY.	TOTAL	ITEM	QTY.	TOTAL
Desk Lamp	2	$ 51.98	34″ square Fold-up Table	1	$ 16.99
Bentwood Rocker	3	293.97	40″ round Fold-up Table	2	43.98
Roll-top Desk	2	273.98	Oak Curio Cabinet	1	165.99
5-piece Dinette	1	339.99	Brass Serving Cart	1	156.99
Maple Serving Cart	1	119.99	Recliner	2	321.98

Find the total sales for the first hour.

Adding Decimals

MODEL

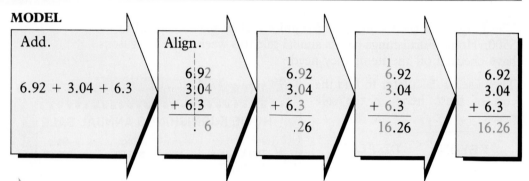

Add.	Align.			
6.92 + 3.04 + 6.3	6.92 3.04 + 6.3 6	1 6.92 3.04 + 6.3 .26	1 6.92 3.04 + 6.3 16.26	6.92 3.04 + 6.3 16.26

 GET READY.

1. 26.0 + 43.8	**2.** 84.3 + 13.5	**3.** 4.56 + 2.13	**4.** 8.23 + 1.04	**5.** 6.08 + 3.90

6. 34.2 + 5.38 **7.** 5.47 + 82.2 **8.** 91.7 + 4.22

NOW USE SKILL 7.

9. 2.146 + 3.915	**10.** 12.34 + 38.09	**11.** 82.49 + 45.28	**12.** 671.3 + 548.9	**13.** 73.103 + 9.477

14. 3.79 + 26.34 **15.** 829.4 + 96.2

16. 67.23 + 8.486 **17.** 103.5 + 6.982

18. 6.03 + 306.761 + 49.439 **19.** 474.67 + 9.436 + 7.469

20. 4637.04 + 193.764 + 106.4 **21.** 9.4613 + 27.347 + 4691.9

22. $87.46 + $29.77 **23.** $647.87 + $278.69

24. $476.43 + $9.08 + $47.92 **25.** $3789.47 + $47.03 + $1.84

Add the rows and columns to complete the blanks.

26.	9.6	203.15	?
27.	16	0.03	?
28.	?	?	

29.	$5.98	$6.16	?
30.	$4.89	$28.88	?
31.	?	?	

For more practice: **Skill Bank**, page 408.

Subtracting Decimals

MODEL

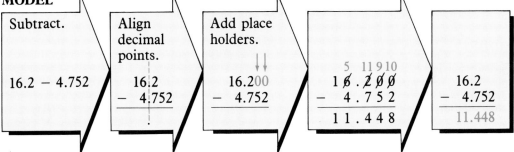

Subtract.	Align decimal points.	Add place holders.		
16.2 − 4.752	16.2 − 4.752	16.200 − 4.752	5 11 9 10 1 6 . 2 0 0 − 4 . 7 5 2 1 1 . 4 4 8	16.2 − 4.752 11.448

 GET READY.

1. 9.28
− 4.16

2. 6.49
− 3.19

3. 33.8
− 12.6

4. 86.8
− 43.5

5. 79.4
− 62.4

6. 39.5 − 18.2   **7.** 89.4 − 56.1   **8.** 11.87 − 10.6

 NOW USE SKILL 8.

9. 87.83
− 12.26

10. 447.5
− 324.8

11. 6.98
− 4.19

12. 38.92
− 16.76

13. 18.05
− 13.92

14. 8.624
− 4.732

15. 63.45
− 41.89

16. 423.7
− 271.9

17. 96.43
− 23.84

18. 6.27
− 2.68

19. 26.32
− 8.157

20. 129.3
− 82.79

21. 478.8
− 9.05

22. 6.348
− 0.899

23. 2.496
− 0.697

24. $29.41 − $18.96   **25.** $46.63 − $4.83   **26.** $304.82 − $19.98

27. 84.091 − 6.9   **28.** 2600 − 159.75   **29.** 26.32 − 8.157

Subtract the rows and columns to complete the blanks.

30.	36.43	12.071	?
31.	27.092	3.81	?
32.	?	?	

33.	197	43.076	?
34.	76.94	26.7	?
35.	?	?	

For more practice: **Skill Bank,** page 408.   23

Estimating Sums and Differences of Decimals

To estimate the sum or difference of two or more decimals, round the
numbers to the same place. Then add or subtract.

MODEL

Estimate the
sum to the
nearest tenth.

8.038
+ 15.75

Round to
the nearest
tenth.
Then add.

Skill 2

$$8.038 \longrightarrow 8.0$$
$$+ 15.75 \longrightarrow + 15.8$$
$$\overline{ 23.8}$$

An estimate of

8.038
+ 15.75

is 23.8.

Estimate the
difference to
the nearest
whole number.

135.405
− 53.873

Round to
the nearest
whole number.
Then subtract.

Skill 2

$$135.405 \longrightarrow 135$$
$$- 53.873 \longrightarrow - 54$$
$$\overline{ 81}$$

An estimate of

135.405
− 53.873

is 81.

 GET READY. Round to the nearest tenth.

1. 0.26   **2.** 1.44   **3.** 3.09   **4.** 2.15   **5.** 7.098

Round to the nearest hundredth.

6. 0.478   **7.** 7.088   **8.** 15.209   **9.** 7.767   **10.** 2.396

Round to the nearest whole number.

11. 1.63   **12.** 5.15   **13.** 4.6   **14.** 30.72   **15.** 8.49

16. 7.06   **17.** 29.95   **18.** 19.86   **19.** 99.6   **20.** 0.48

NOW USE APPLICATION A3. Estimate the sum to the nearest tenth.

21. 0.63 + 0.20   **22.** 1.68 + 0.29   **23.** 7.46 + 7.29

24. 14.26 + 10.80   **25.** 31.85 + 10.95   **26.** 56.74 + 56.25

27. 0.46 + 37.19 + 37   **28.** 291.17 + 14.02 + 1.8   **29.** 7.68 + 9.14 + 4.26

Estimate the sum to the nearest whole number.

30. 86.45 + 9.91   **31.** 42.49 + 82.11   **32.** 6.95 + 98.02

33. $11.45 + $19.39   **34.** $159.95 + $420.59   **35.** $.84 + $9.49

36. 6.18 + 9.82 + 0.90   **37.** $8642.89 + $1010.39   **38.** $49 + $9.81 + $11.25

Estimate the difference to the nearest tenth.

39. 45.019 − 6.728   **40.** 16.42 − 9.099   **41.** 754.950 − 309.777

42. 79 − 8.451   **43.** 160.07 − 79.13   **44.** 8.0539 − 0.4609

Estimate the difference to the nearest whole number.

45. 86.94 − 48.15   **46.** 10.405 − 5.5   **47.** 37.902 − 1.1495

48. 482.5 − 69.721   **49.** 1089.52 − 750   **50.** 9.95 − 3.95

51. $19.95 − $2.43   **52.** $63.42 − $19.80   **53.** $447.49 − $32.03

Estimate the sum or difference.

54. You purchase a mop for $3.29, a can of cleanser for $.59, a box of laundry detergent for $2.29, and a can of floor wax for $1.69. To the nearest dime, estimate the amount of your purchases.

55. The median weight of 10-year-old boys is 69.17 pounds, and of 10-year-old girls is 71.61 pounds. To the nearest tenth, estimate the difference of these median weights.

56. In the first six months of the year, the consumer price index rises from 253.5 to 265.1. To the nearest whole number, estimate how much the consumer price index rose.

57. You want to buy four books that cost $12.95, $9.95, $4.39, and $5.25. You have $25.00 to spend. To the nearest whole dollar, estimate the amount of your purchase. Do you have enough money?

For more practice: **Application Bank,** page 449.

Clarion Rally Coupe

You have $5600 with which to buy a new car. You decide to purchase a 2-door Clarion Rally Coupe. The base price of the car is $4777.45. A dealer has four 2-door models in stock. You look at these file cards showing the factory-installed options on each.

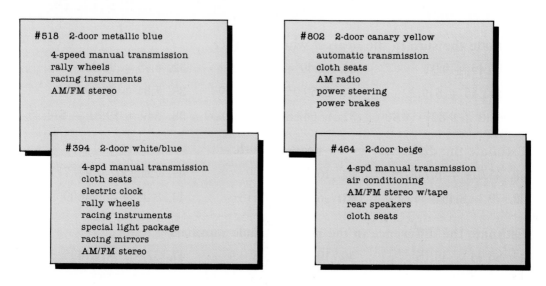

#518   2-door metallic blue

    4-speed manual transmission
    rally wheels
    racing instruments
    AM/FM stereo

#802   2-door canary yellow

    automatic transmission
    cloth seats
    AM radio
    power steering
    power brakes

#394   2-door white/blue

    4-spd manual transmission
    cloth seats
    electric clock
    rally wheels
    racing instruments
    special light package
    racing mirrors
    AM/FM stereo

#464   2-door beige

    4-spd manual transmission
    air conditioning
    AM/FM stereo w/tape
    rear speakers
    cloth seats

PROBLEM: Select the cars that you are able to buy.

Solution Plan: Estimate the amount that you can spend on options. Estimate the cost of options on the cars in stock. Choose the cars with options that cost less than the amount you can spend.

WHAT DO YOU KNOW? You know the amount that you can spend, the base price of the car, and the options available on the cars in stock. Answer these questions.

1. How much can you spend?

2. What is the base price of your car?

3. To the nearest ten dollars, how much can you spend on options?

4. If you want a car with manual transmission, which cars will you consider?

WHAT DO YOU WANT TO KNOW? You want to estimate the cost of options on the cars in stock.

The list of available options on a Clarion Rally Coupe is shown below. To the nearest ten dollars, you estimate that the cost of options on the 2-door, metallic blue model is $570.

Estimate the cost of options to the nearest ten dollars.

4-spd manual:	$135 $\rightarrow$	$140
rally wheels:	93 $\rightarrow$	90
racing instr:	112 $\rightarrow$	110
AM/FM stereo:	+ 232 $\rightarrow$	+ 230
		$570

Application A3

5. Canary yellow model.

6. White-on-blue model.

7. Beige model.

Clarion Rally Coupe
Available Options

Transmission,
  – automatic $359
  – 4-spd manual 135
Air conditioning 562
Power brakes 76
Power steering 159
Cloth seats 23
Electric clock 28
Rally wheels 93
Racing instruments 112
Special light pkg 64
Racing mirrors 43
Radios,
  – AM only 85
  – AM/FM stereo 232
  – AM/FM stereo w/tape . . . 385
Rear speakers 98

NOW SOLVE THE PROBLEM. Use the information that you have found to make your selection.

8. Which cars have options totaling more than the amount you can spend on options?

9. Which cars are you able to buy?

10. Select one of the cars that you can buy. Find the sticker price of the car (base price plus cost of options).

Quiz

Skill 7

1.   73.63
   + 8.71

2. 2.98 + 8.7

Skill 8

3.   1.036
    − 0.359

4. 1.2 − 0.093

Application A3

Estimate to the nearest tenth.

5.   611.31
   + 30.08

6.   200.58
    + 173.54

7.   8.811
    − 0.98

8.   12.26
    − 4.792

Problem Solving

9. You want to buy a new van. The base price is $5398. Estimate the sticker price of the van, with the options shown, to the nearest hundred dollars.

Air conditioning	$659
AM/FM Stereo radio	225
Rear speakers	120
Rally wheels	165
Leather upholstery	815

CONSUMER NOTE

National Camera Company is offering *rebates* on its cameras. When you buy a camera you will receive a rebate form. You mail the form and your sales receipt to the company. They will send you a check for the amount of the rebate.

The M-112 automatic camera usually sells for $54.99. The rebate is $8.50. What is the final cost?

$54.99	Regular Price
− 8.50	Rebate
$46.49	Final Cost

1. The price for an M-432 camera is $27.50, after the rebate. The regular price is $32.00. Find the amount of the rebate.

2. The Camera Corner is reducing the AZ-319 by $20. The manufacturer is also offering a $25 rebate. If the regular price for the camera is $259.99, what is the final cost?

28

Skill 2 **Round each number to the places shown.** (pages 4–5)

		Nearest Hundred	Nearest Ten	Nearest ·Unit	Nearest Tenth
1.	0.875			?	?
2.	4836	?	?		
3.	10.63		?	?	?
4.	1923	?	?		

Skill 4 (page 13)

5. 43,196
    + 18,205

6. $6089
    + 920

7. $408
    + 517

8. 746
    + 39

9. 5186
    + 2705

10. 746
    35
    1149
    + 3206

11. 23,416
    6,409
    + 117,983

12. 97
    48
    76
    + 15

13. 144
    66
    33
    + 99

14. 8,926
    478
    + 60,016

Skill 6 (page 15)

15. 4862
    − 1479

16. 6374
    − 2586

17. 4281
    − 2638

18. 7238
    − 6841

19. 2005
    − 388

Skill 7 (page 22)

20. $18.36 + 7.4$

21. $138.09 + 16.145$

22. $3.08 + 11.64$

23. $298 + $26.30 + $981.80

24. $0.04 + 635.8 + 94.8$

Skill 8 (page 23)

25. $63.28
    − 49.30

26. 372.4
    − 259.6

27. 2.806
    − 0.798

28. $52.39
    − 39.82

29. $239.72 − 49.69$

30. $17.3 − 8.47$

31. $6.05 − 3.886$

Applications Review

Application A1 **Compare the numbers. Which is larger?** (pages 6–7)

1. 323 or 409   **2.** 1477 or 1500   **3.** 17,459 or 17,954

4. 14.19 or 14.09   **5.** 1.1109 or 1.1119   **6.** 34.5 or 34.2957

Application A2 **Estimate to the nearest ten.** (pages 16–17)

7.   66
     + 23

8.   189
     + 145

9.   278
       84
     + 113

10.   535
      138
     + 91

Estimate to the nearest hundred.

11.   563
     − 179

12.   181
     − 98

13.   1357
     − 492

14.   6383
     − 1619

Application A3 **Estimate to the nearest tenth.** (pages 24–25)

15.   0.78
     + 0.34

16.   2.93
     + 0.87

17.   5.8043
     + 2.462

18.   43.036
      6.8174
     + 0.555

Estimate to the nearest whole number.

19.   78.92
      7.019
     + 5.053

20.   83.5
     15.93
     + 7.0

21.   95.42
     12.5
     + 39

22.   63.21
      0.257
     + 12.800

Estimate to the nearest hundredth.

23.   1.019
     − 0.122

24.   31.9935
     − 9.056

25.   109.05
     − 45.909

26.   5.1667
     − 1.8

Estimate to the nearest whole number.

27. $73.15
     − 29.95

28. $259.95
     − 49.19

29. $1095.29
     − 957.50

30. $995.49
     − 199.29

Unit Test

Skill 1

Give the place and value of the underlined digit.

1. 732,003   2. 33.3<u>0</u>8

Skill 2

Round to the given place.

3. Nearest thousand: 65,302
4. Nearest tenth: 607.38

Skill 3

5. $\begin{array}{r} 832 \\ + 151 \\ \hline \end{array}$ 6. $\begin{array}{r} 13,024 \\ + 922 \\ \hline \end{array}$

Skill 4

7. $\begin{array}{r} 6483 \\ + 908 \\ \hline \end{array}$ 8. $\begin{array}{r} 83,510 \\ + 29,491 \\ \hline \end{array}$

Skill 5

9. $\begin{array}{r} 273 \\ - 201 \\ \hline \end{array}$ 10. $\begin{array}{r} 7073 \\ - 6042 \\ \hline \end{array}$

Skill 6

11. $\begin{array}{r} 807 \\ - 762 \\ \hline \end{array}$ 12. $\begin{array}{r} 54,019 \\ - 35,432 \\ \hline \end{array}$

Skill 7

13. $13.2 + 3.8$

14. $11.27 + 4.892 + 15.6$

Skill 8

15. $777.2 - 668.1$

16. $200.01 - 11.808$

Application A1 **Compare the two numbers. Which is larger?**

17. 3010 or 3119   18. 27.015 or 27.0136

Application A2

Estimate to the nearest ten.

19. $\begin{array}{r} 129 \\ + 83 \\ \hline \end{array}$ 20. $\begin{array}{r} 6735 \\ + 882 \\ \hline \end{array}$

Estimate to the nearest thousand

21. $\begin{array}{r} 4800 \\ - 3011 \\ \hline \end{array}$ 22. $\begin{array}{r} 6075 \\ - 1893 \\ \hline \end{array}$

Application A3

Estimate to the nearest hundredth.

23. $\begin{array}{r} 12.2191 \\ + 0.5339 \\ \hline \end{array}$ 24. $\begin{array}{r} 0.6529 \\ + 0.7976 \\ \hline \end{array}$

Estimate to the nearest whole number.

25. $\begin{array}{r} \$7.03 \\ - 5.95 \\ \hline \end{array}$ 26. $\begin{array}{r} 10.9804 \\ - 4.4995 \\ \hline \end{array}$

2

Multiplication

Customary System of Measurement

Problem Solving

Unit Preview

 You will use these skills and applications in this unit. Which do you already know? Work each problem.

Skill 9

1. $\begin{array}{r} 32 \\ \times\ 3 \\ \hline \end{array}$

2. $\begin{array}{r} 43 \\ \times\ 2 \\ \hline \end{array}$

Skill 10

3. $\begin{array}{r} 172 \\ \times\ 4 \\ \hline \end{array}$

4. $\begin{array}{r} 558 \\ \times\ 9 \\ \hline \end{array}$

Skill 11

5. $\begin{array}{r} 4812 \\ \times\ 10 \\ \hline \end{array}$

6. $\begin{array}{r} 637 \\ \times\ 400 \\ \hline \end{array}$

Skill 12

7. $\begin{array}{r} 887 \\ \times\ 28 \\ \hline \end{array}$

8. $\begin{array}{r} 1375 \\ \times\ 64 \\ \hline \end{array}$

Skill 13

9. $\begin{array}{r} 9431 \\ \times\ 311 \\ \hline \end{array}$

10. $\begin{array}{r} 1969 \\ \times\ 324 \\ \hline \end{array}$

Skill 14

11. $\begin{array}{r} 314 \\ \times\ 105 \\ \hline \end{array}$

12. $\begin{array}{r} 6403 \\ \times\ 706 \\ \hline \end{array}$

Skill 15

13. $\begin{array}{r} 15.26 \\ \times\ 6.3 \\ \hline \end{array}$

14. $\begin{array}{r} 1.034 \\ \times\ 35 \\ \hline \end{array}$

15. $\begin{array}{r} 0.00134 \\ \times\ 0.07 \\ \hline \end{array}$

16. $\begin{array}{r} 0.0007 \\ \times\ 0.13 \\ \hline \end{array}$

Application A4

17. 5 feet = __?__ inches

18. 4 yards = __?__ feet

19. 1 ft 6 in. = __?__ in.

20. 10 miles = __?__ feet

Application A5

21. 3 pounds = __?__ ounces

22. 3 tons = __?__ pounds

23. 1 lb 12 oz = __?__ oz

24. 3 lb 2 oz = __?__ oz

Application A6

25. 1 cup = __?__ fluid ounces

26. 2 gallons = __?__ quarts

27. 8 pints = __?__ cups

28. 10 quarts = __?__ pints

Check your answers. If you had difficulty with any skill or application, be sure to study the corresponding lesson in this unit.

Multiplication Without Carrying

MODEL

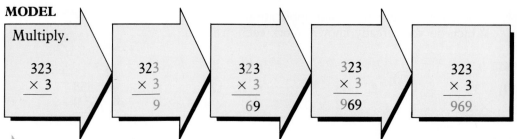

Multiply.

323	323	323	323	323
× 3	× 3	× 3	× 3	× 3
	9	69	969	969

GET READY.

1. 3
   × 4

2. 6
   × 5

3. 8
   × 0

4. 4
   × 2

5. 7
   × 1

6. 3
   × 6

7. 10
   × 3

8. 20
   × 4

9. 30
   × 2

10. 20
    × 3

11. 40
    × 0

12. 30
    × 3

NOW USE SKILL 9.

13. 21
    × 3

14. 34
    × 2

15. 42
    × 2

16. 12
    × 4

17. 25
    × 1

18. 22
    × 4

19. 31
    × 3

20. 41
    × 2

21. 33
    × 2

22. 30
    × 3

23. 202
    × 2

24. 132
    × 3

25. 243
    × 1

26. 121
    × 3

27. 111
    × 2

28. 150
    × 1

29. 131
    × 2

30. 223
    × 2

31. 123
    × 3

32. 121
    × 4

33. 3101
    × 1

34. 2120
    × 2

35. 1412
    × 2

36. 2231
    × 3

37. 3104
    × 2

38. 1232
    × 3

39. 3002
    × 2

40. 4012
    × 2

41. 2030
    × 3

42. 1005
    × 1

For more practice: **Skill Bank,** page 409.

Multiplication With Carrying

MODEL

Multiply.

$$\begin{array}{r} 486 \\ \times\ 3 \\ \hline \end{array}$$

$$\begin{array}{r} {}^{1} \\ 486 \\ \times\ 3 \\ \hline 8 \end{array}$$

$$\begin{array}{r} {}^{2\,1} \\ 486 \\ \times\ 3 \\ \hline 58 \end{array}$$

$$\begin{array}{r} {}^{2\,1} \\ 486 \\ \times\ 3 \\ \hline 1458 \end{array}$$

$$\begin{array}{r} 486 \\ \times\ 3 \\ \hline 1458 \end{array}$$

GET READY.

1. $\begin{array}{r} 32 \\ \times\ 3 \\ \hline \end{array}$	**2.** $\begin{array}{r} 24 \\ \times\ 2 \\ \hline \end{array}$	**3.** $\begin{array}{r} 34 \\ \times\ 1 \\ \hline \end{array}$	**4.** $\begin{array}{r} 56 \\ \times\ 0 \\ \hline \end{array}$	**5.** $\begin{array}{r} 42 \\ \times\ 2 \\ \hline \end{array}$
6. $\begin{array}{r} 213 \\ \times\ 3 \\ \hline \end{array}$	**7.** $\begin{array}{r} 402 \\ \times\ 1 \\ \hline \end{array}$	**8.** $\begin{array}{r} 200 \\ \times\ 4 \\ \hline \end{array}$	**9.** $\begin{array}{r} 321 \\ \times\ 3 \\ \hline \end{array}$	**10.** $\begin{array}{r} 130 \\ \times\ 3 \\ \hline \end{array}$

NOW USE SKILL 10.

11. $\begin{array}{r} 24 \\ \times\ 3 \\ \hline \end{array}$	**12.** $\begin{array}{r} 37 \\ \times\ 2 \\ \hline \end{array}$	**13.** $\begin{array}{r} 25 \\ \times\ 2 \\ \hline \end{array}$	**14.** $\begin{array}{r} 48 \\ \times\ 2 \\ \hline \end{array}$	**15.** $\begin{array}{r} 46 \\ \times\ 2 \\ \hline \end{array}$
16. $\begin{array}{r} 216 \\ \times\ 2 \\ \hline \end{array}$	**17.** $\begin{array}{r} 127 \\ \times\ 3 \\ \hline \end{array}$	**18.** $\begin{array}{r} 314 \\ \times\ 3 \\ \hline \end{array}$	**19.** $\begin{array}{r} 228 \\ \times\ 3 \\ \hline \end{array}$	**20.** $\begin{array}{r} 125 \\ \times\ 2 \\ \hline \end{array}$
21. $\begin{array}{r} 241 \\ \times\ 3 \\ \hline \end{array}$	**22.** $\begin{array}{r} 173 \\ \times\ 2 \\ \hline \end{array}$	**23.** $\begin{array}{r} 352 \\ \times\ 2 \\ \hline \end{array}$	**24.** $\begin{array}{r} 260 \\ \times\ 3 \\ \hline \end{array}$	**25.** $\begin{array}{r} 182 \\ \times\ 3 \\ \hline \end{array}$
26. $\begin{array}{r} 145 \\ \times\ 5 \\ \hline \end{array}$	**27.** $\begin{array}{r} 124 \\ \times\ 6 \\ \hline \end{array}$	**28.** $\begin{array}{r} 236 \\ \times\ 4 \\ \hline \end{array}$	**29.** $\begin{array}{r} 176 \\ \times\ 3 \\ \hline \end{array}$	**30.** $\begin{array}{r} 246 \\ \times\ 3 \\ \hline \end{array}$
31. $\begin{array}{r} 427 \\ \times\ 4 \\ \hline \end{array}$	**32.** $\begin{array}{r} 576 \\ \times\ 3 \\ \hline \end{array}$	**33.** $\begin{array}{r} 226 \\ \times\ 6 \\ \hline \end{array}$	**34.** $\begin{array}{r} 918 \\ \times\ 2 \\ \hline \end{array}$	**35.** $\begin{array}{r} 234 \\ \times\ 9 \\ \hline \end{array}$
36. $\begin{array}{r} 2376 \\ \times\ 2 \\ \hline \end{array}$	**37.** $\begin{array}{r} 1426 \\ \times\ 3 \\ \hline \end{array}$	**38.** $\begin{array}{r} 4672 \\ \times\ 4 \\ \hline \end{array}$	**39.** $\begin{array}{r} 8168 \\ \times\ 2 \\ \hline \end{array}$	**40.** $\begin{array}{r} 6827 \\ \times\ 3 \\ \hline \end{array}$

For more practice: **Skill Bank,** page 409.

Multiplication by a Multiple of 10 or 100

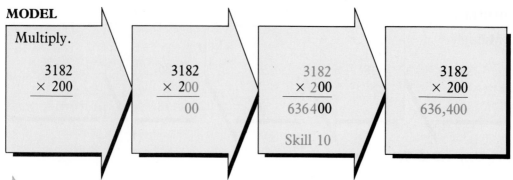

MODEL

Multiply.

3182	3182	3182	3182
× 200	× 200	× 200	× 200
	00	636400	636,400
		Skill 10	

 GET READY.

1. 26 × 10

2. 30 × 10

3. 78 × 10

4. 56 × 10

5. 80 × 10

6. 312 × 100

7. 498 × 100

8. 605 × 100

9. 893 × 100

10. 490 × 100

 NOW USE SKILL 11.

11. 72 × 20

12. 25 × 60

13. 83 × 70

14. 69 × 40

15. 20 × 50

16. 321 × 40

17. 243 × 30

18. 190 × 20

19. 205 × 20

20. 120 × 30

21. 372 × 400

22. 735 × 600

23. 423 × 700

24. 721 × 500

25. 483 × 300

26. 1234 × 10

27. 2196 × 20

28. 4123 × 30

29. 9169 × 20

30. 4711 × 50

31. 4196 × 400

32. 6127 × 600

33. 5941 × 200

34. 5140 × 100

35. 9091 × 300

36. 19,267 × 500

37. 43,268 × 400

38. 49,890 × 800

39. 12,345 × 900

Multiplication by a Multiple of 1000

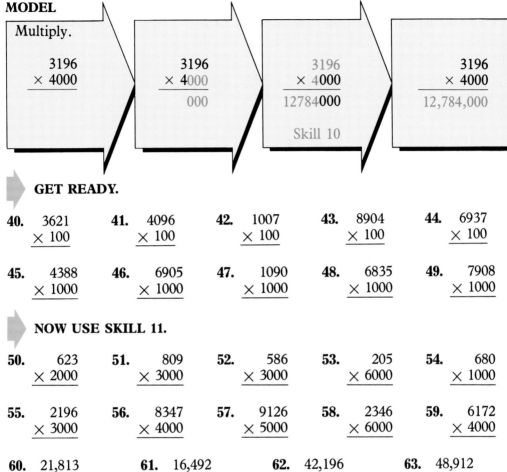

MODEL

Multiply.

3196	3196	3196	3196
× 4000	× 4000	× 4000	× 4000
	000	12784000	12,784,000
		Skill 10	

GET READY.

40. 3621
× 100

41. 4096
× 100

42. 1007
× 100

43. 8904
× 100

44. 6937
× 100

45. 4388
× 1000

46. 6905
× 1000

47. 1090
× 1000

48. 6835
× 1000

49. 7908
× 1000

NOW USE SKILL 11.

50. 623
× 2000

51. 809
× 3000

52. 586
× 3000

53. 205
× 6000

54. 680
× 1000

55. 2196
× 3000

56. 8347
× 4000

57. 9126
× 5000

58. 2346
× 6000

59. 6172
× 4000

60. 21,813
× 3000

61. 16,492
× 5000

62. 42,196
× 2000

63. 48,912
× 2000

64. 12,345
× 4000

65. 32,143
× 4000

66. 16,482
× 8000

67. 25,109
× 6000

68. 44,192
× 7000

69. 52,487
× 5000

70. 46,172
× 6000

71. 24,461
× 6000

72. $19,860
× 3000

73. $72,419
× 7000

74. $86,412
× 8000

75. $43,268
× 9000

For more practice: **Skill Bank**, page 409.

Using U.S. Customary Units of Length

Some common customary units of length and their symbols are shown in the table. To measure shorter distances, use inch or foot. To measure longer distances, use yard or mile.

Unit	Symbol
inch	in. (or ″)
foot	ft (or ′)
yard	yd
mile	mi

To change from one customary unit to another, use the customary equivalent. Some common equivalents are shown below.

1 ft = 12 in.   1 mi = 5280 ft
1 yd = 3 ft   1 mi = 1760 yd

MODEL

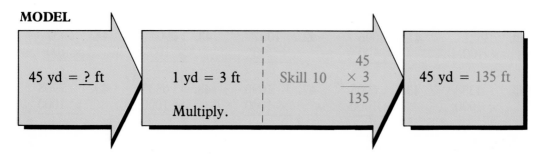

45 yd = ? ft

1 yd = 3 ft

Multiply.

Skill 10

$$\begin{array}{r} 45 \\ \times\ 3 \\ \hline 135 \end{array}$$

45 yd = 135 ft

GET READY.

1. $\begin{array}{r} 12 \\ \times\ 5 \\ \hline \end{array}$
2. $\begin{array}{r} 18 \\ \times\ 3 \\ \hline \end{array}$
3. $\begin{array}{r} 1760 \\ \times\ 9 \\ \hline \end{array}$
4. $\begin{array}{r} 5280 \\ \times\ 10 \\ \hline \end{array}$
5. $\begin{array}{r} 12 \\ \times\ 10 \\ \hline \end{array}$
6. $\begin{array}{r} 5280 \\ \times\ 50 \\ \hline \end{array}$

Complete the statement.

7. 12 in. = __?__ ft

8. 3 ft = __?__ yd

9. 1 ft = __?__ in.

10. __?__ ft = 1 mi

11. __?__ ft = 1 yd

12. 1 mi = __?__ yd

Name the customary unit most commonly used to measure the following.

13. The height of this textbook

14. The length of a football field

15. The width of your classroom

16. The length of the Nile River

17. The length of a paper clip

18. The width of a record

19. The distance between cities

20. The height of a building

NOW USE APPLICATION A4. Find the equivalent measure. Complete the table.

	21.	22.	23.	24.	25.	26.
yards			1	5	10	20
feet	5	10	?	?	?	?
inches	?	?	?	?	?	?

	27.	28.	29.	30.	31.	32.
miles	2	5	9	10	30	70
yards	?	?	?	?	?	?
feet	?	?	?	?	?	?

Complete the statement.

33. 1 ft 4 in. = 1 ft + 4 in.
= __?__ in. + 4 in.
= __?__ in.

34. 1 ft 9 in. = 1 ft + 9 in.
= __?__ in. + 9 in.
= __?__ in.

35. 4 yd 2 ft = 4 yd + 2 ft
= __?__ ft + 2 ft
= __?__ ft

36. 3 yd 8 in. = 3 yd + 8 in.
= __?__ in. + 8 in.
= __?__ in.

37. 1 yd 1 ft 5 in. = 1 yd + 1 ft + 5 in.
= __?__ in. + 12 in. + 5 in.
= __?__ in.

38. 1 yd 2 ft 11 in. = 1 yd + 2 ft + 11 in.
= __?__ in. + __?__ in. + 11 in.
= __?__ in.

39. You have one board that measures 3 ft 7 in. Another board measures 4 ft. How much longer is the second board?

40. The distance from Reading to Hollis is 53 miles. The distance from Hollis to Kettering is 112 miles. What is the distance from Reading to Hollis to Kettering and back?

For more practice: **Application Bank,** page 450.

Route 6

You are a supervisor for the Department of Roads. Your crew has been assigned to place mileposts and paint the lanes for a newly completed section of Route 6. The map shows the highway.

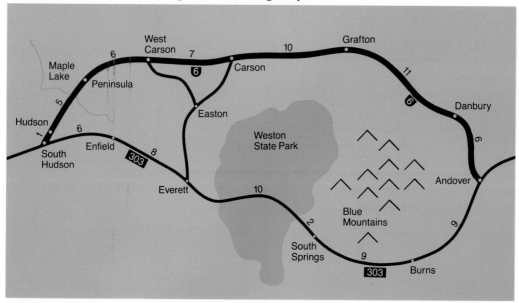

PROBLEM: How much material should you order from the supply depot to complete your job?

Solution Plan: Find the total length of the new section of highway. Determine the number of mileposts you will need. Calculate the amount of paint needed to paint the lanes. Make a list of items you should order from the supply depot.

WHAT DO YOU KNOW? You can use the map to find the length of the new section of Route 6. Answer these questions.

1. The new section of Route 6 begins at Andover. At what town does the section end?

2. How many towns are on the newly completed highway?

3. What is the distance from Andover to Danbury?

4. What is the total length of the new highway?

WHAT DO YOU WANT TO KNOW? You want to determine the number of mileposts needed. Use the map to answer the questions.

For each mile of highway, you need two mileposts, one for each side of the road. For instance, to place mileposts for a 5-mile section of highway, you would need twelve mileposts.

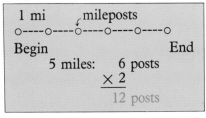

```
 1 mi        ⌐mileposts
o----o----o----o----o----o
Begin                    End
    5 miles:     6 posts
                 × 2
                 12 posts
```

5. What is the length of the new section of Route 6?

6. How many posts are needed for the new section of Route 6?

7. The milepost that begins in Andover will be number 23. What number will be on the post in South Hudson where the new section ends?

WHAT MORE DO YOU WANT? You want to find the amount of paint needed to paint the lanes of the highway.

A paint machine uses about 6 gallons of paint for each mile of highway. To paint the lanes from Andover to Grafton, you will use about 102 gallons of paint.

Andover to Grafton:	17 miles
	× 6 gal/mi
Skill 10	102 gallons

8. What is the length of the new section?

9. How many gallons of paint will you need to order to complete the job?

NOW SOLVE THE PROBLEM. To allow for waste, you always order one extra gallon of paint per mile.

10. How many gallons of paint do you order?

41

Quiz

Skill 9

1. 23
  × 3

2. 231
  × 3

Skill 10

3. 363
  × 4

4. 5039
  × 8

Skill 11

5.   35
  × 70

6.   981
  × 300

Application A4

7. On your sixteenth birthday, you measure 6 feet tall. How many inches is this?

8. One wire measures 3 feet long. Another wire is 4 feet 3 inches long. If you splice the two wires together, what is the total length of the wire in inches?

Problem Solving

9. You find that it takes 367 truckloads of concrete to pave one mile of highway. How many truckloads are needed to pave 50 miles?

CONSUMER NOTE

Energy is consumed when you use your hair dryer, television, or any other electrical appliance. When you watch your favorite television program, the amount of electrical energy consumed is measured in kilowatt-hours. Appliances have a power rating in watts. One kilowatt (kW) equals 1000 watts (W). If your hair dryer is rated 1000 W (1 kW), then it uses 1 kilowatt-hour of energy every hour it is in use. An iron rated at 1300 W (1.3 kW) uses 1.3 kilowatt-hours of energy every hour it is in use.

Find the total number of kilowatt-hours of energy consumed in each room when all the appliances are used for 2 hours.

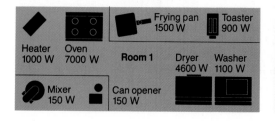

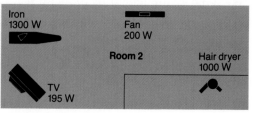

Computers and Their Uses

Computers are everywhere! Although they have been around for only about 30 years, they touch everyone's life. One reason computers are such an important part of society is that they simplify work. With computers you can do many things much more quickly and with greater reliability and accuracy. Computers make life a little easier by giving you more time for other activities.

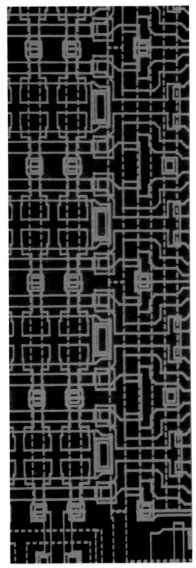

Computers do arithmetic and make decisions. They can answer questions which require a "yes" or "no" answer. For example: Is a number less than 10? Does one name come before another in alphabetizing a list? Computers use electronic circuitry that enables them to work at lightning speed. This speed and the ability to make decisions, to follow a programmer's instructions, and to store information for later use make computers powerful tools. With these features computers can do jobs which, not long ago, were impossible.

Computers are used to:
- guide spacecraft through outer space
- determine your grocery bill at the supermarket
- compute utility bills, such as electricity and telephone
- calculate your savings account balance
- control air traffic
- run machinery in a factory
- compute payrolls and print paychecks

Activities

Write to several computer manufacturers to find out how computers are used in business. What are the names of some computers that are sold for business purposes? How much do they cost? What types of computer programs are available for these computers? How much information can the computers store?

Multiplication by a 2-Digit Number

MODEL

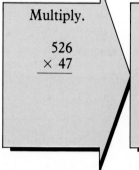

Multiply.

526
× 47

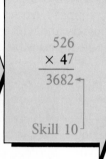

526
× 47
3682

Skill 10

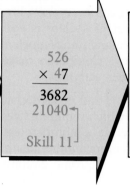

526
× 47
3682
21040

Skill 11

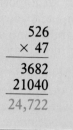

526
× 47
3682
21040
24,722

 GET READY.

1. 231 × 4	**2.** 384 × 6	**3.** 527 × 8	**4.** 280 × 5	**5.** 678 × 7
6. 633 + 8440	**7.** 927 + 3090	**8.** 543 + 3650	**9.** 1,482 + 22,210	**10.** 6,982 + 48,340

NOW USE SKILL 12.

11. 62 × 31	**12.** 51 × 14	**13.** 43 × 22	**14.** 20 × 13	**15.** 40 × 21
16. 314 × 23	**17.** 215 × 12	**18.** 423 × 11	**19.** 320 × 31	**20.** 460 × 42
21. 478 × 64	**22.** 620 × 57	**23.** 824 × 49	**24.** 507 × 84	**25.** 7132 × 42
26. 7609 × 68	**27.** 8123 × 76	**28.** 3906 × 36	**29.** 2097 × 84	**30.** 3009 × 49
31. 7634 × 25	**32.** 2006 × 73	**33.** 7380 × 48	**34.** 3894 × 29	**35.** 9999 × 99

For more practice: **Skill Bank,** page 410.

Multiplication by a 3- or 4-Digit Number

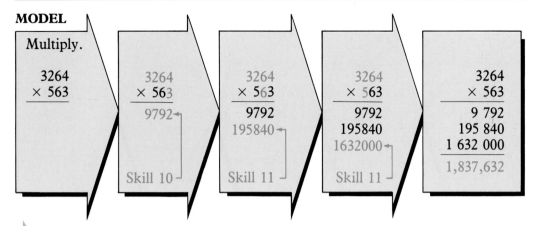

MODEL

Multiply.

3264
× 563

3264
× 563
9792

Skill 10

3264
× 563
9792
195840

Skill 11

3264
× 563
9792
195840
1632000

Skill 11

3264
× 563
9 792
195 840
1 632 000
1,837,632

GET READY.

1. 596 × 43	**2.** 783 × 68	**3.** 469 × 72	**4.** 637 × 54	**5.** 826 × 49

6. 6,798
42,410
+ 679,800

7. 1,596
23,480
+ 594,600

8. 4,106
31,920
+ 638,400

9. 1,240
32,400
+ 961,000

NOW USE SKILL 13.

10. 621 × 413	**11.** 425 × 216	**12.** 524 × 324	**13.** 432 × 214	**14.** 364 × 425
15. 870 × 574	**16.** 736 × 863	**17.** 698 × 782	**18.** 669 × 930	**19.** 649 × 738
20. 2163 × 214	**21.** 4196 × 248	**22.** 1896 × 326	**23.** 2561 × 287	**24.** 3692 × 428
25. 7491 × 2618	**26.** 6475 × 3192	**27.** 4320 × 5432	**28.** 3268 × 4616	**29.** 6619 × 8367

For more practice: **Skill Bank,** page 410.

Multiplication Using Numbers with Zeros

MODEL

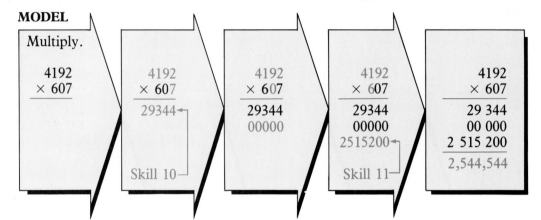

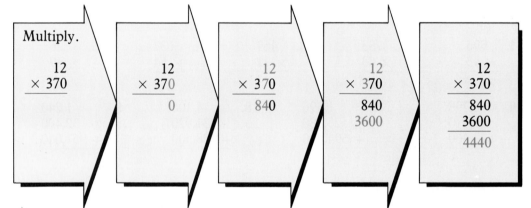

 GET READY.

1.	48 × 20	2.	67 × 30	3.	50 × 60	4.	39 × 70	5.	70 × 40
6.	317 × 40	7.	628 × 60	8.	203 × 70	9.	728 × 50	10.	806 × 40

 NOW USE SKILL 14.

11.	289 × 209	12.	408 × 402	13.	827 × 705	14.	730 × 501	15.	487 × 806

16. 612
      × 204

17. 409
      × 418

18. 920
      × 705

19. 807
      × 436

20. 781
      × 702

21.   18
      × 430

22.   26
      × 503

23.   32
      × 120

24.   62
      × 403

25.   37
      × 240

26.   93
      × 807

27.   75
      × 910

28.   38
      × 509

29.   86
      × 470

30.   52
      × 307

31. 4137
      × 200

32. 8293
      × 600

33. 6309
      × 700

34. 7249
      × 400

35. 3572
      × 300

36. 6101
      × 204

37. 7340
      × 390

38. 4006
      × 609

39. 5020
      × 340

40. 8000
      × 504

41.   84
      × 120

42.   32
      × 230

43.   74
      × 120

44.   67
      × 140

45.   12
      × 130

46.   37
      × 6703

47.   86
      × 5027

48.   76
      × 3140

49.   44
      × 7023

50.   76
      × 4050

51. 2463
      × 304

52. 270
      × 4583

53. 8038
      × 123

54. 6768
      × 260

55. 879
      × 2025

56. 3279
      × 7008

57. 492
      × 4206

58. 6744
      × 9360

59. 284
      × 3150

60. 8934
      × 903

Multiply the rows and columns to complete the blanks.

61.	674	470	?
62.	103	914	?
63.	?	?	

64.	2365	360	?
65.	809	4433	?
66.	?	?	

For more practice: **Skill Bank,** page 411.

Using U.S. Customary Units of Weight

Some common customary units of weight and their symbols are shown at the right. To measure lighter objects, use ounce or pound. To measure heavier objects, use ton.

To change from one customary unit of weight to another, use the following equivalents.

Unit	Symbol
ounce	oz
pound	lb
ton	t

$$1 \text{ lb} = 16 \text{ oz} \qquad 1 \text{ t} = 2000 \text{ lb}$$

MODEL

| 5 lb = ? oz | 1 lb = 16 oz Multiply. | Skill 12 | $\begin{array}{r} 16 \\ \times\ 5 \\ \hline 80 \end{array}$ | 5 lb = 80 oz |

 GET READY.

1. $\begin{array}{r} 16 \\ \times\ 7 \\ \hline \end{array}$ 　　**2.** $\begin{array}{r} 25 \\ \times\ 16 \\ \hline \end{array}$ 　　**3.** $\begin{array}{r} 100 \\ \times\ 16 \\ \hline \end{array}$ 　　**4.** $\begin{array}{r} 2000 \\ \times\ 9 \\ \hline \end{array}$ 　　**5.** $\begin{array}{r} 2000 \\ \times\ 15 \\ \hline \end{array}$

Complete the statement.

6. 16 oz = __?__ lb

7. 1 t = __?__ lb

8. __?__ oz = 1 lb

9. 1 lb = 16 __?__

10. 2000 __?__ = 1 t

11. 16 oz = 1 __?__

Name the customary unit of weight most commonly used to measure the following.

12. Your weight

13. Box of breakfast cereal

14. Large bag of potatoes

15. A load of steel

16. A letter

17. A pencil eraser

18. Shipment of grain

19. A car

NOW USE APPLICATION A5. Find the equivalent measure. Complete the table.

	20.	**21.**	**22.**	**23.**	**24.**	**25.**	**26.**
tons				2	9	15	5
pounds	3	10	25	?	?	?	?
ounces	?	?	?				?

27. 1 lb 5 oz = 1 lb + 5 oz
            = 16 oz + 5 oz
            = __?__ oz

28. 5 lb 6 oz = 5 lb + 6 oz
            = 80 oz + 6 oz
            = __?__ oz

29. 7 lb 10 oz = 7 lb + 10 oz
             = __?__ oz + 10 oz
             = __?__ oz

30. 11 lb 14 oz = 11 lb + 14 oz
               = __?__ oz + 14 oz
               = __?__ oz

Which weighs more?

31. A 2 lb cake or a 28 oz box of cookies

32. A 3 lb rock or a 2 lb 12 oz rock

33. A 1 lb 5 oz box of cereal or a 20 oz box of cereal

34. Two tons of cement or 4200 pounds of cement

35. The directions for mixing a batch of cement call for four parts cement mix to one part gravel. You order 100 pounds of gravel. How many pounds of cement mix should you order?

36. You have three books. One weighs 1 lb 5 oz, another weighs 2 lb 1 oz, and the third weighs 2 lb 13 oz. You put the books in a box that weighs 5 oz. What is the total weight of the package?

37. Your favorite baseball player uses a 1-pound bat. You prefer to use an 18-ounce bat. How much heavier is your bat?

38. You can buy raisins in 1-pound boxes for 89¢ each. You can also buy a package of 8 2-ounce boxes for 98¢. Do these packages contain the same amount of raisins? Which is the better buy?

For more practice: **Application Bank,** page 450.

Harmon Sand & Gravel

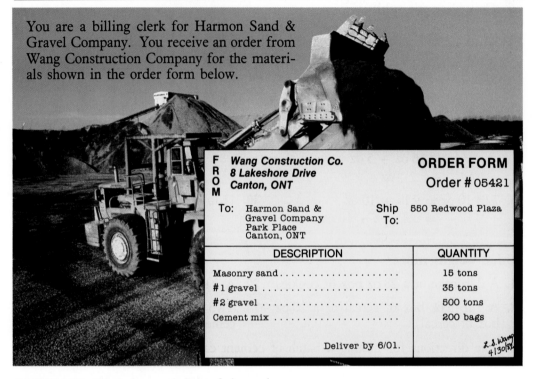

You are a billing clerk for Harmon Sand & Gravel Company. You receive an order from Wang Construction Company for the materials shown in the order form below.

FROM	Wang Construction Co. 8 Lakeshore Drive Canton, ONT	**ORDER FORM** Order # 05421
To:	Harmon Sand & Gravel Company Park Place Canton, ONT	Ship To: 550 Redwood Plaza

DESCRIPTION	QUANTITY
Masonry sand.....................	15 tons
#1 gravel	35 tons
#2 gravel	500 tons
Cement mix	200 bags

Deliver by 6/01.

L. S. Wang
4/30/8?

PROBLEM: Find the total cost of the order.

Solution Plan: Find the materials and the amounts ordered. Use a rate table to find the charges. Figure the delivery charge. Find the total cost.

WHAT DO YOU KNOW? You look at the order to find the materials needed. Answer the questions.

1. Some of the materials are ordered by the ton. Which materials are ordered by the ton?

2. Which material is ordered in bags?

3. How much #1 gravel is ordered?

4. How much #2 gravel is ordered?

5. How many bags of cement mix are ordered?

WHAT DO YOU WANT TO KNOW? You want to find the amount to charge for each material ordered.

You begin with the masonry sand. You use the rate table shown below. At $11 per ton, 15 tons of sand will cost $165.

Masonry sand: $11 per ton
           × 15 tons
              $165

Skill 12

Rates	
Masonry sand	$11/ton
Masonry sand	$2/bag
#1 gravel	$12/ton
#2 gravel	$15/ton
Silica	$4/bag
Cement mix	$3/bag

6. What do you charge for the #1 gravel ordered?

7. What do you charge for the #2 gravel ordered?

8. What do you charge for the cement mix ordered?

WHAT MORE DO YOU WANT? You want to figure the total weight.

You figure that 15 tons of masonry sand is equivalent to 30,000 pounds.

1 t = 2000 lb   15
              × 2000
Application A5   30,000

9. #1 gravel: 35 t = ? lb

10. #2 gravel: 500 t = ? lb

11. One bag of cement mix weighs 85 pounds. What is the total weight of cement mix to be delivered?

12. What is the total weight of all deliveries?

NOW SOLVE THE PROBLEM. Use the information you have found to compute the total cost.

13. What is the total charge for sand, gravel, and cement mix?

14. The delivery charge is $2 for each 1000 pounds delivered. What is the delivery charge?

15. What is the total billing cost to Wang Construction Company?

Quiz

Skill 12

1. $\begin{array}{r} 326 \\ \times\ 17 \\ \hline \end{array}$

2. $\begin{array}{r} 3301 \\ \times\ 13 \\ \hline \end{array}$

Skill 13

3. $\begin{array}{r} 6361 \\ \times\ 211 \\ \hline \end{array}$

4. $\begin{array}{r} 2983 \\ \times\ 621 \\ \hline \end{array}$

Skill 14

5. $\begin{array}{r} 309 \\ \times\ 165 \\ \hline \end{array}$

6. $\begin{array}{r} 8342 \\ \times\ 509 \\ \hline \end{array}$

7. $\begin{array}{r} 827 \\ \times\ 110 \\ \hline \end{array}$

8. $\begin{array}{r} 128 \\ \times\ 2003 \\ \hline \end{array}$

Application A5

9. You buy a 3-pound ham. How many ounces of ham do you have?

10. A shipment of grain weighs 10 tons. How many pounds is this?

Problem Solving

11. You have purchased 50 bags of cement mix. Each bag weighs 85 pounds. Your pickup truck has a load capacity of 3700 pounds. Can you carry all the cement mix in one trip?

EXTRA!

ZIP codes help speed mail delivery. They are used so that machines can sort and route mail. The mail is first sent to one of the 552 major post offices in the country, then to branch offices. The nine-digit ZIP code for Warrenton Junior High School is:

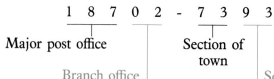

1 8 7 0 2 - 7 3 9 3

Major post office

Branch office

Section of town

Section of street

1. Should mail coded 18719-2191 be routed to the same major post office as mail for Warrenton Junior High?

2. Should mail coded 18702-3158 be routed to the same branch office as mail for Warrenton Junior High?

3. Should mail coded 18702-7893 be routed to the same section of town as mail for Warrenton Junior High?

4. You live in the same section of town as the junior high school. What are the first seven digits of your ZIP code?

Job Applications

Lee Whalen wants to work part-time after school or on weekends. Many businesses will ask Lee to complete an application form when applying for a job.

Please Print.

Name *Whalen   Lee   M.* Age *16* Soc. Sec. No. *000-18-7153*
     Last   First   M.I.

Address *72 Park Blvd., Crescent, OH* Tel. No. *555-6185*

AVAILABILITY
I can work ☐ full-time ☑ part-time at *14* hrs/wk
I can work on ☐ weekdays ☐ weekends ☑ both

SCHOOL MOST RECENTLY ATTENDED

Name *Crescent High Sch.* Address *Crescent, OH* Tel. No. *555-6264*

Teacher/Counselor *Mr. Roland Kendall* Last grade attended *10*

Now enrolled? ☑ yes ☐ no If no, date of graduation _____

School activities *Choir, Intramural Sports*

MOST RECENT JOB

Company *Town of Crescent* Address *Crescent, OH* Tel. No. *555-6444*

Position *Playground Dir.* Supervisor *Mrs. L. Farber*

Salary *$3/hr.* Dates worked: from *6/15* to *8/28*

Reason for leaving *Summer job only*

REFERENCES (other than family members)
Name *Sylvia Conti* Address *78 Park Blvd.* Tel. No. *585-8094*
Name *Bob Barton* Address *10 River Road* Tel. No. *873-9011*

All applications will be kept confidential.

---

1. How much did Lee earn for each 5-hour workday at the playground?

2. Lee will work 2 hours each weekday. At most, how many weekend hours will Lee work?

3. If this job pays $3.85 per hour, how much can Lee expect to earn for 4 weeks, working 14 hours each week?

Skill 15

Multiplying Decimals

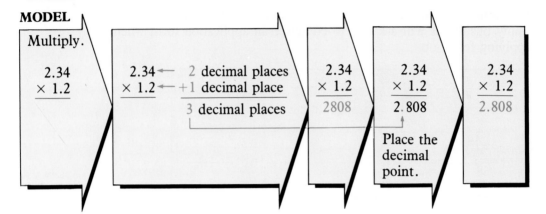

Multiply.

2.34	2.34 ← 2 decimal places
× 1.2	× 1.2 ← +1 decimal place
	3 decimal places

2.34
× 1.2
2808

2.34
× 1.2
2.808

Place the decimal point.

2.34
× 1.2
2.808

> **GET READY. Place the decimal point in the answer.**

1.	8.23 × 2.1 17283	2.	4.37 × 3.4 14858	3.	41.2 × 1.3 5356	4.	16.8 × 2.3 3864	5.	1.06 × 3.2 3392
6.	343 × 3.2 10976	7.	43.96 × 63 276948	8.	237.9 × 3.2 76128	9.	50.81 × 3.7 187997	10.	619.2 × 5.2 321984

> **NOW USE SKILL 15.**

11.	2.15 × 31	12.	9.23 × 12	13.	16.21 × 42	14.	22.38 × 16	15.	34.12 × 45
16.	26.82 × 4.1	17.	67.23 × 5.9	18.	35.09 × 6.2	19.	71.83 × 9.2	20.	33.42 × 8.7
21.	1.892 × 7.8	22.	9.081 × 2.3	23.	8.547 × 3.7	24.	3.549 × 8.6	25.	6.901 × 2.8
26.	16.23 × 1.47	27.	39.65 × 29.3	28.	65.81 × 12.4	29.	89.27 × 45.6	30.	83.15 × 34.9
31.	6.541 × 3.47	32.	9.198 × 6.02	33.	5.014 × 7.13	34.	3.105 × 2.03	35.	8.009 × 5.04

Multiplying Decimals

MODEL

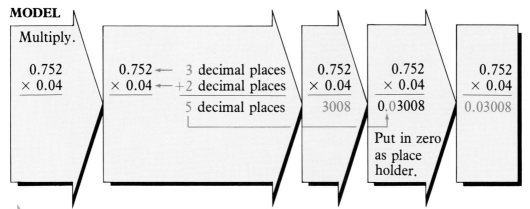

Multiply.

$$
\begin{array}{r} 0.752 \\ \times\ 0.04 \\ \hline \end{array}
$$

$0.752 \leftarrow$	3 decimal places	
$\times\ 0.04 \leftarrow$	$+2$ decimal places	
	5 decimal places	

$$
\begin{array}{r} 0.752 \\ \times\ 0.04 \\ \hline 3008 \end{array}
$$

$$
\begin{array}{r} 0.752 \\ \times\ 0.04 \\ \hline 0.03008 \end{array}
$$

Put in zero as place holder.

$$
\begin{array}{r} 0.752 \\ \times\ 0.04 \\ \hline 0.03008 \end{array}
$$

 GET READY. Place the decimal point in the answer. When necessary, put in zeros as place holders.

36. $\begin{array}{r} 0.5 \\ \times\ 0.6 \\ \hline 30 \end{array}$
37. $\begin{array}{r} 0.7 \\ \times\ 0.1 \\ \hline 7 \end{array}$
38. $\begin{array}{r} 0.03 \\ \times\ 0.2 \\ \hline 6 \end{array}$
39. $\begin{array}{r} 0.04 \\ \times\ 0.9 \\ \hline 36 \end{array}$
40. $\begin{array}{r} 0.003 \\ \times\ 0.4 \\ \hline 12 \end{array}$

41. $\begin{array}{r} 0.28 \\ \times\ 0.3 \\ \hline 84 \end{array}$
42. $\begin{array}{r} 0.19 \\ \times\ 1.2 \\ \hline 228 \end{array}$
43. $\begin{array}{r} 4.03 \\ \times\ 0.06 \\ \hline 2418 \end{array}$
44. $\begin{array}{r} 0.03 \\ \times\ 0.12 \\ \hline 36 \end{array}$
45. $\begin{array}{r} 100.7 \\ \times\ 0.003 \\ \hline 3021 \end{array}$

NOW USE SKILL 15.

46. $\begin{array}{r} 0.28 \\ \times\ 0.6 \\ \hline \end{array}$
47. $\begin{array}{r} 0.23 \\ \times\ 0.7 \\ \hline \end{array}$
48. $\begin{array}{r} 0.81 \\ \times\ 0.9 \\ \hline \end{array}$
49. $\begin{array}{r} 0.51 \\ \times\ 0.4 \\ \hline \end{array}$
50. $\begin{array}{r} 0.91 \\ \times\ 0.7 \\ \hline \end{array}$

51. $\begin{array}{r} 0.47 \\ \times\ 0.2 \\ \hline \end{array}$
52. $\begin{array}{r} 0.3 \\ \times\ 0.26 \\ \hline \end{array}$
53. $\begin{array}{r} 0.03 \\ \times\ 0.3 \\ \hline \end{array}$
54. $\begin{array}{r} 0.94 \\ \times\ 0.02 \\ \hline \end{array}$
55. $\begin{array}{r} 0.87 \\ \times\ 0.05 \\ \hline \end{array}$

56. $\begin{array}{r} 0.147 \\ \times\ 0.91 \\ \hline \end{array}$
57. $\begin{array}{r} 0.641 \\ \times\ 0.52 \\ \hline \end{array}$
58. $\begin{array}{r} 0.149 \\ \times\ 0.78 \\ \hline \end{array}$
59. $\begin{array}{r} 0.432 \\ \times\ 0.31 \\ \hline \end{array}$
60. $\begin{array}{r} 0.547 \\ \times\ 0.51 \\ \hline \end{array}$

61. $\begin{array}{r} 0.241 \\ \times\ 0.002 \\ \hline \end{array}$
62. $\begin{array}{r} 0.583 \\ \times\ 0.003 \\ \hline \end{array}$
63. $\begin{array}{r} 0.178 \\ \times\ 0.02 \\ \hline \end{array}$
64. $\begin{array}{r} 0.104 \\ \times\ 0.12 \\ \hline \end{array}$
65. $\begin{array}{r} 0.873 \\ \times\ 0.006 \\ \hline \end{array}$

66. $\begin{array}{r} 0.1415 \\ \times\ 0.2 \\ \hline \end{array}$
67. $\begin{array}{r} 0.8715 \\ \times\ 0.03 \\ \hline \end{array}$
68. $\begin{array}{r} 0.8001 \\ \times\ 0.01 \\ \hline \end{array}$
69. $\begin{array}{r} 0.0003 \\ \times\ 0.003 \\ \hline \end{array}$
70. $\begin{array}{r} 0.0102 \\ \times\ 0.12 \\ \hline \end{array}$

For more practice: **Skill Bank,** page 411.

Using U.S. Customary Units of Capacity

Some common customary units of capacity and their symbols are shown at the right. To measure smaller amounts, use fluid ounce or cup. To measure larger amounts, use pint, quart, or gallon.

To change from one customary unit of capacity to another, use the following equivalents.

Unit	Symbol
fluid ounce	fl oz
cup	c
pint	pt
quart	qt
gallon	gal

$$1 \text{ c} = 8 \text{ fl oz} \qquad 1 \text{ pt} = 2 \text{ c}$$
$$1 \text{ qt} = 2 \text{ pt} \qquad 1 \text{ gal} = 4 \text{ qt}$$

MODEL

6 gal = ? qt

1 gal = 4 qt

Multiply.

$$\begin{array}{r} 4 \\ \times\, 6 \\ \hline 24 \end{array}$$

6 gal = 24 qt

GET READY. Complete the statement.

1. 1 pt = __?__ c
2. 4 __?__ = 1 gal
3. 1 qt = __?__ pt
4. 8 __?__ = 1 c
5. 2 pt = 1 __?__
6. 1 c = __?__ fl oz
7. __?__ qt = 1 gal
8. 1 __?__ = 2 pt
9. 1 __?__ = 4 qt

Name the customary unit of capacity most commonly used to measure the following.

10. Gasoline purchase
11. Bottle of vinegar
12. Can of lemonade
13. Carton of whipping cream
14. Motor oil
15. Windshield washer fluid
16. Bottle of nail polish
17. Flour
18. Water in swimming pool
19. Bottle of shampoo

NOW USE APPLICATION A6. Find the equivalent measure. Complete the table.

	20.	21.	22.	23.	24.	25.
fluid ounces	?	?	?			
cups	?	?	2	?	?	
pints	3	?	?	?	?	?
quarts		4		?	12	?
gallons				2	?	10

Find the amount.

26. 5 quarts of motor oil.
$1.09 per quart.
Total cost?

27. 15 gallons of gasoline.
$1.669 per gallon.
Total cost?

28. 2 pints of whipping cream.
89¢ per pint.
Total cost?

29. 3 cups of flour.
To double the recipe,
how much flour?

30. A 5-ounce bottle of nail polish costs $1.29. A 10-ounce bottle costs $1.99. How much can you save by purchasing the 10-ounce bottle instead of two 5-ounce bottles?

31. Your car averages about 28 miles per gallon of gasoline. About how far can you drive on 25 gallons of gasoline?

32. You purchase a 1-quart bottle of vinegar. You use 8 ounces in a recipe. How much vinegar is left in the bottle?

33. Each day the cafeteria receives 21 cases of milk. Each case contains 24 half-pint cartons. How many cartons of milk are received daily?

34. The fuel gauge on your car shows empty. You fill the tank with gas, and it takes 16.8 gallons. Your owner's manual says your car has an 18-gallon tank. How much gas do you have in the gas tank when the gauge shows empty?

For more practice: **Application Bank,** page 450.

Clearwater Bottling Company

You are the warehouse supervisor for Clearwater Bottling Company. Your job is to maintain the stock of materials used to bottle Clearwater drinks. You have just received the memo shown below.

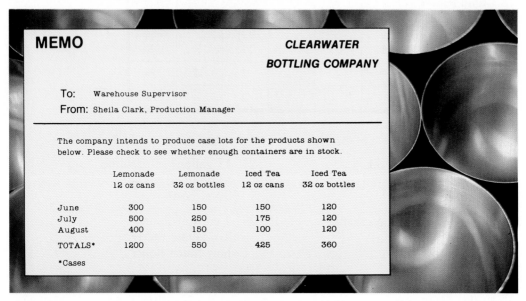

MEMO   **CLEARWATER BOTTLING COMPANY**

To: Warehouse Supervisor

From: Sheila Clark, Production Manager

The company intends to produce case lots for the products shown below. Please check to see whether enough containers are in stock.

	Lemonade 12 oz cans	Lemonade 32 oz bottles	Iced Tea 12 oz cans	Iced Tea 32 oz bottles
June	300	150	150	120
July	500	250	175	120
August	400	150	100	120
TOTALS*	1200	550	425	360

*Cases

PROBLEM: Do you have enough containers in stock?

Solution Plan: Find the number of cans and bottles you will need. Check these amounts against the numbers you have in stock. Determine whether you have the amounts needed to fill production quantities.

WHAT DO YOU KNOW? You know the production quantities for this summer. Use the memo.

1. What is the capacity of the can used for your products?

2. What is the capacity of the bottle used for your products?

3. How many cases will the company produce this summer for the following?

   a. 12 oz cans of lemonade   **b.** 12 oz cans of iced tea

   c. 32 oz bottles of lemonade   **d.** 32 oz bottles of iced tea

WHAT DO YOU WANT TO KNOW? You want to find the number of containers needed.

You need enough cans for 1200 cases of lemonade. There are 24 cans in a case. You estimate that you need 24,000 cans. You figure that you actually need 28,800 cans.

Lemonade, 12 oz cans
Estimated: Actual:

             1200 cases   1200 cases

Skill 2 $\times$ 20 cans per $\times$ 24 cans per

           24,000   case   28,800   case

Complete the table.

                                 4.   **5.**   **6.**

		Lemonade		Iced Tea	
		12 oz cans	32 oz bottles	12 oz cans	32 oz bottles
Production Quantity		1200 cases	?	?	?
Containers per Case		24	12	24	12
Containers Needed	**Estimated**	24,000	?	?	?
	Actual	28,800	?	?	?

WHAT MORE DO YOU WANT? You want to know how many containers you have in stock.

7. You have 263 gross of 12 oz cans in stock. There are 144 cans per gross. How many 12 oz cans do you have in stock?

8. You have 162 gross of 32 oz bottles in stock. There are 144 bottles per gross. How many 32 oz bottles do you have in stock?

NOW SOLVE THE PROBLEM. Use the information you have found to answer the question.

9. How many 12 oz cans do you need altogether?

10. Do you need to order more 12 oz cans? If so, how many?

11. How many 32 oz bottles do you need altogether?

12. Do you need to order more 32 oz bottles? If so, how many?

Skill 15

1. $\begin{array}{r} 1.93 \\ \times\ 7.5 \\ \hline \end{array}$	**2.** $\begin{array}{r} 37.73 \\ \times\ 1.56 \\ \hline \end{array}$	**3.** $\begin{array}{r} 7.25 \\ \times\ 0.04 \\ \hline \end{array}$	**4.** $\begin{array}{r} 0.0109 \\ \times\ 0.0015 \\ \hline \end{array}$

Application A6

5. You purchased 10 pints of different flavors of yogurt. Have you purchased more or less than a gallon of yogurt?

6. A recipe calls for 6 fl oz of condensed milk. Is this more or less than a cup?

Problem Solving

7. You have ordered 15 cases of juice. There are 24 cans per case. If it costs you $.25 per can, and you sell it for $.35 per can, how much profit will you make by selling all 15 cases?

CALCULATOR DISPLAYS

Patterns

Use a calculator to find the product. Watch for a pattern.

$5 \times 1.9 = ?$
$5 \times 1.99 = ?$
$5 \times 1.999 = ?$
$5 \times 1.9999 = ?$

Find the products. If the answer is too large for your calculator, use the pattern to predict the final product.

1. $2 \times 1.3 = ?$
$\quad 2 \times 1.33 = ?$
$\quad 2 \times 1.333 = ?$
$\quad 2 \times 1.33333333 = ?$

2. $6 \times 1.6 = ?$
$\quad 6 \times 1.66 = ?$
$\quad 6 \times 1.666 = ?$
$\quad 6 \times 1.666666666 = ?$

3. $40 \times 1.2 = ?$
$\quad 40 \times 1.22 = ?$
$\quad 40 \times 1.222 = ?$
$\quad 40 \times 1.2222222222 = ?$

4. $20 \times 1.5 = ?$
$\quad 20 \times 1.55 = ?$
$\quad 20 \times 1.555 = ?$
$\quad 20 \times 1.5555555555 = ?$

Skills Review

Skill 10 (page 35)

1. 216
×3

2. 47
×2

3. 847
×8

4. 628
×5

5. 1326
×4

6. 143
×9

7. 476
×2

8. 99
×9

9. 82
×7

10. 6179
×6

Skill 12 (page 44)

11. 26
×13

12. 87
×32

13. 419
×18

14. 632
×34

15. 76
×52

Skill 13 (page 45)

16. 6283
×327

17. 7192
×373

18. 4327
×141

19. 182
×157

20. 2436
×7531

Skill 14 (pages 46–47)

21. 728
×10

22. 309
×23

23. 473
×60

24. 1897
×305

25. 294
×508

26. 843
×200

27. 14
×400

28. 1030
×502

29. 6204
×709

30. 1086
×430

Skill 15 (pages 54–55)

31. 9.21
×3.2

32. 63.2
×7.2

33. 4.97
×3.8

34. 9.82
×6.2

35. 16.38
×4.7

36. 0.02
×0.3

37. 0.063
×0.08

38. 0.017
×0.32

39. 26.07
×0.19

40. 4.093
×0.4

Application A4 Complete the table. (pages 38–39)

	1.	2.	3.	4.	5.	6.
Inches	?	?			?	
Feet	3	?		?	?	?
Yards		5	?	?	20	?
Miles			4	5		10

Complete the statement.

7. 5 ft 8 in. = __?__ in.   **8.** 6 ft 2 in. = __?__ in.   **9.** 4 ft 11 in. = __?__ in.

Application A5 Complete the table. (pages 48–49)

	10.	11.	12.	13.	14.	15.
Ounces	?	?	?			
Pounds	4	7	4000	?	?	?
Tons			?	3	6	17

Complete the statement.

16. 9 lb 7 oz = __?__ oz   **17.** 5 lb 11 oz = __?__ oz   **18.** 4 t = __?__ lb

Application A6 Complete the table. (pages 56–57)

	19.	20.	21.	22.	23.	24.
Fluid Ounces	?	?		?		
Cups	2	?	?	?		?
Pints		5	?	?	?	?
Quarts			4	?	?	?
Gallons				1	5	25

Unit Test

1. 31
× 2

2. 212
× 4

3. 245
× 3

4. 416
× 6

Skill 11

5. 94
× 10

6. 527
× 500

Skill 12

7. 279
× 81

8. 414
× 67

Skill 13

9. 112
× 131

10. 6415
× 928

Skill 14

11. 654
× 903

12. 579
× 306

Skill 15

13. 1.15
× 2.3

14. 91.05
× 6.93

15. 0.0045
× 0.9

16. 0.0009
× 0.015

Application A4 **Complete the statement.**

17. 8 yards = __?__ feet

18. 2 miles = __?__ feet

19. 5 ft 9 in. = __?__ in.

20. 15 feet = __?__ inches

Application A5 **Complete the statement.**

21. 2 pounds = __?__ ounces

22. 5 tons = __?__ pounds

23. 1 lb 10 oz = __?__ oz

24. 8 lb 12 oz = __?__ oz

Application A6 **Complete the statement.**

25. 2 cups = __?__ fluid ounces

26. 4 quarts = __?__ pints

27. 3 pints = __?__ fluid ounces

28. 10 gallons = __?__ quarts

29. 2 gallons = __?__ pints

30. 5 pints = __?__ cups

3

Visitor Center — 1.6 km / 1.0 mi

→ Race Point — 3.7 km / 2.3 mi

↑ Herring Cove — 5.0 km / 3.1 mi

Division

Metric System

Problem Solving

Unit Preview

 You will use these skills and applications in this unit. Which do you already know? Work each problem.

Skill 16

1. $9\overline{)468}$ **2.** $6\overline{)480}$

Skill 17 Write remainders with R.

3. $5\overline{)386}$ **4.** $4\overline{)237}$

Skill 18 Write remainders with R.

5. $47\overline{)309}$ **6.** $30\overline{)6593}$ **7.** $121\overline{)5138}$ **8.** $452\overline{)78,345}$

Skill 19 Write remainders as fractions.

9. $8\overline{)905}$ **10.** $28\overline{)899}$ **11.** $49\overline{)919}$ **12.** $325\overline{)5517}$

Skill 20 Divide to hundredths. Round to the nearest tenth.

13. $22\overline{)208}$ **14.** $147\overline{)8073}$

Skill 21 Divide to hundredths. Round to the nearest tenth.

15. $0.09\overline{)6.42}$ **16.** $1.56\overline{)54.7}$

Application A7

17. Arrange in order from the longest to the shortest.
   63 km   0.112 km   70,123 m   2200 m

Application A8

18. How much longer is a piece of string that measures 60 cm than one that measures 522 mm?

Application A9

19. 105 g = __?__ kg **20.** 0.78 g = __?__ kg
21. 0.782 kg = __?__ g **22.** 23.5 kg = __?__ g

Application A10

23. Arrange in order from the smallest to the largest amount.
   98.3 mL   0.098 L   0.731 L   9813 mL

Check your answers. If you had difficulty with any skill or application, be sure to study the corresponding lesson in this unit.

Dividing Without Remainders

MODEL

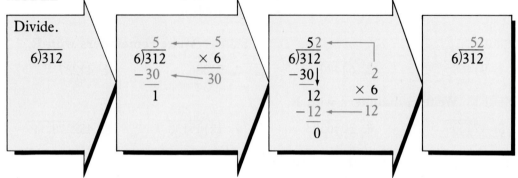

 GET READY.

1. 7)14   **2.** 6)30   **3.** 9)63   **4.** 5)25   **5.** 9)54

6. 8)80   **7.** 4)80   **8.** 2)40   **9.** 3)90   **10.** 5)50

 NOW USE SKILL 16.

11. 2)62   **12.** 5)55   **13.** 3)69   **14.** 2)48   **15.** 2)84

16. 3)123   **17.** 5)305   **18.** 7)420   **19.** 8)728   **20.** 4)240

21. 7)672   **22.** 8)592   **23.** 3)192   **24.** 6)468   **25.** 4)268

26. 5)630   **27.** 6)864   **28.** 3)345   **29.** 9)927   **30.** 2)552

31. 4)3480   **32.** 3)4215   **33.** 8)1088   **34.** 2)1470

35. 6)4686   **36.** 7)2506   **37.** 9)1422   **38.** 6)2664

For more practice: **Skill Bank,** page 412.

Dividing with Remainders

MODEL

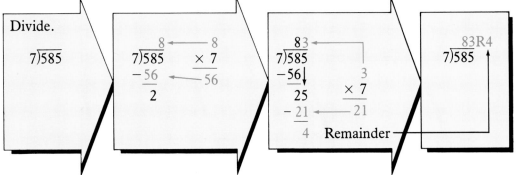

Divide.

$7\overline{)585}$

GET READY.

1. $8\overline{)54}$　　　**2.** $9\overline{)23}$　　　**3.** $6\overline{)28}$　　　**4.** $9\overline{)70}$　　　**5.** $4\overline{)34}$

6. $3\overline{)19}$　　　**7.** $8\overline{)33}$　　　**8.** $4\overline{)26}$　　　**9.** $5\overline{)34}$　　　**10.** $3\overline{)28}$

NOW USE SKILL 17.

11. $4\overline{)85}$　　　**12.** $2\overline{)29}$　　　**13.** $3\overline{)68}$　　　**14.** $7\overline{)73}$　　　**15.** $9\overline{)94}$

16. $8\overline{)169}$　　　**17.** $4\overline{)209}$　　　**18.** $2\overline{)149}$　　　**19.** $6\overline{)429}$　　　**20.** $5\overline{)157}$

21. $4\overline{)246}$　　　**22.** $9\overline{)880}$　　　**23.** $7\overline{)166}$　　　**24.** $8\overline{)691}$　　　**25.** $2\overline{)113}$

26. $8\overline{)506}$　　　**27.** $5\overline{)457}$　　　**28.** $7\overline{)593}$　　　**29.** $4\overline{)315}$　　　**30.** $6\overline{)569}$

31. $4\overline{)2470}$　　　**32.** $3\overline{)1645}$　　　**33.** $5\overline{)3623}$　　　**34.** $2\overline{)1879}$

35. $6\overline{)3287}$　　　**36.** $9\overline{)7527}$　　　**37.** $8\overline{)5395}$　　　**38.** $7\overline{)2480}$

For more practice: **Skill Bank,** page 412.

Dividing by 2- and 3-Digit Numbers

MODEL

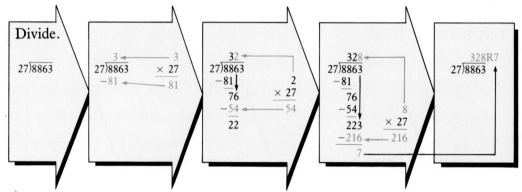

 GET READY.

1.	3 × 6	**2.**	4 × 2	**3.**	9 × 7	**4.**	6 × 8	**5.**	5 × 9	**6.**	8 × 4

7.	23 × 1	**8.**	14 × 7	**9.**	61 × 5	**10.**	72 × 9	**11.**	11 × 6	**12.**	56 × 4

13.	243 × 9	**14.**	427 × 2	**15.**	259 × 7	**16.**	638 × 4	**17.**	917 × 3	**18.**	172 × 5

NOW USE SKILL 18.

19. $23\overline{)88}$ **20.** $16\overline{)92}$ **21.** $34\overline{)72}$ **22.** $12\overline{)88}$ **23.** $27\overline{)69}$

24. $32\overline{)795}$ **25.** $23\overline{)604}$ **26.** $19\overline{)816}$ **27.** $68\overline{)753}$ **28.** $52\overline{)983}$

29. $78\overline{)326}$ **30.** $58\overline{)427}$ **31.** $42\overline{)376}$ **32.** $98\overline{)562}$ **33.** $23\overline{)186}$

34. $42\overline{)5237}$ **35.** $68\overline{)7926}$ **36.** $72\overline{)8092}$ **37.** $53\overline{)9206}$

38. $32\overline{)2168}$ **39.** $86\overline{)4863}$ **40.** $48\overline{)2657}$ **41.** $76\overline{)6089}$

42. 58$\overline{)4090}$ **43.** 21$\overline{)1906}$ **44.** 92$\overline{)5006}$ **45.** 65$\overline{)4308}$

MODEL

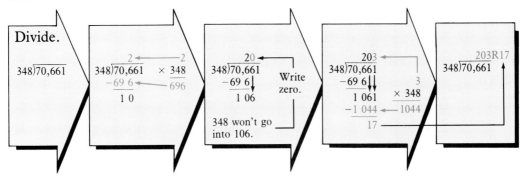

 GET READY.

46. 100$\overline{)500}$ **47.** 300$\overline{)600}$ **48.** 400$\overline{)800}$ **49.** 500$\overline{)500}$ **50.** 200$\overline{)600}$

51. 200$\overline{)8000}$ **52.** 300$\overline{)9000}$ **53.** 200$\overline{)4000}$ **54.** 300$\overline{)6000}$ **55.** 800$\overline{)8000}$

 NOW USE SKILL 18.

56. 168$\overline{)870}$ **57.** 301$\overline{)940}$ **58.** 250$\overline{)868}$ **59.** 192$\overline{)665}$

60. 481$\overline{)967}$ **61.** 742$\overline{)750}$ **62.** 362$\overline{)796}$ **63.** 641$\overline{)917}$

64. 241$\overline{)6849}$ **65.** 325$\overline{)7204}$ **66.** 416$\overline{)6509}$ **67.** 632$\overline{)9863}$

68. 802$\overline{)3210}$ **69.** 670$\overline{)4027}$ **70.** 754$\overline{)6789}$ **71.** 283$\overline{)1136}$

72. 395$\overline{)37,928}$ **73.** 210$\overline{)92,401}$ **74.** 614$\overline{)89,650}$

75. 824$\overline{)61,809}$ **76.** 509$\overline{)29,536}$ **77.** 732$\overline{)46,126}$

78. 578$\overline{)482,632}$ **79.** 118$\overline{)420,219}$ **80.** 873$\overline{)609,360}$

For more practice: **Skill Bank,** page 413.

Using Metric Units of Length

The kilometer (km) is a unit of length in the metric system. It is used to measure long distances. For shorter distances, use meters (m).

To change from one unit of length to another, use the metric equivalents shown below.

$$1 \text{ km} = 1000 \text{ m} \qquad 1 \text{ m} = 0.001 \text{ km}$$

MODEL

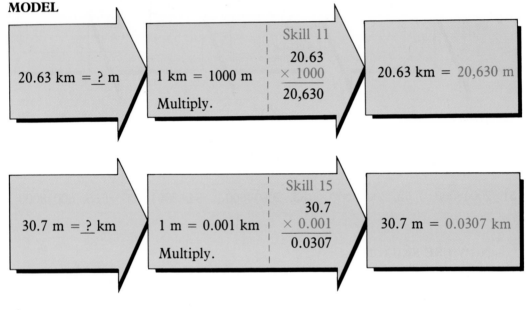

GET READY.

1.	2.	3.	4.	5.
0.3 × 1000	7.93 × 1000	40,000 × 0.001	25 × 0.001	12,476 × 0.001

Name the metric unit of length most commonly used to measure the following.

6. The distance from Chicago to New York

7. The length of a rope

8. The length of a city block

9. The distance in a marathon

10. The length of a highway

11. The height of a house

NOW USE APPLICATION A7. Complete the table.

	12.	13.	14.	15.	16.	17.	18.
km	0.01	7.56	5	0.9	30.86	150	0.0468
m	?	?	?	?	?	?	?

	19.	20.	21.	22.	23.	24.	25.
m	1500	693	14,896	7840	3430	170,680	4413
km	?	?	?	?	?	?	?

	26.	27.	28.	29.	30.	31.	32.
km	8.631	?	3.325	?	0.736	0.042	?
m	?	500	?	12,740	?	?	20,910

33. Which measurement is longer?

   a. 7600 m or 12.8 km   **b.** 7.43 km or 7450 m

34. Which measurement is shorter?

   a. 230 km or 1500 m   **b.** 17.3 km or 17.3 m

35. Arrange in order of length, starting with the longest.
84 km   1900 m   114,683 m   2.6 km

36. Arrange in order of length, starting with the shortest.
200 m   19 km   7670 m   4.43 km

37. A plane travels 150 meters per second. How many kilometers does it travel in 20 minutes?

38. Jan Tremont lives 2 km from her office. She rides her bike to and from work every day. How far does Jan ride in a 5-day work week?

39. There are 3 windows in the office. Each window requires 8.6 m of fabric for curtains. The cost of the fabric is $15.65 per meter. How much would it cost to purchase the needed fabric?

For more practice: **Application Bank,** page 451.

Meadville Marathon

You are organizing the Meadville Marathon, a 20 km race. To be an approved race, it must meet the requirements shown at the right. You have the names of 340 people who have volunteered to work at the race.

Required Stations
Medical tent every 1000 m and at finish
Water station every 250 m and at finish
Time checkpoint every 500 m and at finish
Crowd monitor every 200 m and at finish

PROBLEM: Do you have enough volunteers to assign to each station to meet marathon requirements?

Solution Plan: Find out what the marathon requirements are. Determine the number of stations of each type needed. Check the number of volunteers you have. Determine whether you have enough.

WHAT DO YOU KNOW? You know the marathon requirements. You know the length of the race. Answer the questions.

1. What is the length of the marathon?   2. What is this in meters?

3. What is the distance between stations?

    a. Medical tents   **b.** Water stations

    c. Crowd monitors   **d.** Time checkpoints

WHAT DO YOU WANT TO KNOW? You want to calculate the number of stations needed.

You need a medical tent every 1000 m. The length of the race is 20 km, or 20,000 m. You figure that you will need 20 medical tents in all.

> 20 km = 20,000 m   Application A7
> $$\frac{20 \text{ stations}}{1000\,\overline{)20{,}000}}$$ Skill 18

How many of each are needed?

4. Water stations   **5.** Crowd monitors   **6.** Time checkpoints

WHAT MORE DO YOU WANT? You want to determine whether you have enough volunteers to assign to the various stations.

You need 20 medical tents. You want to assign 3 volunteers to each medical tent. You figure that you will need 60 volunteers for the medical tents.

> ```
>    20   tents
>  × 3   volunteers per tent
>  ───
>    60   volunteers
> ```

7. You want one volunteer per time checkpoint. How many volunteers do you need?

8. You want two volunteers per water station. How many volunteers do you need?

9. You need one crowd monitor per station. How many volunteers will you need to work as crowd monitors?

NOW SOLVE THE PROBLEM.
Use the information you have found to answer the questions.

10. In all, how many volunteers do you need?

11. How many volunteers do you have now?

12. Do you have enough volunteers? If not, how many more will you have to find?

Skill 16

1. $7\overline{)252}$ **2.** $8\overline{)1728}$

Skill 17

3. $6\overline{)389}$ **4.** $6\overline{)2387}$

Skill 18

5. $57\overline{)873}$ **6.** $429\overline{)3206}$ **7.** $653\overline{)58,224}$

Application A7

8. Which is longer: 1500 m or 1.6 km?

Problem Solving

9. The Walkathon is 30 km long. You need a refreshment station every 2500 m and at the end. There will be 3 volunteers at each refreshment station. How many volunteers do you need?

CONSUMER NOTE

Leslie R. Duncan	5/13	40	217.85		217.85	14.49	36.20	8.95	6.85		66.49	151.36
Employee's Name	Period Ending	Time Worked	Regular Earnings	O.T.	Gross Earnings	F.I.C.A.	Fed. With.	State Tax	Med. Ins.		Total Deductions	Net Pay
Payee's Record of Earnings or Payments				**Earnings**				**Deductions**				

HMC Company
Gainsville, FL 32601

63-068
631

DATE _____

82868

PAY TO THE
ORDER OF ___LESLIE R. DUNCAN___ $ 151.36

___ONE HUNDRED FIFTY-ONE & 36/100 DOLLARS___

1 FIRST
CITY
BANK

TREASURER, HMC COMPANY

⑈0631⑈ ⑈0068⑈ ⑈0100 565851⑈

Leslie Duncan earns $217.85 per week. However, Leslie's *take-home pay* is less than that amount. Part of Leslie's pay is withheld for social security (FICA), federal and state taxes, and medical insurance.

$$\text{Earnings} - \text{Deductions} = \text{Take-Home Pay}$$

$$\$217.85 - \$66.49 = \$151.36$$

Find the take-home pay: Earnings $338.85; Federal tax $54.70; State tax $16.94; FICA $23.53; Union dues $5.50; Take-home pay ___?___.

Temperature Scales

There are different scales to measure temperature. The two most common scales are the Fahrenheit and the Celsius.

The Metric System uses the Celsius scale. On this scale water boils at 100°C and freezes at 0°C.

The Customary System uses the Fahrenheit scale. On this scale water boils at 212°F and freezes at 32°F.

Use the thermometer to answer the following questions.

What is the corresponding temperature?

1. 5°C = __?__ °F   **2.** 140°F = __?__ °C

3. 212°F = __?__ °C   **4.** 68°F = __?__ °C

5. 0°C = __?__ °F   **6.** 32°C = __?__ °F

Which temperature is higher?

7. 150°F, 70°C   **8.** 80°F, 20°C

9. 100°F, 100°C   **10.** 145°F, 65°C

List from the lowest to the highest temperature.

11. 100°F, 20°C, 60°C, 80°F   **12.** 40°F, 90°C, 90°F, 75°C

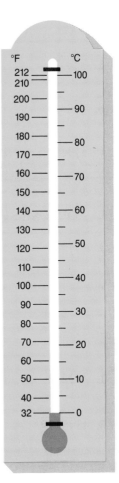

75

Skill 19

Dividing with Fraction Remainders

MODEL

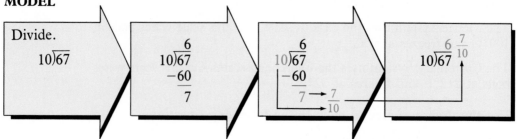

 GET READY. Write remainders as fractions.

1. $5\overline{)6}$ **2.** $3\overline{)8}$ **3.** $4\overline{)9}$ **4.** $7\overline{)9}$ **5.** $2\overline{)7}$

6. $7\overline{)44}$ **7.** $9\overline{)75}$ **8.** $8\overline{)20}$ **9.** $3\overline{)16}$ **10.** $4\overline{)75}$

11. $5\overline{)341}$ **12.** $2\overline{)301}$ **13.** $7\overline{)540}$ **14.** $6\overline{)335}$ **15.** $4\overline{)381}$

 NOW USE SKILL 19. Write remainders as fractions.

16. $40\overline{)92}$ **17.** $31\overline{)83}$ **18.** $26\overline{)94}$ **19.** $69\overline{)90}$ **20.** $42\overline{)76}$

21. $40\overline{)96}$ **22.** $23\overline{)87}$ **23.** $19\overline{)97}$ **24.** $54\overline{)57}$ **25.** $19\overline{)59}$

26. $26\overline{)934}$ **27.** $34\overline{)750}$ **28.** $60\overline{)659}$ **29.** $54\overline{)909}$ **30.** $51\overline{)317}$

31. $73\overline{)222}$ **32.** $61\overline{)318}$ **33.** $32\overline{)740}$ **34.** $59\overline{)653}$ **35.** $30\overline{)408}$

36. $45\overline{)5099}$ **37.** $51\overline{)4310}$ **38.** $31\overline{)6922}$ **39.** $84\overline{)4906}$

40. $78\overline{)9709}$ **41.** $89\overline{)8993}$ **42.** $66\overline{)4900}$ **43.** $30\overline{)9206}$

MODEL

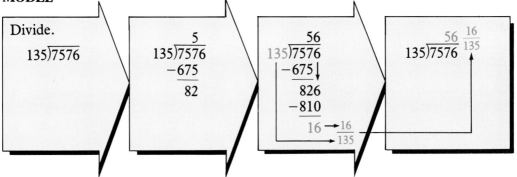

Divide.

135)7576

$$\begin{array}{r} 5 \\ 135\overline{)7576} \\ -675 \\ \hline 82 \end{array}$$

$$\begin{array}{r} 56 \\ 135\overline{)7576} \\ -675\downarrow \\ \hline 826 \\ -810 \\ \hline 16 \end{array}$$

$$135\overline{)7576}\;\; 56\frac{16}{135}$$

 GET READY. Write remainders as fractions.

44. 100)501 **45.** 100)631 **46.** 400)813 **47.** 300)973

48. 120)247 **49.** 222)445 **50.** 160)323 **51.** 250)577

52. 160)1603 **53.** 500)1067 **54.** 350)3511 **55.** 423)8466

 NOW USE SKILL 19. Write remainders as fractions.

56. 527)844 **57.** 107)600 **58.** 352)809 **59.** 744)894

60. 300)2709 **61.** 803)6430 **62.** 540)4323 **63.** 395)3557

64. 946)7573 **65.** 391)2776 **66.** 191)7700 **67.** 764)2300

68. 475)9520 **69.** 832)9160 **70.** 427)9488 **71.** 847)6779

72. 216)13,397 **73.** 541)17,327 **74.** 753)93,409

75. 231)15,822 **76.** 142)30,561 **77.** 323)69,472

For more practice: **Skill Bank,** page 413.

Skill 20

Dividing with Decimal Remainders

MODEL

| Divide to hundredths. Round answer to the nearest tenth. $63\overline{)225}$ | Align the decimal points. Write two zeros after the decimal. $63\overline{)225.00}$ | 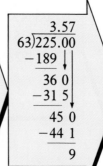 | To the nearest tenth, the answer is 3.6. Skill 2 |

 GET READY. Round to the nearest tenth.

1. 4.06 **2.** 7.34 **3.** 2.59 **4.** 11.42 **5.** 20.89

6. 7.26 **7.** 9.66 **8.** 3.14 **9.** 2.23 **10.** 8.16

 NOW USE SKILL 20. Divide to hundredths. Round to the nearest tenth.

11. $8\overline{)58}$ **12.** $4\overline{)11}$ **13.** $2\overline{)27}$ **14.** $6\overline{)63}$ **15.** $9\overline{)74}$

16. $3\overline{)259}$ **17.** $5\overline{)433}$ **18.** $4\overline{)181}$ **19.** $6\overline{)299}$ **20.** $4\overline{)145}$

21. $25\overline{)412}$ **22.** $18\overline{)642}$ **23.** $79\overline{)932}$ **24.** $42\overline{)869}$ **25.** $41\overline{)789}$

26. $64\overline{)408}$ **27.** $25\overline{)205}$ **28.** $48\overline{)210}$ **29.** $32\overline{)148}$ **30.** $78\overline{)529}$

31. $26\overline{)137}$ **32.** $58\overline{)541}$ **33.** $63\overline{)600}$ **34.** $74\overline{)379}$ **35.** $94\overline{)787}$

36. $86\overline{)2640}$ **37.** $68\overline{)1708}$ **38.** $22\overline{)1780}$ **39.** $37\overline{)1867}$

40. $21\overline{)2055}$ **41.** $42\overline{)4025}$ **42.** $81\overline{)1200}$ **43.** $49\overline{)3085}$

MODEL

| Divide to thousandths. Round answer to the nearest hundredth.

34)54 | Align the decimal points. Write three zeros after the decimal.

34)54.000 | 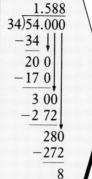 | To the nearest hundredth, the answer is 1.59.

Skill 2 |

 GET READY. Round to the nearest hundredth.

44. 9.853 **45.** 6.529 **46.** 1.807 **47.** 1.796 **48.** 4.363

 NOW USE SKILL 20. Divide to thousandths. Round to the nearest hundredth.

49. 7)9 **50.** 3)7 **51.** 8)9 **52.** 6)1 **53.** 3)4

54. 8)29 **55.** 7)83 **56.** 9)84 **57.** 6)43 **58.** 8)59

59. 34)90 **60.** 67)71 **61.** 82)36 **62.** 49)36 **63.** 96)99

64. 23)140 **65.** 15)484 **66.** 37)308 **67.** 29)179

68. 56)513 **69.** 83)358 **70.** 67)389 **71.** 93)589

72. 317)315 **73.** 487)935 **74.** 732)598 **75.** 647)739

76. 231)1391 **77.** 649)1951 **78.** 598)2403 **79.** 376)3218

80. 740)2930 **81.** 486)2517 **82.** 869)6161 **83.** 548)3704

For more practice: **Skill Bank,** page 414.

Using Metric Units of Length

To measure short lengths, use centimeters (cm) or millimeters (mm).
The thickness of a dime is about a millimeter.

To change from one unit of length to another, use the metric equivalents
shown below.

$$100 \text{ cm} = 1 \text{ m} \qquad 1000 \text{ mm} = 1 \text{ m}$$
$$1 \text{ cm} = 10 \text{ mm} \qquad 1 \text{ mm} = 0.1 \text{ cm}$$
$$1 \text{ cm} = 0.01 \text{ m} \qquad 1 \text{ mm} = 0.001 \text{ m}$$

MODEL

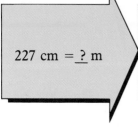

227 cm = _?_ m

1 cm = 0.01 m

Multiply.

Skill 15

$$\begin{array}{r} 227 \\ \times\ 0.01 \\ \hline 2.27 \end{array}$$

227 cm = 2.27 m

8.03 m = _?_ cm

1 m = 100 cm

Multiply.

Skill 11

$$\begin{array}{r} 8.03 \\ \times\ 100 \\ \hline 803 \end{array}$$

8.03 m = 803 cm

 GET READY.

1.	0.478	2.	9456	3.	77.48	4.	2.391	5.	87.12
	× 100		× 0.01		× 0.1		× 1000		× 10

 NOW USE APPLICATION A8. Complete the table.

	6.	**7.**	**8.**	**9.**	**10.**
m	6	4	7.3	0.25	0.08
cm	?	?	?	?	?

	11.	**12.**	**13.**	**14.**	**15.**
mm	23	70	18	123	4.3
cm	?	?	?	?	?

	16.	17.	18.	19.	20.
cm	325	603	250	42	101.2
m	?	?	?	?	?

	21.	22.	23.	24.	25.
cm	2.5	0.7	19.2	50	76
mm	?	?	?	?	?

	26.	27.	28.	29.	30.	31.	32.
mm	1	?	?	?	176.25	?	7
cm	?	750	?	4389	?	?	?
m	?	?	0.927	?	?	0.0367	?

Use this metric ruler to answer the following questions.

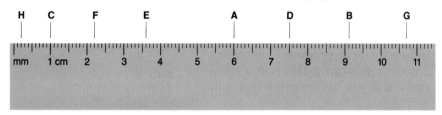

33. How many centimeters are indicated by the following letters?

    a. E   **b.** C   **c.** H   **d.** G   **e.** A

34. How many millimeters are indicated by the following letters?

    a. D   **b.** F   **c.** B   **d.** E   **e.** H

35. Measure.

    a. __?__ mm   **b.** __?__ mm   **c.** __?__ mm

36. Which measurement is more reasonable?

    Car key: 50 mm or 50 m?   Paper clip: 3 m or 3 cm?

    Your height: 180 cm or 180 mm?   Fabric: 5 m or 5 mm?

37. How much longer is a stick that measures 13.4 cm than one that measures 27.60 mm?

38. If the thickness of a dime is about one millimeter, how thick is a stack of 25 dimes?

For more practice: **Application Bank**, page 451.

Norwood Office Suppliers

You are the supervisor for Norwood Office Suppliers. You have a request for 2500 boxes of small paper clips, 1500 boxes of jumbo clips, and 1000 boxes of butterfly clips to restock inventory. The amount of wire needed to produce each size clip is shown at the right.

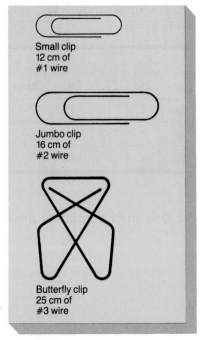

Small clip
12 cm of
#1 wire

Jumbo clip
16 cm of
#2 wire

Butterfly clip
25 cm of
#3 wire

PROBLEM: How many spools of wire will you need to produce the clips?

Solution Plan: You know the amounts to produce. Take the number of clips per box. Compute the total number of clips of each size to produce. Determine the number of clips that you can make from a spool of wire. Then find out if you have enough spools of wire.

WHAT DO YOU KNOW? Use the information above to answer the questions.

1. How many boxes of each size of paper clips are to be produced?
   a. Small clips   **b.** Jumbo clips   **c.** Butterfly clips

2. What length of wire is needed to manufacture each size of clip?
   a. Small clip   **b.** Jumbo clip   **c.** Butterfly clip

WHAT DO YOU WANT TO KNOW? You want to find the total number of clips that you must manufacture. Complete the table.

		Small	Jumbo	Butterfly
3.	**Boxes of Clips**	?	?	?
	Number of Clips per Box	100	100	50
4.	**Total Number to Produce**	?	?	?

WHAT MORE DO YOU WANT? Find the number of clips that can be produced from a spool of wire.

A spool of #1 wire has 500 m, or 50,000 cm, of wire. You estimate that you can make 5000 small clips from one spool of wire. You figure that you can actually make 4166 small clips from one spool, with 8 cm of wire wasted.

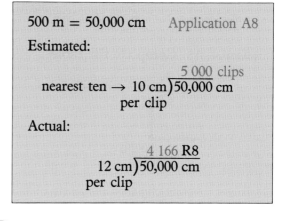

500 m = 50,000 cm   Application A8

Estimated:

$$\begin{array}{r} 5\ 000 \text{ clips} \\ \text{nearest ten} \rightarrow 10 \text{ cm} \overline{)50,000 \text{ cm}} \\ \text{per clip} \end{array}$$

Actual:

$$\begin{array}{r} 4\ 166 \textbf{ R8} \\ 12 \text{ cm} \overline{)50,000 \text{ cm}} \\ \text{per clip} \end{array}$$

5. A spool of #2 wire has 450 m of wire. How many centimeters of wire is this?

6. How many jumbo paper clips can be produced from one spool of #2 wire?
   a. estimated   **b.** actual

7. A spool of #3 wire has 400 m of wire. How many centimeters of wire is this?

8. How many butterfly clips can be produced from one spool of #3 wire?
   a. estimated   **b.** actual

NOW SOLVE THE PROBLEM. Calculate the number of spools of wire needed to produce the clips.

One spool of #1 wire produces 4166 small paper clips. You want to produce 250,000 small clips. You have 60 spools of #1 wire. This is not enough.

$$\begin{array}{r} 4166 \text{ clips per spool} \\ \times\ 60 \text{ spools} \\ \hline 249,960 \text{ clips} \end{array}$$

Skill 11

9. You have 55 spools of #2 wire. Is this enough to produce the jumbo clips needed?

10. You have 30 spools of #3 wire. Is this enough to produce the butterfly clips needed?

Quiz

Skill 19 **Divide. Write remainders as fractions.**

1. $73 \overline{)5306}$ **2.** $651 \overline{)51,217}$

Skill 20 **Divide to hundredths. Round to the nearest tenth.**

3. $75 \overline{)4328}$ **4.** $376 \overline{)4229}$

Application A8 **Complete the tables.**

     5. **6.** **7.**

	5.	6.	7.
m	4	?	7.2
cm	?	250	?

     8. **9.** **10.**

	8.	9.	10.
mm	240	?	?
cm	?	12.3	25

Problem Solving

11. A heavyweight clip requires 18 cm of #4 wire. A spool of #4 wire has 350 m of wire. How many heavyweight clips can be produced from a spool of #4 wire?

EXTRA!

Chronological Expressions

Chronological expressions tell you the number of times an event occurs during a given period. The table on the right shows some common expressions.

Common Chronological Expressions (Number of occurrences per year)		
	weekly (52)	biweekly (26)
semimonthly (24)	monthly (12)	bimonthly (6)
	quarterly (4)	
semiannually (2)	annually (1)	

Christina Bemko has received weekly paychecks from her employer for the past 12 years. How many paychecks has she received so far?

Number of times per year × Number of years = Number of paychecks

         52   ×   12   =   624

Christina has received a total of 624 paychecks.

Use the table to find the number of occurrences.

1. Annually for 13 years   **2.** Bimonthly for 3 years
3. Biweekly for 6 years   **4.** Weekly for 2 years
5. Quarterly for 9 years   **6.** Semimonthly for $1\frac{1}{2}$ years
7. Monthly for 27 years   **8.** Semiannually for $3\frac{1}{2}$ years

Fractional Remainders

The calculator expresses the remainder in division as a decimal. You can, however, use the calculator to express the remainder as a fraction. For example, divide 359 by 24.

KEY-IN	**DISPLAY**
359	*359*
$\div$	*359*
24	*24*
$=$	*14.958333* ← The answer shows a decimal remainder.

From the calculator display, you know that the whole number part of the answer is 14. Now use the calculator to find the remainder.

KEY-IN   **DISPLAY**

Multiply the whole number part of the answer by the divisor.

	KEY-IN	DISPLAY	
	14	*14*	
	$\times$	*14*	14
	24	*24*	$24\overline{)359}$
	$=$	*336* $\longrightarrow$	336

Subtract the product from 359.

	KEY-IN	DISPLAY	
	359	*359*	14
	$-$	*359*	$24\overline{)359}$
	336	*336*	336
	$=$	*23* $\rightarrow$ Remainder $\rightarrow$	23

Write the answer in fractional form: $14\frac{23}{24}$.

Divide. Write the remainder as a fraction.

1. $9\overline{)302}$ 2. $7\overline{)521}$

3. $6\overline{)413}$ 4. $23\overline{)645}$

5. $27\overline{)829}$ 6. $49\overline{)2165}$

7. $38\overline{)8742}$ 8. $366\overline{)7895}$

Dividing Decimals

MODEL

Divide to hundredths. Round answer to the nearest tenth.

$$3.51\overline{)9.2}$$

Change divisor to a whole number. Multiply by 100.

$$3.51\overline{)9.20}$$

Move two decimal places. Add a zero.

$$351\overline{)920}$$

Align the decimal points. Write two zeros after the decimal point.

$$351\overline{)920.00}$$

$$\begin{array}{r} 2.62 \\ 351\overline{)920.00} \\ -702 \\ \hline 218\ 0 \\ -210\ 6 \\ \hline 7\ 40 \\ -7\ 02 \\ \hline 38 \end{array}$$

To the nearest tenth the answer is 2.6.

Skill 2

 GET READY.

1. $\begin{array}{r} 7.4 \\ \times\ 10 \\ \hline \end{array}$
2. $\begin{array}{r} 3.6 \\ \times\ 10 \\ \hline \end{array}$
3. $\begin{array}{r} 0.18 \\ \times\ 10 \\ \hline \end{array}$
4. $\begin{array}{r} 71 \\ \times\ 10 \\ \hline \end{array}$
5. $\begin{array}{r} 0.04 \\ \times\ 10 \\ \hline \end{array}$
6. $\begin{array}{r} 36 \\ \times\ 10 \\ \hline \end{array}$

7. $\begin{array}{r} 9.4 \\ \times\ 100 \\ \hline \end{array}$
8. $\begin{array}{r} 7.93 \\ \times\ 100 \\ \hline \end{array}$
9. $\begin{array}{r} 0.91 \\ \times\ 100 \\ \hline \end{array}$
10. $\begin{array}{r} 0.045 \\ \times\ 100 \\ \hline \end{array}$
11. $\begin{array}{r} 49.7 \\ \times\ 100 \\ \hline \end{array}$

Round to the nearest tenth.

12. 0.37　　13. 7.08　　14. 5.84　　15. 41.52　　16. 3.96

 NOW USE SKILL 21. Divide to hundredths. Round to the nearest tenth.

17. $3.1\overline{)2.48}$　　18. $1.2\overline{)8.29}$　　19. $6.3\overline{)8.56}$　　20. $4.9\overline{)40.14}$

21. $3.2\overline{)7.89}$　　22. $4.1\overline{)5.697}$　　23. $7.2\overline{)38.88}$　　24. $6.1\overline{)55.39}$

25. $2.95\overline{)6.98}$　　26. $5.26\overline{)74.6}$　　27. $3.82\overline{)55.9}$　　28. $2.73\overline{)6.103}$

29. $4.62\overline{)9.27}$　　30. $1.85\overline{)37.94}$　　31. $3.19\overline{)29.75}$　　32. $4.35\overline{)67.88}$

33. $18.2\overline{)3.735}$　　34. $9.33\overline{)81.4}$　　35. $1.104\overline{)12.68}$　　36. $38.7\overline{)201.35}$

MODEL

| Divide to thousandths. Round answer to the nearest hundredth. $0.19\overline{)0.066}$ | Change divisor to a whole number. Multiply by 100. $0.19\overline{)0.06.6}$ Move two decimal places. | $19\overline{)6.6}$ Align the decimal points. Write two zeros as place holders. $19\overline{)6.600}$ | $\begin{array}{r} 0.347 \\ 19\overline{)6.600} \\ -57 \\ \hline 90 \\ -76 \\ \hline 140 \\ -133 \\ \hline 7 \end{array}$ | To the nearest hundredth the answer is 0.35. Skill 2 |

 GET READY.

37. $\begin{array}{r} 0.802 \\ \times\ 100 \\ \hline \end{array}$

38. $\begin{array}{r} 2.386 \\ \times\ 100 \\ \hline \end{array}$

39. $\begin{array}{r} 0.743 \\ \times\ 100 \\ \hline \end{array}$

40. $\begin{array}{r} 0.8468 \\ \times\ 100 \\ \hline \end{array}$

41. $\begin{array}{r} 0.018 \\ \times\ 1000 \\ \hline \end{array}$

42. $\begin{array}{r} 0.5807 \\ \times\ 1000 \\ \hline \end{array}$

43. $\begin{array}{r} 3.21 \\ \times\ 1000 \\ \hline \end{array}$

44. $\begin{array}{r} 77.47 \\ \times\ 1000 \\ \hline \end{array}$

NOW USE SKILL 21. Divide to hundredths. Round to the nearest tenth.

45. $0.6\overline{)1.8}$

46. $0.4\overline{)2.08}$

47. $0.7\overline{)5.07}$

48. $0.8\overline{)2.56}$

49. $0.02\overline{)26}$

50. $0.08\overline{)96}$

51. $0.06\overline{)72}$

52. $0.03\overline{)45}$

53. $0.038\overline{)0.8132}$

54. $2.41\overline{)0.7807}$

55. $0.068\overline{)16.96}$

56. $0.526\overline{)74.983}$

57. $1.15\overline{)0.3707}$

58. $0.392\overline{)69.032}$

Divide to thousandths. Round to the nearest hundredth.

59. $0.3\overline{)2.2}$

60. $0.9\overline{)3.04}$

61. $0.07\overline{)1.5}$

62. $0.38\overline{)0.02}$

63. $0.76\overline{)1.83}$

64. $1.08\overline{)0.027}$

65. $0.42\overline{)2.839}$

66. $0.26\overline{)17.8425}$

For more practice: **Skill Bank,** page 414.

Using Metric Units of Mass

The kilogram (kg) and gram (g) are units of mass in the metric system. For larger objects use kilograms. For smaller objects use grams. To change from one unit of mass to another, use the metric equivalents shown below.

$$1 \text{ kg} = 1000 \text{ g} \qquad 1 \text{ g} = 0.001 \text{ kg}$$

MODEL

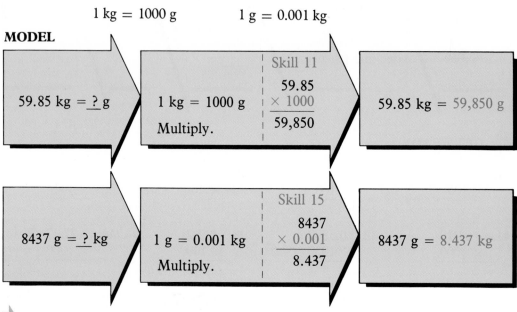

		Skill 11	
59.85 kg = _?_ g	1 kg = 1000 g Multiply.	59.85 × 1000 59,850	59.85 kg = 59,850 g

		Skill 15	
8437 g = _?_ kg	1 g = 0.001 kg Multiply.	8437 × 0.001 8.437	8437 g = 8.437 kg

NOW USE APPLICATION A9.

Which is more?

1. A 200 g box or a 0.4 kg box
2. A 700 kg crate or an 800,000 g crate
3. A 0.5 kg baseball bat or a 550 g bat
4. A 0.7 kg book or a 900 g book

Complete the statement.

5. 75 kg = _?_ g
6. 17,289 g = _?_ kg
7. 6.398 kg = _?_ g
8. 9783 g = _?_ kg

Complete the table.

	9.	10.	11.	12.	13.	14.
g	?	296	?	?	18,374	77
kg	55.6	?	0.49	1.763	?	?

For more practice: **Application Bank,** page 451.

Using Metric Units of Capacity

The liter (L) and the milliliter (mL) are units of capacity in the metric
system. For smaller amounts, use milliliters. For larger amounts, use
liters. A drop of liquid from an eye dropper is about one milliliter. To
change from one unit of capacity to another, use the metric equivalents
shown below.

$$1 \text{ L} = 1000 \text{ mL} \qquad 1 \text{ mL} = 0.001 \text{ L}$$

MODEL

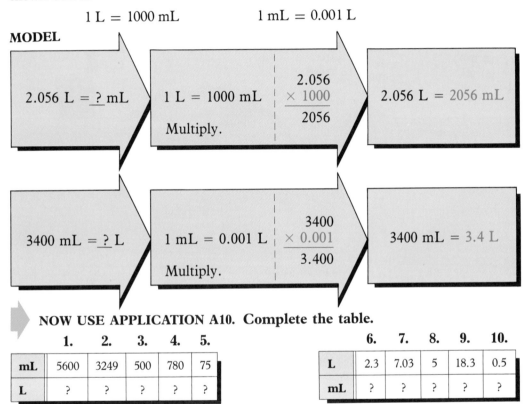

| 2.056 L = ? mL | 1 L = 1000 mL Multiply. | $\begin{array}{r} 2.056 \\ \times\ 1000 \\ \hline 2056 \end{array}$ | 2.056 L = 2056 mL |

| 3400 mL = ? L | 1 mL = 0.001 L Multiply. | $\begin{array}{r} 3400 \\ \times\ 0.001 \\ \hline 3.400 \end{array}$ | 3400 mL = 3.4 L |

▶ **NOW USE APPLICATION A10. Complete the table.**

	1.	**2.**	**3.**	**4.**	**5.**
mL	5600	3249	500	780	75
L	?	?	?	?	?

	6.	**7.**	**8.**	**9.**	**10.**
L	2.3	7.03	5	18.3	0.5
mL	?	?	?	?	?

11. Arrange in order from the largest to the smallest amount.
   67.8 mL   0.483 L   15 L   14,360 mL

12. Arrange in order from the smallest to the largest amount.
   100 mL   4.3 L   2000 mL   0.368 L

13. Bottled water costs $7.50 per 5-L bottle. If you average 3 bottles a
   month, how much is your cost per year?

14. A recipe calls for 5 mL of almond extract, 15 mL of water, and
   0.25 L of milk. How many milliliters of liquids do you need in all?

For more practice: **Application Bank,** page 451.

Augusta Bridge Club

ZESTY CHICKEN (4 servings)

70 g	butter or margarine
1.5 kg	chicken breast
125 mL	dark corn syrup
40 g	raisins
10 mL	grated lemon peel
15 mL	lemon juice
20 mL	seasoned salt
	pepper to taste

You are a member of the Augusta Bridge Club. Your club is having a banquet to honor those members who have earned 100 master points. There will be 36 club members at the banquet. You plan to use a special recipe you got while traveling overseas last summer.

PROBLEM: Prepare a shopping list of the ingredients you need to buy, and find out how much they will cost.

Solution Plan: Determine from the recipe how much of each ingredient is needed for four servings. Calculate how much will be needed for 36 servings. Find how many packages of each ingredient you will need. Figure the cost.

WHAT DO YOU KNOW? The recipe shows the amount needed for each ingredient. Answer the questions.

1. How many people does the recipe serve?

2. How many people do you plan to serve at the banquet?

3. How much of each main ingredient does the recipe call for?

    a. margarine   **b.** chicken   **c.** corn syrup   **d.** raisins

WHAT DO YOU WANT TO KNOW? You want to find how much you need for 36 people.

The recipe serves 4 people, and you want to serve 36. You find the recipe must be increased 9 times.

Since the recipe calls for 70 g of butter or margarine, you will need 630 g altogether.

$$\begin{array}{r} 9 \\ 4\overline{)36} \end{array}$$

Skill 16

$$\begin{array}{r} 70 \text{ g for } 4 \\ \times\ 9 \\ \hline 630 \text{ g for } 36 \end{array}$$

Skill 10

Complete the table.

	Butter or margarine	4. Chicken	5. Corn syrup	6. Raisins	7. Lemon peel	8. Lemon juice	9. Seasoned salt
Amount for 4	70 g	?	?	?	?	?	?
Amount for 36	630 g	?	?	?	?	?	?

WHAT MORE DO YOU WANT TO KNOW? You want to know how many packages or containers to buy of each ingredient.

After shopping and checking ads, you find the package sizes and prices of the items.

Margarine: 450-g box for $.89
Corn syrup: 400-mL bottle for $.79
Lemons: $.18 each (a lemon yields 50 mL of peel and 150 mL of juice)

Frozen chicken breasts: 2-kg box for $4.49
Raisins: 400-g box for $1.39
Seasoned salt: 100-mL jar for $.89

You need 90 mL of lemon peel and 135 mL of lemon juice. Since one lemon yields about 50 mL of peel, you will need two lemons. From one of the lemons, you will get enough juice for the recipe.

$$\begin{array}{r} 1.8 \\ 50\overline{)90.0} \end{array}$$

Skill 20

How many packages or containers do you buy of each ingredient?

10. margarine **11.** chicken **12.** corn syrup

13. raisins **14.** seasoned salt

NOW SOLVE THE PROBLEM. Use the information you have found to answer the questions.

How much do you spend on each ingredient?

15. margarine **16.** chicken **17.** corn syrup

18. raisins **19.** seasoned salt **20.** lemons

21. How much will the ingredients cost altogether?

Quiz

Skill 21 **Divide to hundredths. Round to the nearest tenth.**

1. $5.7\overline{)12.86}$ **2.** $4.13\overline{)8.425}$ **3.** $0.3\overline{)3.83}$ **4.** $2.03\overline{)0.485}$

Application A9

5. Complete the statement:
$13{,}725\text{ g} = \underline{\ ?\ }\text{ kg}$

6. Which is more, a 2.1 kg melon or a 2150 g melon?

Application A10

7. Arrange in order from the smallest to the largest amount.
2000 mL   0.42 L   2.13 L   385 mL

Problem Solving

Complete the table.

	8. Ground beef	**9.** Dried beans	**10.** Tomatoes	**11.** Chili powder
Amount for 8	500 g	250 g	650 g	30 mL
Amount for 40	?	?	?	?

CONSUMER NOTE

Miles Per Gallon

To find your car's fuel efficiency rating, calculate the number of miles traveled per gallon of gasoline. First, fill up the tank and record the mileage. The next time you fill your tank, record both the number of gallons of gasoline and the new mileage. Subtract the two mileage readings to find the number of miles traveled. Then divide:

$$\frac{\text{Miles traveled}}{\text{Gallons of gas used}} = \text{Miles per gallon (MPG)}$$

Find the MPG rating for each car. Round to the nearest tenth.

1. Miles traveled   423
     Gallons of gas used   18

2. Miles traveled   143
     Gallons of gas used   9.2

3. Miles traveled   325
     Gallons of gas used   12

4. Miles traveled   540
     Gallons of gas used   16

Skills Review

Skill 16 (page 66)

1. $6\overline{)96}$ 2. $8\overline{)312}$ 3. $5\overline{)320}$ 4. $7\overline{)1659}$ 5. $9\overline{)7776}$

Skill 18 **Write remainders with R.** (pages 68–69)

6. $38\overline{)896}$ 7. $49\overline{)522}$ 8. $26\overline{)4890}$ 9. $82\overline{)9403}$ 10. $52\overline{)8399}$

11. $603\overline{)8976}$ 12. $491\overline{)8902}$ 13. $206\overline{)1890}$ 14. $899\overline{)5475}$ 15. $730\overline{)6075}$

Skill 19 **Write the remainder as a fraction.** (pages 76–77)

16. $26\overline{)470}$ 17. $82\overline{)908}$ 18. $108\overline{)832}$ 19. $492\overline{)986}$

20. $391\overline{)842}$ 21. $270\overline{)861}$ 22. $797\overline{)6470}$ 23. $395\overline{)4900}$

Skill 20 **Divide to hundredths. Round to the nearest tenth.** (pages 78–79)

24. $24\overline{)183}$ 25. $47\overline{)396}$ 26. $63\overline{)3096}$ 27. $44\overline{)6002}$

Divide to thousandths. Round to the nearest hundredth.

28. $385\overline{)854}$ 29. $203\overline{)980}$ 30. $732\overline{)1146}$ 31. $508\overline{)8050}$

Skill 21 **Divide to hundredths. Round to the nearest tenth.** (pages 86–87)

32. $4.2\overline{)30.66}$ 33. $8.1\overline{)81.81}$ 34. $9.3\overline{)9.88}$ 35. $3.19\overline{)4.37}$

36. $0.06\overline{)24}$ 37. $0.09\overline{)54}$ 38. $0.05\overline{)20}$ 39. $0.15\overline{)90}$

Divide to thousandths. Round to the nearest hundredth.

40. $6.42\overline{)102.9}$ 41. $3.4\overline{)282.35}$ 42. $10.6\overline{)100.2}$ 43. $2.51\overline{)0.061}$

44. $0.05\overline{)66861}$ 45. $0.073\overline{)83.001}$ 46. $0.032\overline{)29.344}$ 47. $0.021\overline{)15.309}$

Applications Review

Application A7 **Complete the table.** (pages 70–71)

	1.	2.	3.	4.	5.	6.	7.
km	0.951	67.04	75	20.98	0.02	250	0.398
m	?	?	?	?	?	?	?

	8.	9.	10.	11.	12.	13.	14.
m	1962	325	25.7	900	5321	6200	1
km	?	?	?	?	?	?	?

15. Arrange in order from the longest to the shortest.

   19 km   23,323 m   9.6 km   3572 m

Application A8 **Complete the table.** (pages 80–81)

	16.	17.	18.	19.	20.	21.	22.
mm	3	?	?	?	235.14	?	?
cm	?	6	?	7352	?	5.81	?
m	?	?	0.831	?	?	?	0.006

23. Wen-Ying says her height is 170 cm. Orlando says his height is 1.6 m. Who is taller?

Application A9 **Complete the statement.** (page 88)

24. 95.7 kg = __?__ g   **25.** 0.007 kg = __?__ g

26. 52.8 kg = __?__ g   **27.** 65,727 g = __?__ kg

28. 732 g = __?__ kg   **29.** 1900 g = __?__ kg

Application A10 **Complete the table.** (page 89)

	30.	31.	32.	33.	34.	35.	36.	37.	38.
mL	7500	1300	?	?	3209	?	350.4	?	?
L	?	?	2.7	0.7	?	0.68	?	0.3751	8.893

39. You are refilling 50-mL bottles of glue from a 0.5-L bottle. How many 50-mL bottles can you fill?

Unit Test

Skill 16

1. $2\overline{)22}$ **2.** $7\overline{)392}$

Skill 17 Write remainders with R.

3. $4\overline{)138}$ **4.** $3\overline{)475}$

Skill 18 Write remainders with R.

5. $13\overline{)8777}$ **6.** $81\overline{)8753}$ **7.** $521\overline{)9317}$ **8.** $137\overline{)8077}$

Skill 19 Write remainders as fractions.

9. $23\overline{)786}$ **10.** $25\overline{)633}$ **11.** $143\overline{)721}$ **12.** $467\overline{)6186}$

Skill 20 Divide to hundredths. Round to the nearest tenth.

13. $55\overline{)443}$ **14.** $271\overline{)588}$

Skill 21 Divide to hundredths. Round to the nearest tenth.

15. $8.1\overline{)6.42}$ **16.** $0.062\overline{)2.71}$

Application A7 Which is longer?

17. 9.01 km or 888 m

18. 63.01 m or 0.666 km

19. Judy Wood drives her automobile at 50 kilometers per hour for 3.5 hours. Her destination is 240 km away from her starting point. How much farther does Judy have to travel to reach her destination?

Application A8

20. How much longer is a steel rod that measures 85 cm than one that measures 752 mm?

21. Hank Sanchez throws the discus 54.94 m. The school record is 55.02 m. How many centimeters is Hank short of the record?

Application A9 Which is more?

22. a 531 g book or a 0.673 kg book

23. a 531 g box or a 0.5 kg box

24. a 65 g bracelet or a 0.05 kg bracelet

25. a 3000 g basket or a 3.1 kg basket

Application A10

26. Arrange in order from the smallest to the largest amount.
783 mL   6.2 L   316 mL   0.3 L

Cumulative Review: Units 1–3

Skills

Skill 1 **Give the place and value of the underlined digit.** (pages 2–3)

1. 4<u>3</u>
2. <u>7</u>85
3. 31.<u>3</u>2
4. 184.6<u>8</u>

Skill 2 **Round to the given place value.** (pages 4–5)

5. to the nearest ten: 145
6. to the nearest hundred: 593
7. to the nearest tenth: 0.54
8. to the nearest hundredth: 75.637

Skill 3 (page 12)

9.
```
  58
+ 21
```

10.
```
  17
+ 42
```

11.
```
  582
+ 311
```

12.
```
  1094
+ 603
```

Skill 4 (page 13)

13.
```
  33
+ 69
```

14.
```
  94
+ 57
```

15.
```
  792
+ 658
```

16.
```
  8676
+ 7515
```

Skill 5 (page 14)

17.
```
  65
- 14
```

18.
```
  184
- 13
```

19.
```
  536
- 123
```

20.
```
  9197
- 3103
```

Skill 6 (page 15)

21.
```
  276
- 95
```

22.
```
  337
- 48
```

23.
```
  8320
- 650
```

24.
```
  5080
- 792
```

Skill 7 (page 22)

25.
```
  33.8
+ 2.9
```

26.
```
  63.07
+ 43.08
```

27. 198.5 + 567.276

28. 8674.23 + 1773.2

Skill 8 (page 23)

29.
```
  25.7
- 0.8
```

30.
```
  51.97
- 43.08
```

31. 784.03 − 567.276

32. 3864.16 − 1773.2

Skill 9 (page 34)

33. 43
× 2

34. 52
× 4

35. 223
× 3

36. 7243
× 2

Skill 10 (page 35)

37. 683
× 3

38. 728
× 5

39. 284
× 6

40. 5492
× 9

Skill 11 (pages 36–37)

41. 3186
× 10

42. 606
× 30

43. 718
× 400

44. 3481
× 8000

Skill 12 (page 44)

45. 881
× 36

46. 475
× 81

47. 772
× 25

48. 7681
× 29

Skill 13 (page 45)

49. 575
× 413

50. 395
× 213

51. 6282
× 862

52. 8506
× 2197

Skill 14 (pages 46–47)

53. 961
× 102

54. 5666
× 700

55. 7078
× 608

56. 4106
× 7083

Skill 15 (pages 54–55)

57. 6.88
× 65

58. 7.948
× 3.6

59. 0.672
× 0.05

60. 0.814
× 0.034

Skill 16 (page 66)

61. $3\overline{)339}$

62. $6\overline{)672}$

63. $5\overline{)1780}$

64. $8\overline{)8424}$

Cumulative Review: Units 1–3

Skill 17 **Write remainders with R.** (page 67)

65. $4\overline{)897}$ **66.** $7\overline{)145}$ **67.** $9\overline{)152}$ **68.** $6\overline{)832}$

Skill 18 **Write remainders with R.** (pages 68–69)

69. $77\overline{)816}$ **70.** $64\overline{)7357}$ **71.** $510\overline{)16,974}$ **72.** $698\overline{)48,865}$

Skill 19 **Write remainders as fractions.** (pages 76–77)

73. $36\overline{)95}$ **74.** $35\overline{)448}$ **75.** $16\overline{)7920}$ **76.** $105\overline{)9626}$

Skill 20 **Divide to hundredths. Round to the nearest tenth.** (pages 78–79)

77. $7\overline{)30}$ **78.** $23\overline{)955}$ **79.** $595\overline{)917}$ **80.** $197\overline{)2245}$

Skill 21 **Divide to hundredths. Round to the nearest tenth.** (pages 86–87)

81. $2.7\overline{)5.9}$ **82.** $8.4\overline{)18.93}$ **83.** $0.97\overline{)547.3}$ **84.** $0.072\overline{)8.484}$

Applications

Application A1 **Compare the two numbers. Which is larger?** (pages 6–7)

1. 472 or 462 **2.** 1070 or 1007 **3.** 8.6031 or 8.6036 **4.** 0.002 or 0.003

Application A2 (pages 16–17)

Estimate to the nearest ten.

5. 306
+ 34

6. 787
+ 258

Estimate to the nearest thousand.

7. 3459
− 1226

8. 8738
− 4160

Application A3 (pages 24–25)

Estimate to the nearest tenth.

9. 14.57
+ 0.82

10. 0.598
+ 0.111

Estimate to the nearest whole number.

11. $6.19
− 1.81

12. 6.31
− 4.82

Application A4 (pages 38–39)

13. 4 yd = __?__ ft

14. 2.5 mi = __?__ ft

15. 10 ft 3 in. = __?__ in.

16. 20 yd 6 ft = __?__ ft

Application A5 (pages 48–49)

17. 3 lb = __?__ oz

18. 7 t = __?__ lb

19. 1 lb 6 oz = __?__ oz

20. 5 lb 8 oz = __?__ oz

Application A6 (pages 56–57)

21. 3 c = __?__ fl oz

22. 6 qt = __?__ pt

23. __?__ qt = 7 gal

24. 6 gal = __?__ pt

Application A7 **Which measurement is longer?** (pages 70–71)

25. 5.3 km or 6000 m

26. 763.6 m or 0.7 km

27. You take a train from Boston to New York City. The trip takes 6 hours. If the train travels at 60 kilometers per hour, how far is it between the two cities?

Application A8 (pages 80–81)

28. Idris Bilal has two plastic sticks. One measures 65 mm and the other measures 7.5 cm. How much will he have to cut from the longer stick to make the two sticks the same length?

29. You have 3 m of adhesive tape. How many packages 50 cm around can you wrap with the tape? Assume you tape each package once around with no overlapping.

Application A9 **Which is more?** (page 88)

30. a 103 g sausage or a 0.13 kg sausage

31. a 30 g box or a 0.35 kg box

32. 9,000 g of sand or 10 kg of sand

33. a 10 g pellet or a 0.013 kg pellet

Application A10 (page 89)

34. Arrange in order from the smallest to the largest amount.
6 L   7650 mL   0.10 L   3.6 mL

4

Basic Fraction Skills

Geometric Definitions

Problem Solving

Unit Preview

 You will use these skills and applications in this unit.
Which do you already know? Work each problem.

Skill 22

1. List the factors of 28.

Skill 23

2. Write the GCF of 24 and 18.

Skill 24

3. Write $\frac{12}{32}$ in lowest terms.

Skill 25

4. Write in higher terms: $\frac{3}{5} = \frac{?}{25}$

Skill 26

5. Write the first four multiples of 8.

Skill 27

6. Write the LCM of 6 and 10.

Skill 28

7. Write the LCD of $\frac{2}{3}$ and $\frac{1}{5}$.

Skill 29

8. Write as fractions with the LCD:
$\frac{1}{8}, \frac{5}{6}$.

Skill 30

9. Write $\frac{207}{30}$ as a mixed number in lowest terms.

Skill 31

10. Write $6\frac{1}{3}$ as a fraction.

Application A11

11. Name the line segments.

Application A12

12. Is the angle right, acute, or obtuse?

Application A13

13. In triangle ABC, what is the measure of $\angle A$?

Application A14

14. If O is the center of the circle, what is $\overline{OP}$ called?

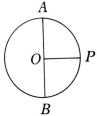

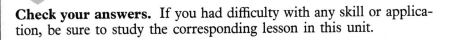

Check your answers. If you had difficulty with any skill or application, be sure to study the corresponding lesson in this unit.

101

Finding Factors of a Number

MODEL

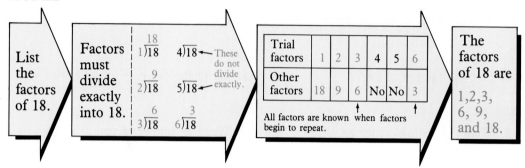

 GET READY. Complete the table. List the factors.

1. Factors of 12

Trial factors	1	2	3	4
Other factors	12	?	?	3

2. Factors of 15

Trial factors	1	2	3	4	5
Other factors	?	?	?	?	3

3. Factors of 24

Trial factors	1	2	3	4	5	6
Other factors	?	?	?	?	?	4

4. Factors of 36

Trial factors	1	2	3	4	5	6
Other factors	?	?	?	?	?	6

 NOW USE SKILL 22. List the factors of each number.

5. 3   **6.** 4   **7.** 5   **8.** 1   **9.** 2

10. 6   **11.** 7   **12.** 19   **13.** 15   **14.** 28

15. 10   **16.** 14   **17.** 12   **18.** 13   **19.** 24

20. 30   **21.** 23   **22.** 26   **23.** 37   **24.** 32

25. 27   **26.** 20   **27.** 50   **28.** 45   **29.** 59

30. 75   **31.** 80   **32.** 81   **33.** 98   **34.** 100

For more practice: **Skill Bank,** page 415.

Finding the Greatest Common Factor (GCF)

MODEL

Write the greatest common factor (GCF) of 16 and 12.

Factors of 16: 1,2,4,8,16

Factors of 12: 1,2,3,4,6,12

Skill 22

Common factors: 1,2,4

The largest common factor is 4.

The GCF is 4.

 GET READY. The factors of two numbers are given. List the common factors. Then write the greatest common factor (GCF).

1. 6: 1, 2, 3, 6
15: 1, 3, 5, 15

2. 8: 1, 2, 4, 8
20: 1, 2, 4, 5, 10, 20

3. 12: 1, 2, 3, 4, 6, 12
16: 1, 2, 4, 8, 16

4. 15: 1, 3, 5, 15
18: 1, 2, 3, 6, 9, 18

5. 20: 1, 2, 4, 5, 10, 20
45: 1, 3, 5, 9, 15, 45

6. 24: 1, 2, 3, 4, 6, 8, 12, 24
30: 1, 2, 3, 5, 6, 10, 15, 30

 NOW USE SKILL 23. Write the GCF.

7. 2 and 4   **8.** 6 and 8   **9.** 7 and 5   **10.** 9 and 3

11. 12 and 8   **12.** 9 and 18   **13.** 6 and 15   **14.** 10 and 20

15. 14 and 16   **16.** 18 and 20   **17.** 45 and 18   **18.** 21 and 42

19. 9 and 15   **20.** 16 and 24   **21.** 30 and 14   **22.** 35 and 21

23. 45 and 30   **24.** 72 and 108   **25.** 65 and 169   **26.** 81 and 54

27. 6, 24, and 36   **28.** 7, 35, and 49   **29.** 5, 25, and 50

30. 30, 60, and 90   **31.** 15, 45, and 60   **32.** 30, 45, and 75

33. 19, 33, and 47   **34.** 50, 100, and 75   **35.** 40, 64, and 72

For more practice: **Skill Bank,** page 415.

Writing Fractions in Lowest Terms

MODEL

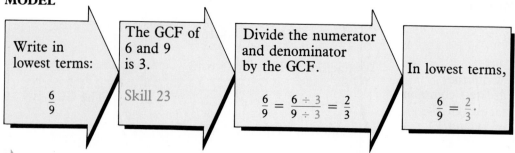

| Write in lowest terms: $\frac{6}{9}$ | The GCF of 6 and 9 is 3. Skill 23 | Divide the numerator and denominator by the GCF. $\frac{6}{9} = \frac{6 \div 3}{9 \div 3} = \frac{2}{3}$ | In lowest terms, $\frac{6}{9} = \frac{2}{3}.$ |

 GET READY. Find the GCF.

1. 3 and 6   **2.** 9 and 4   **3.** 6 and 8   **4.** 15 and 10   **5.** 8 and 12

Write in lowest terms.

6. $\frac{6}{8} = \frac{6 \div 2}{8 \div 2} = ?$ **7.** $\frac{6}{18} = \frac{6 \div 6}{18 \div 6} = ?$ **8.** $\frac{9}{12} = \frac{9 \div 3}{12 \div 3} = ?$

 NOW USE SKILL 24. Use the GCF to write in lowest terms.

9. $\frac{2}{4}$ **10.** $\frac{3}{9}$ **11.** $\frac{4}{8}$ **12.** $\frac{9}{18}$ **13.** $\frac{6}{12}$

14. $\frac{7}{21}$ **15.** $\frac{6}{15}$ **16.** $\frac{8}{16}$ **17.** $\frac{9}{21}$ **18.** $\frac{8}{36}$

19. $\frac{10}{12}$ **20.** $\frac{14}{16}$ **21.** $\frac{18}{32}$ **22.** $\frac{16}{24}$ **23.** $\frac{21}{30}$

24. $\frac{25}{60}$ **25.** $\frac{13}{91}$ **26.** $\frac{54}{81}$ **27.** $\frac{34}{85}$ **28.** $\frac{64}{72}$

29. $\frac{60}{105}$ **30.** $\frac{120}{144}$ **31.** $\frac{48}{132}$ **32.** $\frac{105}{135}$ **33.** $\frac{108}{117}$

34. $4\frac{2}{8}$ **35.** $7\frac{2}{6}$ **36.** $5\frac{6}{8}$ **37.** $7\frac{4}{6}$ **38.** $9\frac{2}{4}$

39. $4\frac{4}{10}$ **40.** $5\frac{9}{12}$ **41.** $6\frac{8}{16}$ **42.** $9\frac{9}{21}$ **43.** $8\frac{10}{20}$

44. $12\frac{24}{60}$ **45.** $14\frac{24}{30}$ **46.** $20\frac{18}{32}$ **47.** $24\frac{50}{75}$ **48.** $40\frac{60}{72}$

For more practice: **Skill Bank,** page 416.

Writing Fractions in Higher Terms

MODEL

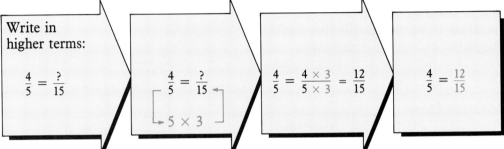

Write in higher terms:

$$\frac{4}{5} = \frac{?}{15}$$

$$\frac{4}{5} = \frac{?}{15}$$

$$5 \times 3$$

$$\frac{4}{5} = \frac{4 \times 3}{5 \times 3} = \frac{12}{15}$$

$$\frac{4}{5} = \frac{12}{15}$$

 GET READY. Find the missing factor.

1. $6 \times ? = 18$ **2.** $5 \times ? = 30$ **3.** $7 \times ? = 21$ **4.** $9 \times ? = 36$

5. $10 \times ? = 30$ **6.** $15 \times ? = 90$ **7.** $12 \times ? = 36$ **8.** $11 \times ? = 44$

Write the fraction in higher terms.

9. $\frac{6}{10} = \frac{6 \times 3}{10 \times 3} = ?$ **10.** $\frac{7}{12} = \frac{7 \times 3}{12 \times 3} = ?$ **11.** $\frac{9}{11} = \frac{9 \times 4}{11 \times 4} = ?$

 NOW USE SKILL 25. Write the fraction in higher terms.

12. $\frac{1}{4} = \frac{?}{8}$ **13.** $\frac{1}{9} = \frac{?}{18}$ **14.** $\frac{1}{5} = \frac{?}{25}$ **15.** $\frac{1}{3} = \frac{7}{?}$

16. $\frac{4}{9} = \frac{?}{27}$ **17.** $\frac{5}{8} = \frac{45}{?}$ **18.** $\frac{6}{9} = \frac{?}{45}$ **19.** $\frac{8}{9} = \frac{?}{36}$

20. $\frac{10}{18} = \frac{40}{?}$ **21.** $\frac{20}{24} = \frac{40}{?}$ **22.** $\frac{21}{32} = \frac{?}{64}$ **23.** $\frac{30}{50} = \frac{?}{100}$

24. $\frac{14}{16} = \frac{42}{?}$ **25.** $\frac{38}{46} = \frac{?}{230}$ **26.** $\frac{27}{52} = \frac{?}{208}$ **27.** $\frac{75}{80} = \frac{450}{?}$

28. $5\frac{3}{4} = 5\frac{?}{8}$ **29.** $8\frac{3}{8} = 8\frac{?}{24}$ **30.** $9\frac{4}{9} = 9\frac{?}{36}$ **31.** $5\frac{6}{15} = 5\frac{30}{?}$

For more practice: **Skill Bank,** page 416.

Point, Line, and Plane

A *point* is a position in space. In the diagram, *A* and *B* are points. A *line* is straight and extends endlessly in opposite directions. The line in the diagram is designated $\overleftrightarrow{AB}$ (read "line AB"). A *plane* is a flat surface extending endlessly.

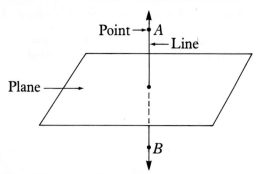

Reference Chart

	Description	Examples
Line segment	A part of a line with two endpoints and all the points between them.	$A \bullet\!\!-\!\!-\!\!-\!\!-\!\!-\!\!-\!\!\bullet B$ $\overline{AB}$ (read "line segment *AB*")
Ray	A part of a line having one endpoint and extending endlessly in one direction.	$A \qquad B$ $\overrightarrow{AB}$ (read "ray *AB*")
Intersecting lines	Lines that meet at a common point.	
Parallel lines	Lines in a plane that never meet. They have no points in common.	
Polygon	A plane figure formed by line segments intersecting at their endpoints.	

 NOW USE APPLICATION A11. Draw each of the following.

1. $\overline{KM}$ **2.** $\overrightarrow{EF}$ **3.** $\overleftrightarrow{YO}$ **4.** $\overline{DC}$

5. two parallel lines   **6.** a polygon   **7.** two intersecting lines

Use the diagram to name the following.

8. Two parallel lines   **9.** Two intersecting lines

10. The corners of the polygon

11. The point where $\overrightarrow{AB}$ and $\overrightarrow{AC}$ intersect

12. Two rays that intersect at point D

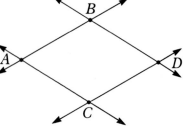

For more practice: **Application Bank,** page 452.

Types of Angles

An **angle** is formed by two rays that have the same endpoint. The common endpoint of the rays is called the **vertex** of the angle. Some special types of angles are shown in the chart below.

Reference Chart

Type of Angle	Description	Examples
Right angle	An angle whose measure is 90°.	90° 90°
Acute angle	An angle whose measure is less than 90°.	30° 50°
Obtuse angle	An angle whose measure is greater than 90° but less than 180°.	140° 130°
Vertical angles	Opposite angles, with equal measures, formed by intersecting lines.	120° 120°

 NOW USE APPLICATION A12. Is the angle right, acute or obtuse?

1. 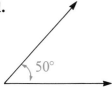   50°

2.    115°

3.   90°

4. 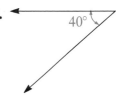 40°

Is the angle right, acute, or obtuse? You may use a protractor or a corner of a sheet of paper to test for right angles.

5.

6.

7.

8.

For more practice: **Application Bank,** page 452.

Universal Computer Services

You are a computer maintenance specialist. To use your time best, you must know the shortest route between the service center and your customer's computers. You use the map below.

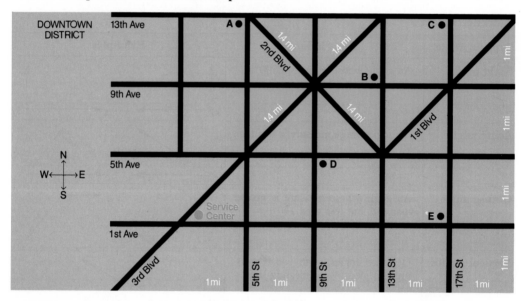

PROBLEM: About how long does it take you to travel between the service center and your customers?

Solution Plan: The distance between two locations multiplied by the time it takes to travel one mile gives your estimated travel time.

WHAT DO YOU KNOW? You know the locations of your customers. Use the map.

1. In which direction does 9th Street run?

2. In which direction do the avenues run?

3. Name the intersection where the service center is located.

4. Name the intersection where the customer is located.

    a. Customer A   **b.** Customer B   **c.** Customer C

    d. Customer D   **e.** Customer E

WHAT DO YOU WANT TO KNOW? You want to know the shortest distance between locations.

You use the map to find the location of Customer A. You determine that the shortest distance between the service center and Customer A is 3.4 miles.

$$\begin{array}{r} 1.4 \\ 1.0 \\ + \; 1.0 \\ \hline 3.4 \end{array} \leftarrow \text{Skill 7}$$

Use the map to find the shortest distance between the two locations.

5. The service center and Customer B.

6. The service center and Customer C.

7. Customer A and Customer B.

8. Customer A and Customer E.

NOW SOLVE THE PROBLEM. Estimate the travel time.

The shortest distance between the service center and Customer A is 3.4 miles. You know that it will take you about 6 minutes to travel 1 mile in the downtown area. You figure it will take you about 20 minutes to travel from the service center to Customer A.

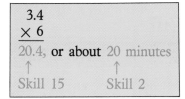

$$\begin{array}{r} 3.4 \\ \times \; 6 \\ \hline 20.4 \end{array}, \textbf{ or about } 20 \text{ minutes}$$

Skill 15   Skill 2

Complete the table. You can travel 1 mile in about 6 minutes.

	Route	Shortest Distance	Approximate Travel Time
	Service Center to Customer A	3.4 miles	20 minutes
9.	Service Center to Customer B	?	23 minutes
10.	Service Center to Customer C	5.2 miles	?
11.	Customer E to Service Center	?	24 minutes
12.	Customer A to Customer E	?	?
13.	Customer A to Customer B	?	?
14.	Service Center to Customer A to Customer B	5.8 miles	?

109

Quiz

Skill 22

1. List the factors of 56.

Skill 23

2. Write the GCF of 60 and 75.

Skill 24

3. Write $\frac{18}{63}$ in lowest terms.

Skill 25

4. Write in higher terms: $\frac{5}{7} = \frac{?}{21}$

Application A11

5. Name the line segments.

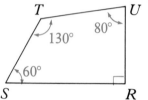

Application A12

6. Is $\angle S$ acute, obtuse, or right?

Problem Solving

7. Find the shortest distance for a single trip beginning and ending in Jacksonville and stopping in Tampa and Miami.

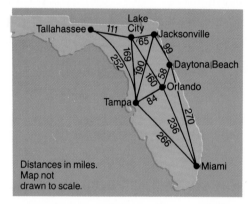

Distances in miles.
Map not
drawn to scale.

CONSUMER NOTE

Comparing Earnings

You receive job offers from two companies. Apex Insurance Company pays an hourly wage of $6.50. Browning Incorporated pays an annual salary of $14,850. Both jobs require a 35-hour work week. To find which one pays more, compare the weekly incomes.

Apex: Hourly wage × hours per week = weekly income
              $6.50   ×   35   = $227.50

Browning: annual salary ÷ weeks per year = weekly income
              $14,850   ÷   52   = $285.58   Browning pays more.

Find the weekly income for each job based on a 35-hour work week. Round to the nearest cent.

1. Annual salary: $11,440

2. Hourly wage: $21.25

3. Hourly wage: $4.25

4. Annual salary: $15,750

Meteorologist

Meteorologists, or weather forecasters, predict the weather. They use information collected by weather stations around the world. They also use satellite pictures to get the "big view."

The weather map shows the forecast for a day in January. The temperatures are shown in degrees Fahrenheit.

Use the map to answer the following questions.

1. What is the highest temperature shown?

2. What is the lowest temperature shown?

3. What is the difference between the highest and the lowest temperatures?

4. In what area is rain predicted?

5. How many cold fronts are predicted?

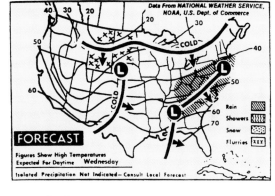

6. In which directions are the cold fronts moving?

Finding Multiples

MODEL

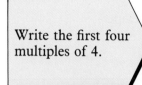

Multiples of 4			
4 × 1	4 × 2	4 × 3	4 × 4
4	8	12	16

The first four multiples of 4 are

4, 8, 12, 16.

 GET READY. Complete the table.

1.

Multiples of 7			
7 × 1	7 × 2	7 × 3	7 × 4
7	?	?	?

2.

Multiples of 10			
10 × 1	10 × 2	10 × 3	10 × 4
?	?	?	?

Complete the list of multiples.

3. 4: 4, 8, 12, ?, ?, ?

4. 6: 6, 12, 18, ?, ?, ?

5. 12: 12, 24, 36, ?, ?, ?

6. 20: 20, 40, 60, ?, ?, ?

NOW USE SKILL 26. Write the first four multiples of the number.

7. 1	**8.** 2	**9.** 5	**10.** 8	**11.** 9
12. 11	**13.** 12	**14.** 13	**15.** 16	**16.** 18
17. 20	**18.** 21	**19.** 22	**20.** 25	**21.** 26
22. 35	**23.** 40	**24.** 45	**25.** 50	**26.** 65
27. 32	**28.** 34	**29.** 36	**30.** 42	**31.** 51
32. 100	**33.** 300	**34.** 102	**35.** 203	**36.** 232
37. 310	**38.** 309	**39.** 299	**40.** 333	**41.** 250

For more practice: **Skill Bank**, page 417.

Finding the Least Common Multiple (LCM)

MODEL

Write the least common multiple (LCM) of 6 and 8.

Multiples of 6: 6,12,18,24,30,...

Multiples of 8: 8,16,24,32,40,...

Skill 26

The smallest common multiple is 24.

The LCM is 24.

 GET READY. The multiples of two or three numbers are given. Write the least common multiple (LCM).

1. 2: 2, 4, 6, 8, 10, . . .
3: 3, 6, 9, 12, 15, . . .

2. 4: 4, 8, 12, 16, 20, . . .
6: 6, 12, 18, 24, 30, . . .

3. 3: 3, 6, 9, 12, 15, . . .
4: 4, 8, 12, 16, 20, . . .
6: 6, 12, 18, 24, 30, . . .

4. 15: 15, 30, 45, 60, 75, . . .
20: 20, 40, 60, 80, 100, . . .
30: 30, 60, 90, 120, 150, . . .

NOW USE SKILL 27. Write the LCM.

5. 3 and 6   **6.** 4 and 8   **7.** 9 and 15   **8.** 5 and 6

9. 4 and 9   **10.** 8 and 12   **11.** 4 and 16   **12.** 8 and 18

13. 12 and 20   **14.** 16 and 24   **15.** 25 and 10   **16.** 20 and 30

17. 12 and 15   **18.** 90 and 60   **19.** 18 and 24   **20.** 12 and 52

21. 36 and 48   **22.** 36 and 24   **23.** 21 and 28   **24.** 30 and 25

25. 6, 12, and 18   **26.** 8, 5, and 20   **27.** 9, 5, and 15

28. 10, 15, and 18   **29.** 12, 15, and 20   **30.** 10, 15, and 30

31. 7, 12, and 3   **32.** 25, 8, and 4   **33.** 12, 32, and 4

For more practice: **Skill Bank,** page 417.   113

Finding the Least Common Denominator (LCD)

MODEL

Write the least common denominator (LCD) of $\frac{1}{3}$, $\frac{1}{4}$, and $\frac{1}{6}$.

The LCD of the fractions is the LCM of the denominators.

Skill 27

Denominators: 3,4,6.

The LCM is 12.

The LCD of $\frac{1}{3}$, $\frac{1}{4}$, and $\frac{1}{6}$ is 12.

 GET READY. Write the first four multiples.

1. 6   **2.** 8   **3.** 14   **4.** 15   **5.** 17

6. 19   **7.** 23   **8.** 30   **9.** 36   **10.** 48

Write the LCM.

11. 8 and 6   **12.** 3 and 9   **13.** 14 and 12   **14.** 25 and 30

15. 16 and 20   **16.** 12 and 36   **17.** 48 and 32   **18.** 22 and 55

NOW USE SKILL 28. Write the LCD of the fractions.

19. $\frac{1}{4}, \frac{1}{5}$ **20.** $\frac{1}{8}, \frac{1}{6}$ **21.** $\frac{1}{4}, \frac{1}{6}$ **22.** $\frac{1}{2}, \frac{1}{5}$ **23.** $\frac{1}{6}, \frac{1}{9}$

24. $\frac{7}{8}, \frac{11}{12}$ **25.** $\frac{3}{4}, \frac{9}{14}$ **26.** $\frac{5}{6}, \frac{13}{16}$ **27.** $\frac{4}{9}, \frac{7}{12}$ **28.** $\frac{3}{7}, \frac{6}{11}$

29. $\frac{5}{12}, \frac{7}{16}$ **30.** $\frac{13}{18}, \frac{8}{15}$ **31.** $\frac{11}{18}, \frac{3}{14}$ **32.** $\frac{7}{15}, \frac{5}{12}$ **33.** $\frac{13}{18}, \frac{7}{12}$

34. $\frac{7}{18}, \frac{14}{19}$ **35.** $\frac{9}{16}, \frac{3}{14}$ **36.** $\frac{8}{15}, \frac{9}{25}$ **37.** $\frac{11}{18}, \frac{19}{24}$ **38.** $\frac{13}{18}, \frac{11}{16}$

39. $\frac{1}{6}, \frac{1}{4}, \frac{1}{2}$ **40.** $\frac{3}{4}, \frac{5}{7}, \frac{3}{8}$ **41.** $\frac{4}{9}, \frac{1}{6}, \frac{7}{12}$

42. $\frac{4}{9}, \frac{7}{12}, \frac{11}{18}$ **43.** $\frac{3}{4}, \frac{6}{7}, \frac{13}{14}$ **44.** $\frac{4}{7}, \frac{5}{9}, \frac{16}{21}$

45. $\frac{7}{20}, \frac{8}{25}, \frac{9}{10}$ **46.** $\frac{7}{12}, \frac{11}{18}, \frac{23}{24}$ **47.** $\frac{5}{12}, \frac{13}{21}, \frac{9}{14}$

       For more practice: **Skill Bank,** page 418.

Writing Fractions with the LCD

MODEL

Write as fractions with the LCD.

$$\frac{1}{2}, \frac{2}{3}, \frac{3}{4}$$

The LCD is 12.

$$\frac{1}{2} = \frac{?}{12}$$

$$\frac{2}{3} = \frac{?}{12}$$ Skill 28

$$\frac{3}{4} = \frac{?}{12}$$

Find the numerators.

$$\frac{1}{2} = \frac{1 \times 6}{2 \times 6} = \frac{6}{12}$$

$$\frac{2}{3} = \frac{2 \times 4}{3 \times 4} = \frac{8}{12}$$ Skill 25

$$\frac{3}{4} = \frac{3 \times 3}{4 \times 3} = \frac{9}{12}$$

$$\frac{1}{2} = \frac{6}{12}$$

$$\frac{2}{3} = \frac{8}{12}$$

$$\frac{3}{4} = \frac{9}{12}$$

GET READY. Write the numerators.

1. $\frac{2}{4} = \frac{2 \times 2}{4 \times 2} = \frac{?}{8}$

$\frac{3}{8} = \frac{3 \times 1}{8 \times 1} = \frac{?}{8}$

2. $\frac{2}{5} = \frac{2 \times 6}{5 \times 6} = \frac{?}{30}$

$\frac{4}{6} = \frac{4 \times 5}{6 \times 5} = \frac{?}{30}$

3. $\frac{2}{3} = \frac{2 \times 4}{3 \times 4} = \frac{?}{12}$

$\frac{5}{6} = \frac{5 \times 2}{6 \times 2} = \frac{?}{12}$

4. $\frac{4}{5} = \frac{?}{10}$

$\frac{1}{2} = \frac{?}{10}$

5. $\frac{6}{8} = \frac{?}{40}$

$\frac{3}{10} = \frac{?}{40}$

6. $\frac{6}{7} = \frac{?}{63}$

$\frac{4}{9} = \frac{?}{63}$

7. $\frac{7}{12} = \frac{?}{48}$

$\frac{5}{16} = \frac{?}{48}$

NOW USE SKILL 29. Write as fractions with the LCD.

8. $\frac{3}{8}, \frac{5}{6}$

9. $\frac{1}{2}, \frac{3}{5}$

10. $\frac{3}{4}, \frac{3}{7}$

11. $\frac{5}{6}, \frac{7}{9}$

12. $\frac{3}{4}, \frac{1}{6}$

13. $\frac{3}{4}, \frac{9}{14}$

14. $\frac{1}{6}, \frac{7}{15}$

15. $\frac{5}{8}, \frac{11}{12}$

16. $\frac{5}{6}, \frac{19}{21}$

17. $\frac{4}{7}, \frac{9}{14}$

18. $\frac{7}{12}, \frac{16}{21}$

19. $\frac{11}{16}, \frac{5}{14}$

20. $\frac{11}{12}, \frac{13}{15}$

21. $\frac{5}{24}, \frac{3}{16}$

22. $\frac{1}{18}, \frac{8}{21}$

23. $\frac{5}{12}, \frac{7}{18}$

24. $\frac{8}{15}, \frac{8}{21}$

25. $\frac{7}{16}, \frac{5}{12}$

26. $\frac{3}{14}, \frac{4}{21}$

27. $\frac{5}{24}, \frac{13}{18}$

28. $\frac{5}{6}, \frac{3}{8}, \frac{5}{12}$

29. $\frac{2}{3}, \frac{3}{5}, \frac{8}{15}$

30. $\frac{2}{9}, \frac{1}{4}, \frac{7}{12}$

31. $\frac{5}{8}, \frac{4}{7}, \frac{3}{14}$

For more practice: **Skill Bank**, page 418.

Triangles

A **triangle** is a polygon that has three sides. The sum of the measures of the angles of a triangle is 180°.

$$\angle A + \angle B + \angle C = 180°$$

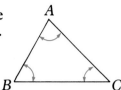

MODEL

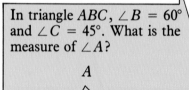

In triangle ABC, $\angle B = 60°$ and $\angle C = 45°$. What is the measure of $\angle A$?

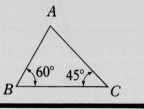

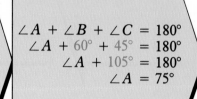

$$\angle A + \angle B + \angle C = 180°$$
$$\angle A + 60° + 45° = 180°$$
$$\angle A + 105° = 180°$$
$$\angle A = 75°$$

The measure of $\angle A$ is 75°.

NOW USE APPLICATION A13. What is the measure of $\angle A$?

1.

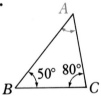

2.

3.

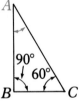

4.

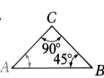

5.

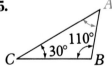

6.

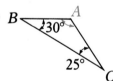

In triangle ABC, what is the measure of $\angle C$?

7. $\angle A = 60°$
$\angle B = 80°$

8. $\angle A = 53°$
$\angle B = 72°$

9. $\angle A = 90°$
$\angle B = 25°$

10. $\angle A = 35°$
$\angle B = 35°$

11. $\angle A = 40.5°$
$\angle B = 60.5°$

12. $\angle A = 26°$
$\angle B = \angle C$

13. $\angle A = 40°$
$\angle A = \angle B$

14. $\angle A = \angle B$
$\angle B = \angle C$

Some triangles have special names because of the measures of their angles or the number of equal sides they have.

Type of Triangle	Description	Examples
Right Triangle	One right angle.	*B* 90° *C* *A*
Equilateral Triangle	All sides are equal. All angles are equal.	*B* *A* *C*
Isosceles Triangle	Two equal sides. Two equal angles.	*B* *A* *C*

What type of triangle is shown?

15. 45° 45°

16. 10 / 10 \ 10

17. 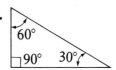 60° 90° 30°

Use the diagram to answer the question.

18. What type of triangle is triangle *DEF*?

19. What is the measure of ∠*F*?

20. What is the measure of ∠*E*?

E 6 6 70.5° *D* 4 *F*

Answer the question.

21. Can an equilateral triangle be a right triangle?

22. Each side of a triangle is 5 cm long. What is the measure of an angle of the triangle?

23. Two angles of a triangle measure 20° and 70°. What is the measure of the third angle? What type of triangle is this triangle?

24. In an isosceles triangle, one angle measures 80° and another angle measures 50°. What is the measure of the third angle?

For more practice: **Application Bank,** page 453.

Wilderness Club

You are the president of the Wilderness Club. You have hired the Frazier Fence Company to enclose the river camp with a fence. You provide the supervisor with the diagram of the camp shown.

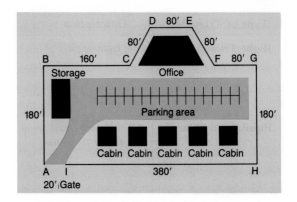

PROBLEM: How much will the fencing job cost?

Solution Plan: Find the cost to build the fence along the borders of the camp. Add the cost of the gate.

WHAT DO YOU KNOW? You know the length of the boundaries of the camp. Use the diagram to complete the table.

	1.	2.	3.	4.	5.	6.	7.	8.	
Side	AB	BC	CD	DE	EF	FG	GH	HI	IA
Length	180'	?	?	?	?	?	?	?	?

WHAT DO YOU WANT TO KNOW? You want the cost for the fence. The Frazier Fence Company states that the cost of labor and materials to build a 4-foot high log fence is $6.25 per foot. Labor and materials to construct a 6-foot high chain-link fence cost $8.50 per foot. A chain-link gate costs $16.50 per foot.

9. You want to use chain-link fence along sides *AB*, *GH*, and *HI*. What is the total length of these three sides?

10. What is the cost per foot to construct a 6-foot high chain-link fence?

11. What is the total cost to construct a 6-foot high chain-link fence along sides *AB*, *GH*, and *HI*?

12. You want to use log fence along sides *CD* and *EF*. What is the total cost to fence these sides?

13. How wide is the gate?

14. What is the cost to construct this chain-link gate?

118

WHAT MORE DO YOU WANT TO KNOW? You want the cost for the remaining sides.

You decide that sides *BC* and *FG* can be either log or chain-link. If the difference in cost is under $500, you will use chain-link.

15. What is the total length of these sides?

16. What is the cost to fence sides *BC* and *FG* with log fence?

17. What is the cost to fence these sides with chain-link fence?

18. Is the difference in cost less than $500?

19. Which type of fence will you use for sides *BC* and *FG*?

NOW SOLVE THE PROBLEM. Find the total cost to fence the camp. Complete the table.

		Type	Cost
20.	*AB, GH,* and *HI*	Chain-link	?
21.	*CD, DE,* and *EF*	Log	?
22.	*BC* and *FG*	?	?
23.	Gate	Chain-link	?
24.	TOTAL COST		?

25. Estimate the cost if the entire fence (1220 ft) were made of 6-foot high chain-link. Round the length of the fence to the nearest hundred feet and the cost per foot to the nearest whole dollar. Do not include the cost of the gate.

Quiz

Skill 26

1. Write the first five multiples of 14.

Skill 28

3. Write the LCD of $\frac{5}{9}$ and $\frac{7}{15}$.

Application A13

5. In triangle *ABC*, what is the measure of $\angle C$?

Problem Solving

6. You are building a fence around your yard. You will need a post at each corner, and every 8 feet around the yard. Treated posts cost $4.50 each. Untreated posts cost $2.75 each. How much more do you spend if you use all treated posts instead of untreated posts?

Skill 27

2. Write the LCM of 6, 9, and 12.

Skill 29

4. Write as fractions with the LCD: $\frac{3}{8}$ and $\frac{5}{12}$.

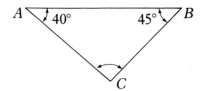

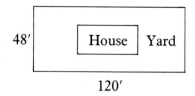

EXTRA!

Write the LCD. Then write as fractions with the LCD.

1.

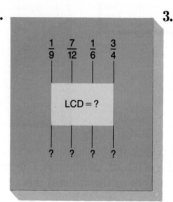

$\frac{1}{2}$ $\frac{3}{5}$ $\frac{5}{6}$ $\frac{2}{3}$

LCD = ?

? ? ? ?

2.

$\frac{1}{9}$ $\frac{7}{12}$ $\frac{1}{6}$ $\frac{3}{4}$

LCD = ?

? ? ? ?

3.

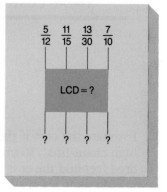

$\frac{5}{12}$ $\frac{11}{15}$ $\frac{13}{30}$ $\frac{7}{10}$

LCD = ?

? ? ? ?

Types of Computer Systems

Computers have come a long way in a relatively short time. In 1946 scientists completed the ENIAC computer. It weighed 30 tons, contained 18,000 vacuum tubes and 6000 switches, and filled a room 30 feet by 50 feet. It cost about $400,000 to build the ENIAC. It was used to calculate army ballistic tables, to forecast the weather, and to perform scientific studies. Today computers have changed drastically. You can still spend $400,000 (and even more) to purchase massive computer systems, but they have capabilities far beyond those of ENIAC.

Large computer systems, called *mainframes,* can process large amounts of data at very fast speeds. These systems are used by organizations such as large insurance companies, the government, banks, and airlines, which must process great amounts of information quickly.

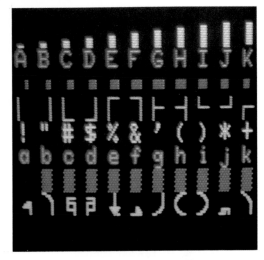

Many jobs originally done by large-scale computer systems are now performed by *minicomputers,* which are much smaller than mainframes. Minicomputers process data at fast speeds but are slower than mainframes. The price of minicomputers, however, makes them much more affordable to smaller businesses.

Even smaller and less expensive are *microcomputers.* Microcomputers usually have a CRT (cathode ray tube) terminal and a keyboard for entering information into the computer. The microprocessor, or "computer on a chip," is the processing unit of the microcomputer.

Today's microcomputer, with its one-quarter-inch-square microprocessor chip, is more powerful than the massive ENIAC of the 40's with its 18,000 vacuum tubes. The $25 microprocessor chip has replaced the $400,000 monster.

Activity

Research the history of the following computers: Mark I, ENIAC, UNIVAC I, IBM 360. Make a chart showing the name of the computer, the name(s) of its inventor(s), its physical features (size, components), and its purpose.

121

Writing Fractions as Whole or Mixed Numbers

MODEL

Write as a mixed number in lowest terms or as a whole number:

$$\frac{9}{6}$$

$$\frac{12}{4}$$

Divide the numerator by the denominator.

$$1\frac{3}{6} = 1\frac{1}{2}$$

$$6\overline{)9}$$

$$4\overline{)12}$$ $$\frac{3}{}$$

Skill 19

$$\frac{9}{6} = 1\frac{1}{2}$$

$$\frac{12}{4} = 3$$

 GET READY. Write in lowest terms.

1. $\frac{8}{12}$ 2. $\frac{24}{48}$ 3. $\frac{36}{108}$ 4. $\frac{75}{145}$ 5. $\frac{27}{81}$ 6. $\frac{32}{64}$

Divide. Write the remainder as a fraction in lowest terms.

7. $7\overline{)32}$ 8. $6\overline{)38}$ 9. $12\overline{)158}$ 10. $9\overline{)219}$ 11. $7\overline{)130}$

NOW USE SKILL 30. Write as a whole number or a mixed number in lowest terms.

12. $\frac{3}{1}$ 13. $\frac{8}{4}$ 14. $\frac{7}{7}$ 15. $\frac{12}{9}$ 16. $\frac{15}{6}$ 17. $\frac{9}{3}$

18. $\frac{21}{17}$ 19. $\frac{13}{13}$ 20. $\frac{6}{2}$ 21. $\frac{68}{20}$ 22. $\frac{8}{8}$ 23. $\frac{15}{5}$

24. $\frac{8}{2}$ 25. $\frac{4}{4}$ 26. $\frac{7}{3}$ 27. $\frac{12}{12}$ 28. $\frac{6}{1}$ 29. $\frac{9}{2}$

30. $\frac{6}{5}$ 31. $\frac{47}{47}$ 32. $\frac{10}{1}$ 33. $\frac{8}{6}$ 34. $\frac{20}{13}$ 35. $\frac{34}{16}$

36. $\frac{72}{51}$ 37. $\frac{25}{5}$ 38. $\frac{60}{8}$ 39. $\frac{16}{16}$ 40. $\frac{38}{38}$ 41. $\frac{76}{9}$

42. $\frac{64}{64}$ 43. $\frac{90}{8}$ 44. $\frac{150}{12}$ 45. $\frac{225}{15}$ 46. $\frac{112}{56}$ 47. $\frac{86}{86}$

48. $\frac{46}{7}$ 49. $\frac{63}{3}$ 50. $\frac{84}{7}$ 51. $\frac{29}{29}$ 52. $\frac{58}{58}$ 53. $\frac{102}{47}$

For more practice: **Skill Bank,** page 419.

Writing Whole or Mixed Numbers as Fractions

MODEL

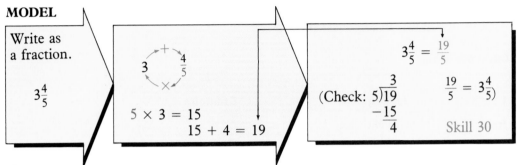

Write as a fraction.

$3\frac{4}{5}$

$3 \overset{+}{\underset{\times}{\circlearrowright}} \frac{4}{5}$

$5 \times 3 = 15$
$15 + 4 = 19$

(Check: $5\overline{)19}$
$\quad\quad\ -15$
$\quad\quad\ \ \ 4$

$3\frac{4}{5} = \frac{19}{5}$

$\frac{19}{5} = 3\frac{4}{5}$)

Skill 30

 GET READY.

1. $5 + \frac{3}{4} = \frac{?}{4} + \frac{3}{4}$

2. $3 + \frac{1}{2} = \frac{?}{2} + \frac{1}{2}$

3. $6 + \frac{7}{9} = \frac{?}{9} + \frac{7}{9}$

4. $2 + \frac{11}{15} = \frac{?}{15} + \frac{11}{15}$

5. $7 + \frac{3}{10} = \frac{?}{10} + \frac{3}{10}$

6. $9 + \frac{4}{5} = \frac{?}{5} + \frac{4}{5}$

 NOW USE SKILL 31. Write as a fraction.

7. $8\frac{1}{5}$

8. $4\frac{1}{3}$

9. $7\frac{1}{8}$

10. 8

11. $3\frac{1}{4}$

12. $9\frac{1}{6}$

13. $4\frac{2}{3}$

14. 7

15. $2\frac{4}{5}$

16. $7\frac{5}{9}$

17. 14

18. $8\frac{3}{7}$

19. $2\frac{3}{5}$

20. $6\frac{6}{7}$

21. $9\frac{3}{8}$

22. $4\frac{4}{9}$

23. $5\frac{5}{7}$

24. $8\frac{3}{4}$

25. $5\frac{3}{7}$

26. $4\frac{8}{9}$

27. $8\frac{5}{6}$

28. 44

29. $3\frac{9}{11}$

30. $2\frac{7}{12}$

31. 23

32. $6\frac{10}{13}$

33. $9\frac{5}{21}$

34. $8\frac{17}{30}$

35. $4\frac{1}{17}$

36. $5\frac{13}{18}$

37. $16\frac{6}{27}$

38. $23\frac{14}{19}$

39. 109

40. $25\frac{19}{30}$

41. $20\frac{6}{25}$

42. $19\frac{13}{20}$

43. 417

44. $8\frac{15}{32}$

45. $5\frac{36}{41}$

46. $16\frac{27}{50}$

47. 38

48. $26\frac{29}{42}$

49. $12\frac{16}{43}$

50. 66

51. $24\frac{31}{50}$

52. $30\frac{9}{32}$

53. $29\frac{3}{40}$

54. $37\frac{6}{43}$

55. $40\frac{5}{52}$

56. $38\frac{3}{61}$

57. $22\frac{2}{43}$

58. $10\frac{7}{52}$

59. $48\frac{3}{20}$

60. $33\frac{5}{28}$

For more practice: **Skill Bank,** page 419.

Quadrilaterals and Circles

A **quadrilateral** is a polygon that has four sides. Some types of quadrilaterals are shown in the chart below.

Reference Chart

Type of Quadrilateral	Description	Examples
Irregular	no parallel sides	
Trapezoid	1 pair of parallel sides	
Parallelogram	2 pairs of parallel sides	
Rectangle	4 right angles	Rectangles are also parallelograms.
Square	4 right angles 4 equal sides	Squares are also rectangles and parallelograms

 GET READY. Name the type of quadrilateral.

1.

2.

3.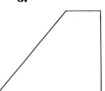

4.

5.

6.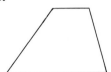

7.

8.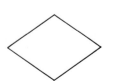

A **circle** is a closed curve in a plane whose points are all at equal distances from the center.

The length of the diameter is two times the length of the radius.

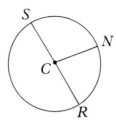

Center C
Radius $\overline{CN}$
Diameter $\overline{SR}$

 GET READY. Is the segment a radius or a diameter?

9. 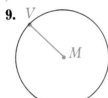 **10.** 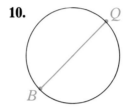 **11.** 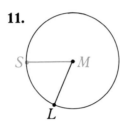 **12.**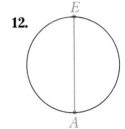

 NOW USE APPLICATION A14. List the quadrilaterals in the diagram that match the description in the table.

	Quadrilateral	Sections
13.	Irregular	?
14.	Trapezoid	?
15.	Parallelogram	?
16.	Rectangle	?
17.	Square	?

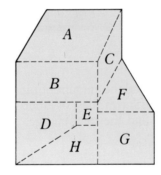

18. How many parallelograms can be identified in this diagram?

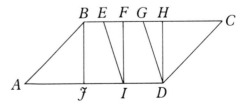

19. If the radius of a circle is 2.5 cm, what is the length of the diameter?

20. If the diameter of a circle is 8 cm, what is the length of the radius?

For more practice: **Application Bank,** page 453.

Beverly's Greenhouse

You are installing an elevated sprinkling system at Beverly's Greenhouse. Beverly provides you with a sketch of the display areas. The sketch below shows the layout of the pipes for the watering system.

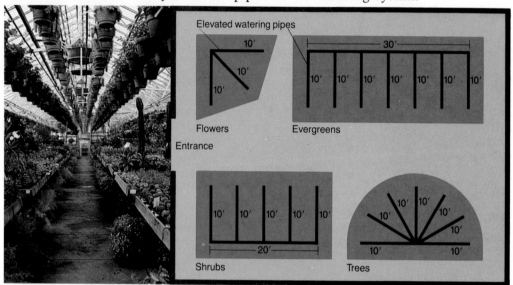

PROBLEM: How much will you charge to install the sprinkling system?

Solution Plan: Find the amount of pipe needed. Determine the cost of materials. Determine the cost of labor. Add the cost of materials to the cost of labor to find the total charge.

WHAT DO YOU KNOW? You know the number of pipes needed in each display. Use the sketch to find the amount of pipe needed.

1. How many 10-foot sections are needed in the flowers display?

2. How many 10-foot sections are needed in the trees display?

3. How many 10-foot sections are needed in the evergreens display?

4. How many 10-foot sections are needed in the shrubs display?

5. How many 10-foot sections are needed in all?

WHAT DO YOU WANT TO KNOW? You want the cost of materials.

Pipes are available in 10-foot sections. Each section costs $10.40. You know that it will cost $31.20 for the pipes in the flowers display alone.

$$\begin{array}{r} \$10.40 \\ \times\ 3 \\ \hline \$31.20 \end{array} \leftarrow \text{Skill 15}$$

6. a. Round the cost per 10-foot section to the nearest whole dollar to estimate the total cost for pipes.

   b. At $10.40 per 10-foot section, what is the actual total cost for pipes?

7. The pipes in the shrubs and evergreens areas are connected with elbow or T joints. How many joints are needed in these areas?

8. At $2.25 per joint, what is the total cost for the joints?

9. A custom joint connects all the pipes in the flower area. Another custom joint connects the pipes in the tree area. At $58.50 each, what is the total cost for the custom joints?

NOW SOLVE THE PROBLEM. Prepare the bill for the job.

Bill to: Beverly's Greenhouse 13 Lake Drive Long Lake, MI	From: Atwater Sprinkling Systems 2315 Hargrove Avenue Somerville, MI

	********** WORK DESCRIPTION **********	COST
10. Pipes:	__?__ 10-foot sections @ $10.40 each	?
11. Joints:	__?__ Elbow or T joints @ $2.25 each	?
12.	__?__ Custom joints @ $58.50 each	?
13.	Total Materials Cost	?
14. Labor:	32 hours @ $11.75 an hour	?
15.	BILLING TOTAL	?
	Terms: 30 days, in full	

Quiz

1. Write $\frac{171}{21}$ as a mixed number in lowest terms.

2. Write $9\frac{17}{32}$ as a fraction.

3. Draw a trapezoid and a square.

4. The layout of a nursery watering system is shown. Pipe can be purchased only in 10-foot sections. How many 10-foot sections will be needed? At $10.40 per 10-foot section, how much will the pipes cost?

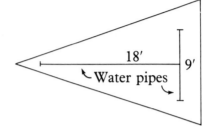

18' 9' Water pipes

CAREER CLIPPINGS

A resume summarizes your work experience, your education, and any other responsibilities you have had. Your resume provides important information to help employers decide whether or not to hire you. It is also your way to tell the employer why *you* are the best person for the job.

NAME: Frances Draper
ADDRESS: 148 First Street, Apt. 12B, Kenton, MN 55400
TELEPHONE: (218) 555-1234

EDUCATION   Kenton High School, Kenton, MN 55400
               27 in class of 143, graduated in 1980

EXPERIENCE
1980 - present Stenographer, Plastic Products Corporation
                       Levitt Industrial Park,
                       Levitt, MN 55412
               Responsibilities: transcribing notes for minutes
                       of meetings, letters to clients

1979 - 1980   Volunteer, Kenton General Hospital
               Responsibilities: assisted in admission of
                       emergency patients

Prepare a copy of your resume.

Skill 24 Write in lowest terms. (page 104)

1. $\frac{2}{4}$ 2. $\frac{4}{8}$ 3. $\frac{9}{12}$ 4. $\frac{14}{16}$ 5. $\frac{15}{18}$ 6. $\frac{36}{81}$

7. $5\frac{3}{9}$ 8. $6\frac{2}{8}$ 9. $4\frac{10}{12}$ 10. $7\frac{16}{24}$ 11. $9\frac{18}{32}$ 12. $8\frac{15}{35}$

Skill 25 Write in higher terms. (page 105)

13. $\frac{1}{2} = \frac{?}{6}$ 14. $\frac{2}{4} = \frac{?}{8}$ 15. $\frac{4}{9} = \frac{16}{?}$ 16. $\frac{12}{16} = \frac{?}{32}$

17. $\frac{20}{40} = \frac{?}{80}$ 18. $\frac{2}{3} = \frac{14}{?}$ 19. $\frac{4}{8} = \frac{12}{?}$ 20. $4\frac{6}{9} = 4\frac{?}{27}$

21. $9\frac{4}{12} = 9\frac{?}{48}$ 22. $12\frac{7}{50} = 12\frac{?}{100}$

Skill 28 Write the LCD of the fractions. (page 114)

23. $\frac{1}{4}, \frac{1}{8}$ 24. $\frac{5}{6}, \frac{3}{4}$ 25. $\frac{4}{5}, \frac{1}{7}$ 26. $\frac{7}{8}, \frac{5}{12}$ 27. $\frac{1}{4}, \frac{11}{18}$

28. $\frac{3}{4}, \frac{6}{13}$ 29. $\frac{4}{9}, \frac{11}{15}$ 30. $\frac{13}{18}, \frac{17}{24}$ 31. $\frac{5}{12}, \frac{5}{21}$ 32. $\frac{13}{15}, \frac{21}{25}$

Skill 29 Write as fractions with the LCD. (page 115)

33. $\frac{1}{2}, \frac{1}{4}$ 34. $\frac{1}{3}, \frac{1}{4}$ 35. $\frac{3}{8}, \frac{2}{3}$ 36. $\frac{1}{6}, \frac{3}{4}$ 37. $\frac{2}{9}, \frac{1}{4}$

38. $\frac{3}{4}, \frac{1}{2}, \frac{2}{3}$ 39. $\frac{1}{2}, \frac{1}{3}, \frac{1}{7}$ 40. $\frac{1}{6}, \frac{4}{9}, \frac{5}{36}$

Skill 30 Write as a whole number or a mixed number in lowest terms. (page 122)

41. $\frac{12}{5}$ 42. $\frac{38}{8}$ 43. $\frac{13}{13}$ 44. $\frac{30}{7}$ 45. $\frac{45}{6}$ 46. $\frac{21}{1}$

47. $\frac{19}{3}$ 48. $\frac{147}{24}$ 49. $\frac{52}{1}$ 50. $\frac{300}{14}$ 51. $\frac{153}{9}$ 52. $\frac{270}{24}$

Skill 31 Write as a fraction. (page 123)

53. $3\frac{1}{7}$ 54. 9 55. $4\frac{3}{8}$ 56. $7\frac{2}{5}$ 57. $6\frac{2}{3}$ 58. $12\frac{4}{9}$

59. $33\frac{3}{4}$ 60. 406 61. $7\frac{11}{13}$ 62. $30\frac{3}{14}$ 63. $8\frac{4}{21}$ 64. $47\frac{9}{10}$

Applications Review

Application A11 **Use the diagram to find the answer.** (page 106)

1. Name two parallel lines.

2. Name two intersecting lines.

3. Name a line segment.

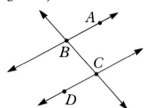

Application A12 **Is the angle right, acute, or obtuse?** (page 107)

4. 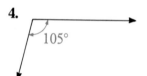 105°

5. 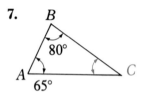 90°

6. 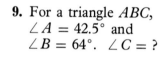 45°

Application A13 **What is the measure of** $\angle C$**?** (pages 116–117)

7.
   B
   80°
   A 65° *C*

8. *B* 90°
   35°
   A *C*
   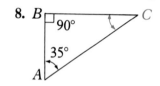

9. For a triangle ABC, $\angle A = 42.5°$ and $\angle B = 64°$. $\angle C = ?$

10. *B*
    68°
    A *C*
    Right angle
    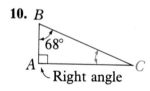

11. *B*
    6" 6"
    48°
    A *C*
    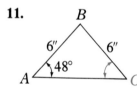

12. *C*
    8 cm 8 cm
    A *B*
    8 cm
    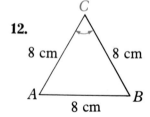

Application A14 **Use the diagram to find the answer.** (pages 124–125)

13. Name the type of quadrilateral.

14. Name the type of quadrilateral.

15. If *O* is the center and $\overline{AO}$ is 3 cm long, how long is $\overline{OC}$? How long is $\overline{CB}$?

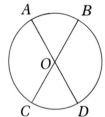

130

Unit Test

Skill 22 List the factors.

1. 20   **2.** 35

Skill 23 Write the GCF.

3. 24 and 32   **4.** 8, 12, and 30

Skill 24 Write in lowest terms.

5. $\frac{15}{25}$ **6.** $\frac{15}{18}$

Skill 25 Write in higher terms.

7. $\frac{8}{9} = \frac{?}{36}$ **8.** $\frac{3}{7} = \frac{48}{?}$

Skill 26 Write the first four multiples.

9. 6   **10.** 15

Skill 27 Write the LCM.

11. 6 and 9   **12.** 12 and 15

Skill 28
Write the LCD.

13. $\frac{3}{4}$ and $\frac{1}{7}$ **14.** $\frac{4}{9}$ and $\frac{11}{12}$

Skill 29
Write as fractions with the LCD.

15. $\frac{1}{6}$ and $\frac{4}{5}$ **16.** $\frac{7}{8}$ and $\frac{2}{3}$

Skill 30 Write as a whole number or a mixed number in lowest terms.

17. $\frac{84}{14}$ **18.** $\frac{220}{24}$

Skill 31 Write as a fraction.

19. $5\frac{4}{13}$ **20.** 78

Application A11

21. Name the line segments.

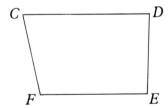

Application A12

22. Is $\angle A$ acute, obtuse, or right?

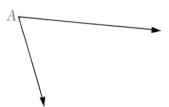

Application A13

23. What is the measure of $\angle C$?

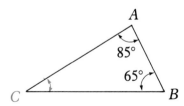

Application A14

24. If the diameter of a circle is 15 in., what is the length of the radius?

25. What type of quadrilateral is shown in the diagram below?

131

5

Add and Subtract Fractions

Pythagorean Theorem, Perimeter

Problem Solving

Unit Preview

 You will use these skills and applications in this unit. Which do you already know? Work each problem.

Skill 32

1. $\dfrac{4}{9}$
   $+\dfrac{7}{9}$

2. $\dfrac{3}{8}$
   $+\dfrac{1}{8}$

Skill 33

3. $4\dfrac{4}{9}$
   $+7\dfrac{4}{9}$

4. $5\dfrac{5}{7}$
   $+8\dfrac{6}{7}$

Skill 34

5. $\dfrac{5}{12}$
   $+\dfrac{7}{10}$

6. $\dfrac{3}{10}$
   $+\dfrac{14}{15}$

Skill 35

7. $7\dfrac{1}{6}$
   $+4\dfrac{1}{9}$

8. $3\dfrac{5}{8}$
   $+7\dfrac{7}{10}$

Skill 36

9. $\dfrac{4}{5}$
   $-\dfrac{1}{5}$

Skill 37

10. $2\dfrac{4}{7}$
    $-1\dfrac{2}{7}$

Skill 38

11. 8
    $-\dfrac{3}{7}$

Skill 39

12. $3\dfrac{2}{9}$
    $-1\dfrac{4}{9}$

Skill 40

13. $\dfrac{1}{4}$
    $-\dfrac{1}{9}$

14. $\dfrac{5}{8}$
    $-\dfrac{4}{9}$

Skill 41

15. $4\dfrac{4}{7}$
    $-3\dfrac{4}{5}$

16. $19\dfrac{13}{15}$
    $-12\dfrac{9}{10}$

Application A15

17. $7^2 = ?$

18. $\sqrt{225} = ?$

Application A16

19. A triangle has sides 15 in., 36 in., and 39 in. long. Is it a right triangle?

Application A17

20. Find the perimeter of a square with each side 5 ft long.

Application A18 Find the circumference of the circle. ($\pi = 3.14$)

21. $d = 5$ in.

22. $r = 11$ yd

 Check your answers. If you had difficulty with any skill or application, be sure to study the corresponding lesson in this unit.

133

Skill 32

Adding Fractions, Same Denominator

MODEL

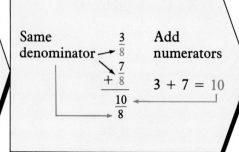

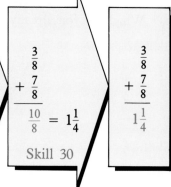

Add.

$$\frac{3}{8}$$
$$+\frac{7}{8}$$

Same denominator → $\frac{3}{8}$
$+\frac{7}{8}$

Add numerators

$3 + 7 = 10$

$$\frac{10}{8}$$

$\frac{3}{8}$
$+\frac{7}{8}$
$\frac{10}{8} = 1\frac{1}{4}$

Skill 30

$\frac{3}{8}$
$+\frac{7}{8}$
$1\frac{1}{4}$

 GET READY. Write in lowest terms.

1. $\frac{6}{8}$ 2. $\frac{4}{6}$ 3. $\frac{6}{9}$ 4. $\frac{4}{16}$ 5. $\frac{8}{10}$ 6. $\frac{7}{14}$

Write as a mixed number in lowest terms or a whole number.

7. $\frac{6}{1}$ 8. $\frac{8}{4}$ 9. $\frac{8}{6}$ 10. $\frac{9}{6}$ 11. $\frac{10}{6}$ 12. $\frac{16}{10}$

13. $1\frac{2}{4}$ 14. $2\frac{3}{9}$ 15. $4\frac{4}{6}$ 16. $2\frac{6}{18}$ 17. $5\frac{5}{10}$ 18. $4\frac{12}{15}$

 NOW USE SKILL 32. Write the answer in lowest terms.

19. $\frac{1}{5}$ $+\frac{1}{5}$ 20. $\frac{2}{7}$ $+\frac{4}{7}$ 21. $\frac{1}{8}$ $+\frac{3}{8}$ 22. $\frac{5}{6}$ $+\frac{1}{6}$ 23. $\frac{2}{9}$ $+\frac{4}{9}$ 24. $\frac{2}{5}$ $+\frac{2}{5}$

25. $\frac{6}{7}$ $+\frac{1}{7}$ 26. $\frac{5}{8}$ $+\frac{7}{8}$ 27. $\frac{5}{6}$ $+\frac{5}{6}$ 28. $\frac{2}{3}$ $+\frac{1}{3}$ 29. $\frac{4}{5}$ $+\frac{3}{5}$ 30. $\frac{8}{9}$ $+\frac{7}{9}$

31. $\frac{9}{10}$ $+\frac{7}{10}$ 32. $\frac{6}{11}$ $+\frac{10}{11}$ 33. $\frac{2}{15}$ $+\frac{8}{15}$ 34. $\frac{17}{18}$ $+\frac{11}{18}$ 35. $\frac{11}{24}$ $+\frac{7}{24}$ 36. $\frac{9}{30}$ $+\frac{11}{30}$

For more practice: **Skill Bank,** p. 420.

Adding Mixed Numbers, Same Denominator

MODEL

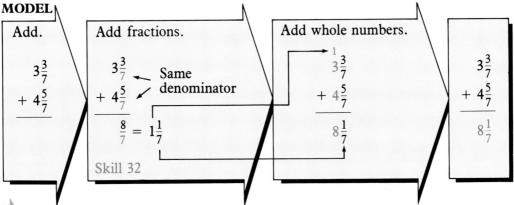

 GET READY. Write in lowest terms.

1. $\frac{4}{6}$ 2. $\frac{3}{9}$ 3. $\frac{2}{8}$ 4. $\frac{5}{10}$ 5. $\frac{3}{15}$ 6. $\frac{4}{16}$

7. $2\frac{4}{8}$ 8. $1\frac{3}{9}$ 9. $6\frac{2}{4}$ 10. $4\frac{9}{12}$ 11. $7\frac{4}{16}$ 12. $10\frac{12}{18}$

Write as a mixed number in lowest terms.

13. $\frac{4}{3}$ 14. $\frac{6}{4}$ 15. $\frac{8}{6}$ 16. $\frac{9}{5}$ 17. $\frac{7}{3}$ 18. $\frac{5}{2}$

19. $\frac{12}{9}$ 20. $\frac{16}{5}$ 21. $\frac{15}{7}$ 22. $\frac{10}{8}$ 23. $\frac{16}{12}$ 24. $\frac{14}{12}$

 NOW USE SKILL 33. Write the answer in lowest terms.

25. $\begin{array}{r} 2\frac{3}{4} \\ + 3\frac{1}{4} \\ \hline \end{array}$ 26. $\begin{array}{r} 5\frac{5}{6} \\ + 2\frac{1}{6} \\ \hline \end{array}$ 27. $\begin{array}{r} 1\frac{5}{9} \\ + 3\frac{8}{9} \\ \hline \end{array}$ 28. $\begin{array}{r} 1\frac{7}{8} \\ + 7\frac{5}{8} \\ \hline \end{array}$ 29. $\begin{array}{r} 6\frac{1}{6} \\ + 2\frac{7}{6} \\ \hline \end{array}$

30. $\begin{array}{r} 7\frac{2}{5} \\ + 6\frac{4}{5} \\ \hline \end{array}$ 31. $\begin{array}{r} 8\frac{5}{8} \\ + 9\frac{3}{8} \\ \hline \end{array}$ 32. $\begin{array}{r} 5\frac{5}{6} \\ + 6\frac{1}{6} \\ \hline \end{array}$ 33. $\begin{array}{r} 5\frac{4}{7} \\ + 8\frac{6}{7} \\ \hline \end{array}$ 34. $\begin{array}{r} 8\frac{1}{6} \\ + 7\frac{5}{6} \\ \hline \end{array}$

35. $2\frac{4}{5} + 4\frac{3}{5} + 1\frac{2}{5}$ 36. $7\frac{8}{9} + 5\frac{7}{9} + 3\frac{5}{9}$

For more practice: **Skill Bank,** p. 420.

Adding Fractions, Different Denominators

MODEL

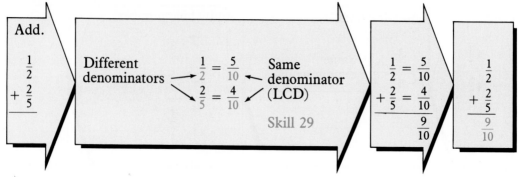

 GET READY. Write the LCD.

1. $\frac{1}{2}, \frac{3}{4}$ 2. $\frac{3}{7}, \frac{1}{6}$ 3. $\frac{3}{5}, \frac{2}{15}$ 4. $\frac{5}{6}, \frac{9}{10}$

Write as fractions with the LCD.

5. $\frac{1}{2}, \frac{5}{6}$ 6. $\frac{2}{3}, \frac{4}{7}$ 7. $\frac{3}{4}, \frac{11}{16}$ 8. $\frac{5}{6}, \frac{7}{9}$

 NOW USE SKILL 34. Write the answer in lowest terms.

9. $\frac{1}{3}$
$+\frac{5}{6}$

10. $\frac{5}{8}$
$+\frac{3}{4}$

11. $\frac{7}{9}$
$+\frac{5}{6}$

12. $\frac{1}{2}$
$+\frac{7}{8}$

13. $\frac{2}{3}$
$+\frac{8}{9}$

14. $\frac{1}{2}$
$+\frac{9}{10}$

15. $\frac{13}{15}$
$+\frac{4}{9}$

16. $\frac{5}{6}$
$+\frac{23}{24}$

17. $\frac{5}{21}$
$+\frac{25}{28}$

18. $\frac{3}{8}$
$+\frac{29}{32}$

19. $\frac{2}{3}$
$+\frac{8}{11}$

20. $\frac{9}{10}$
$+\frac{2}{3}$

21. $\frac{5}{16}$
$+\frac{5}{6}$

22. $\frac{3}{8}$
$+\frac{5}{6}$

23. $\frac{4}{5}$
$+\frac{5}{12}$

24. $\frac{1}{2} + \frac{5}{6} + \frac{2}{3}$ 25. $\frac{3}{5} + \frac{9}{10} + \frac{1}{2}$ 26. $\frac{15}{16} + \frac{7}{8} + \frac{3}{4}$

For more practice: **Skill Bank,** p. 421.

Adding Mixed Numbers, Different Denominators

MODEL

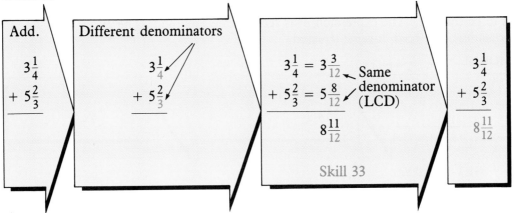

 GET READY. Write as fractions with the LCD.

1. $\frac{5}{7}$, $\frac{1}{2}$ **2.** $\frac{4}{5}$, $\frac{3}{4}$ **3.** $\frac{3}{5}$, $\frac{8}{15}$ **4.** $\frac{5}{6}$, $\frac{3}{8}$

Write as a mixed number in lowest terms.

5. $4\frac{4}{6}$ **6.** $12\frac{6}{9}$ **7.** $\frac{7}{4}$ **8.** $\frac{40}{15}$

 NOW USE SKILL 35. Write the answer in lowest terms.

9. $2\frac{1}{4}$ $+\ 4\frac{5}{6}$

10. $5\frac{5}{6}$ $+\ 7\frac{4}{9}$

11. $3\frac{4}{5}$ $+\ 3\frac{1}{7}$

12. $4\frac{3}{4}$ $+\ 8\frac{5}{6}$

13. $3\frac{8}{9}$ $+\ 5\frac{5}{7}$

14. $7\frac{5}{6}$ $+\ 3\frac{11}{14}$

15. $5\frac{1}{4}$ $+\ 6\frac{9}{13}$

16. $9\frac{7}{8}$ $+\ 8\frac{5}{12}$

17. $8\frac{1}{6}$ $+\ 4\frac{13}{21}$

18. $7\frac{3}{5}$ $+\ 2\frac{6}{11}$

19. $5\frac{9}{14}$ $+\ 3\frac{11}{18}$

20. $7\frac{13}{15}$ $+\ 11\frac{11}{18}$

21. $9\frac{7}{12}$ $+\ 6\frac{3}{14}$

22. $4\frac{5}{18}$ $+\ 13\frac{7}{24}$

23. $14\frac{5}{12}$ $+\ 9\frac{4}{15}$

For more practice: **Skill Bank,** p. 421.

Squares and Square Roots

When you multiply a number by itself, the answer is called the **square** of the number. A raised "2" is the symbol used to show that a number is to be squared. The expression 3^2 is read "three squared." The expression 12^2 is read "twelve squared."

MODEL

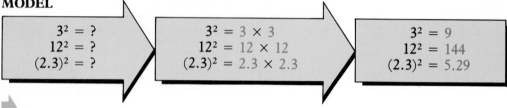

$3^2 = ?$
$12^2 = ?$
$(2.3)^2 = ?$

$3^2 = 3 \times 3$
$12^2 = 12 \times 12$
$(2.3)^2 = 2.3 \times 2.3$

$3^2 = 9$
$12^2 = 144$
$(2.3)^2 = 5.29$

 GET READY.

1.	4	**2.**	5	**3.**	7	**4.**	9	**5.**	10	**6.**	11
	$\times 4$		$\times 5$		$\times 7$		$\times 9$		$\times 10$		$\times 11$

 NOW USE APPLICATION A15. Complete the statement.

7. $6^2 = ?$ **8.** $8^2 = ?$ **9.** $17^2 = ?$ **10.** $20^2 = ?$ **11.** $(4.5)^2 = ?$

Because $9 = 3^2$, you can say that the **square root** of 9 is 3. The symbol for square root is $\sqrt{}$. The expression $\sqrt{9}$ is read "the square root of 9." You write $\sqrt{9} = 3$.

MODEL

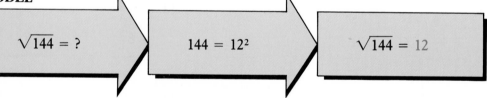

$\sqrt{144} = ?$

$144 = 12^2$

$\sqrt{144} = 12$

 GET READY. Complete the statement.

12. $16 = 4^2, \sqrt{16} = ?$ **13.** $25 = 5^2, \sqrt{25} = ?$ **14.** $49 = 7^2, \sqrt{49} = ?$

 NOW USE APPLICATION A15.

15. $\sqrt{36} = ?$ **16.** $\sqrt{64} = ?$ **17.** $\sqrt{81} = ?$ **18.** $\sqrt{100} = ?$

19. $\sqrt{121} = ?$ **20.** $\sqrt{196} = ?$ **21.** $\sqrt{400} = ?$ **22.** $\sqrt{900} = ?$

For more practice: **Application Bank,** p. 454.

Right Triangles

If the sum of the squares of the two shorter sides equals the square of the long side, the triangle is a right triangle.

MODELS

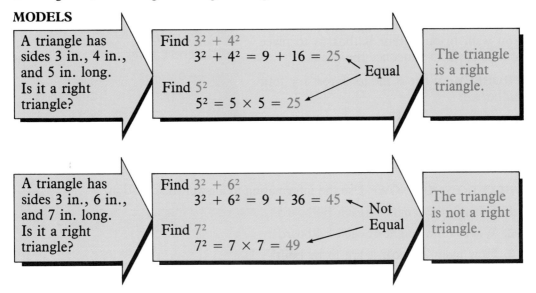

This method of finding out whether a triangle is a right triangle is based on the observations of Pythagoras, a Greek mathematician of the sixth century B.C.

Pythagorean Theorem

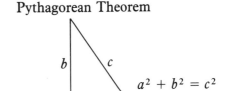

$$a^2 + b^2 = c^2$$

Right angle

 GET READY. Complete the statement.

1. $1^2 + 2^2 = ?$ **2.** $2^2 + 4^2 = ?$ **3.** $8^2 + 10^2 = ?$ **4.** $12^2 + 3^2 = ?$

 NOW USE APPLICATION A16. The lengths of the sides of a triangle are given. Is the triangle a right triangle?

5. 10 ft, 24 ft, 26 ft   **6.** 5 cm, 12 cm, 13 cm   **7.** 9 in., 13 in., 17 in.

8. 15 m, 20 m, 30 m   **9.** 8 yd, 15 yd, 17 yd   **10.** 4 mi, 9.2 mi, 11 mi

For more practice: **Application Bank,** p. 454.   139

Carver High Booster Club

You are the president of the Carver High School Booster Club. You want to raise money for the marching band. You appoint a committee to organize an arts and crafts fair. You want to build four billboards to place around town to announce the event.

PROBLEM: What is the total cost for four billboards?

Solution Plan: Find the materials needed for the billboards and for the supports. Find the total cost.

WHAT DO YOU KNOW? Use the information to answer the question.

1. What is the length of each billboard?

2. What is the height of each billboard?

3. How many billboards are to be built?

WHAT DO YOU WANT TO KNOW? Find the materials needed.

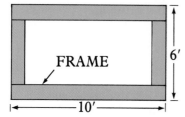

FRAME: Use boards 6″ wide, with no overlap.
FACE: Use 4′ by 6′ sheets of plywood available from a salvage dealer.

First build the frame. Then nail the plywood onto the frame. Make sure the edges of the plywood are flush with the outer frame edges.

4. How many horizontal boards are needed for each frame?

5. Can the two vertical pieces of the frame be cut from one 10-foot length?

6. Six-inch boards are available in 10-foot lengths. How many 10-foot lengths are needed to construct the frame of each billboard?

7. How many 10-foot lengths are needed to construct the frames of four billboards?

8. How many sheets of plywood are needed for the face of a billboard?

9. How many sheets of plywood are needed for four billboards?

WHAT MORE DO YOU WANT? Find the materials needed to support the billboards.

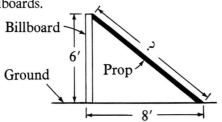

$$6^2 + 8^2 = (\text{length of prop})^2$$

Application A16

10. The billboard forms a right angle with the ground. One end of the prop is 8 feet from the base of the billboard. How long does the prop need to be?

11. Four props are used to support each billboard. How many 10-foot lengths of board are needed to support the four billboards?

NOW SOLVE THE PROBLEM. Find the cost. Fill out your worksheet.

Estimated cost to build and set up four billboards

12. *Frames:* ? *10-foot lengths @ $2.60 each* = $?
13. *Faces:* ? *sheets of plywood @ $13.50 each* = $?
14. *Props:* ? *10-foot lengths @ $1.65 each* = $?
15. *2 pounds of #6 nails @ $2.35 a pound* = $?
16. *Total* = $?

Skill 32	Skill 33	Skill 34	Skill 35
1. $\dfrac{11}{16}$ $+\dfrac{15}{16}$	2. $16\dfrac{3}{7}$ $+13\dfrac{6}{7}$	3. $\dfrac{3}{8}$ $+\dfrac{5}{12}$	4. $9\dfrac{13}{20}$ $+1\dfrac{14}{25}$

Application A15

5. $13^2 = ?$

6. $\sqrt{169} = ?$

Application A16

7. The lengths of the sides of a triangle are 7.5 in., 10 in., and 12.5 in. Is it a right triangle?

Problem Solving

8. You are building a gate for a fence. The gate is 5 ft high and 12 ft wide. For extra strength you add a brace to run from the lower left corner to the upper right corner. How long is the brace?

CAREER CLIPPINGS

To find elapsed time within a 24-hour period, subtract the earlier time from the later time. When necessary, you can "borrow" 60 minutes from the hours.

Kimo was on call from 6:30 A.M. to 10:05 A.M. How long was he on call?

$$\overset{\text{borrow 60 min}}{}$$

later time	10:05 A.M.	9:65 A.M.
— earlier time	— 6:30 A.M.	— 6:30 A.M.
		3:35, or 3 hours 35 minutes

If the time period goes beyond 1:00, add 12 hours to the later time.

Alice saw patients from 7:15 A.M. until 2:25 P.M. How long was this?

$$\overset{\text{add 12 h}}{}$$

later time	2:25 P.M.	14:25 P.M.
— earlier time	— 7:15 A.M.	— 7:15 A.M.
		7:10, or 7 hours 10 minutes

Find the elapsed time.

1. 7:15 A.M. to 11:25 A.M.   2. 9:05 A.M. to 12:20 P.M.   3. 3:20 A.M. to 8:05 A.M.

4. 7:40 P.M. to 11:10 P.M.   5. 10:25 A.M. to 2:30 P.M.   6. 11:40 P.M. to 8:25 A.M.

Pythagorean Theorem

Using a calculator and the Pythagorean theorem, you can check to see if a triangle is a right triangle.

First, recall the formula for the Pythagorean theorem:

$$a^2 + b^2 = c^2$$

Use the calculator to find $a^2 + b^2$. Then compare this sum with c^2.

The lengths of the sides of a triangle are 8 cm, 11 cm, and 17 cm. Is it a right triangle?

KEY-IN	DISPLAY	
8	8	
⊠	8	
8	8	
⊜	64	⟵ a^2
11	11	
⊠	11	
11	11	
⊜	121	⟵ b^2
64	64	
⊞	64	
121	121	
⊜	185	⟵ $a^2 + b^2$
17	17	
⊠	17	Not equal. The triangle
17	17	is not a right triangle.
⊜	289	⟵ c^2

The lengths of the sides of a triangle are given. Is it a right triangle?

1. 9 ft, 12 ft, 15 ft

2. 7 in., 11 in., 13 in.

3. 5 m, 12 m, 13 m

4. 5.25 cm, 5.25 cm, 7.5 cm

5. 6 m, 8 m, 10 m

6. 7 ft, 14 ft, 17 ft

7. 43.9 cm, 36.8 cm, 67 cm

8. 27 in., 36 in., 45 in.

9. 48 yd, 64 yd, 80 yd

10. 723 in., 964 in., 1146 in.

Subtracting Fractions, Same Denominator

MODEL

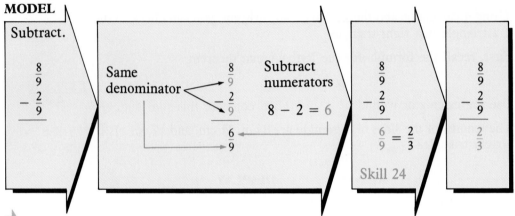

Subtract.

$$\frac{8}{9}$$
$$-\frac{2}{9}$$

Same denominator

$$\frac{8}{9}$$
$$-\frac{2}{9}$$
$$\frac{6}{9}$$

Subtract numerators

$$8 - 2 = 6$$

$$\frac{8}{9}$$
$$-\frac{2}{9}$$
$$\frac{6}{9} = \frac{2}{3}$$

Skill 24

$$\frac{8}{9}$$
$$-\frac{2}{9}$$
$$\frac{2}{3}$$

 GET READY. Write in lowest terms.

1. $\frac{2}{8}$　　**2.** $\frac{4}{10}$　　**3.** $\frac{3}{12}$　　**4.** $\frac{9}{15}$　　**5.** $\frac{10}{20}$　　**6.** $\frac{15}{21}$

7. $\frac{36}{42}$　　**8.** $\frac{13}{39}$　　**9.** $\frac{15}{25}$　　**10.** $\frac{16}{28}$　　**11.** $\frac{48}{100}$　　**12.** $\frac{18}{36}$

 NOW USE SKILL 36. Write the answer in lowest terms.

13. $\frac{2}{3}$ $-\frac{1}{3}$　　**14.** $\frac{3}{4}$ $-\frac{1}{4}$　　**15.** $\frac{5}{6}$ $-\frac{1}{6}$　　**16.** $\frac{7}{8}$ $-\frac{5}{8}$　　**17.** $\frac{3}{5}$ $-\frac{2}{5}$　　**18.** $\frac{8}{9}$ $-\frac{5}{9}$

19. $\frac{7}{9}$ $-\frac{5}{9}$　　**20.** $\frac{3}{7}$ $-\frac{1}{7}$　　**21.** $\frac{5}{9}$ $-\frac{2}{9}$　　**22.** $\frac{4}{5}$ $-\frac{1}{5}$　　**23.** $\frac{5}{8}$ $-\frac{1}{8}$　　**24.** $\frac{4}{5}$ $-\frac{2}{5}$

25. $\frac{6}{7}$ $-\frac{3}{7}$　　**26.** $\frac{4}{5}$ $-\frac{3}{5}$　　**27.** $\frac{11}{12}$ $-\frac{7}{12}$　　**28.** $\frac{10}{11}$ $-\frac{5}{11}$　　**29.** $\frac{13}{15}$ $-\frac{7}{15}$　　**30.** $\frac{11}{12}$ $-\frac{5}{12}$

31. $\frac{11}{14}$ $-\frac{3}{14}$　　**32.** $\frac{15}{16}$ $-\frac{11}{16}$　　**33.** $\frac{13}{18}$ $-\frac{5}{18}$　　**34.** $\frac{21}{25}$ $-\frac{16}{25}$　　**35.** $\frac{17}{18}$ $-\frac{5}{18}$　　**36.** $\frac{17}{21}$ $-\frac{10}{21}$

　　　　For more practice: **Skill Bank,** page 422.

Subtracting Mixed Numbers, Same Denominator

MODEL

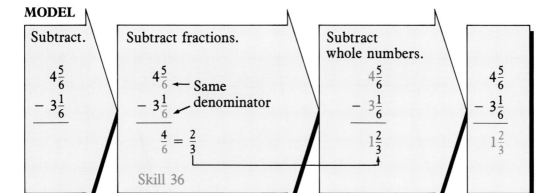

 GET READY. Write in lowest terms.

1. $\frac{2}{4}$ **2.** $\frac{6}{9}$ **3.** $\frac{15}{25}$ **4.** $\frac{9}{18}$ **5.** $\frac{6}{15}$ **6.** $\frac{20}{30}$

7. $3\frac{2}{6}$ **8.** $5\frac{6}{8}$ **9.** $10\frac{14}{20}$ **10.** $4\frac{6}{12}$ **11.** $20\frac{9}{21}$ **12.** $15\frac{85}{100}$

 NOW USE SKILL 37. Write the answer in lowest terms.

13. $4\frac{3}{4} - 3\frac{1}{4}$ **14.** $6\frac{3}{5} - 2\frac{2}{5}$ **15.** $7\frac{5}{8} - 4\frac{3}{8}$ **16.** $9\frac{8}{9} - 8\frac{5}{9}$ **17.** $8\frac{5}{6} - 3\frac{1}{6}$

18. $10\frac{5}{9} - 4\frac{2}{9}$ **19.** $15\frac{7}{8} - 7\frac{1}{8}$ **20.** $18\frac{5}{6} - 9\frac{1}{6}$ **21.** $20\frac{3}{4} - 5\frac{1}{4}$ **22.** $25\frac{5}{8} - 7\frac{3}{8}$

23. $27\frac{11}{16} - 25\frac{9}{16}$ **24.** $16\frac{17}{18} - 11\frac{7}{18}$ **25.** $48\frac{13}{15} - 23\frac{4}{15}$ **26.** $39\frac{17}{20} - 27\frac{3}{20}$ **27.** $68\frac{22}{25} - 55\frac{6}{25}$

28. $8\frac{8}{9} - 2\frac{2}{9}$ **29.** $10\frac{8}{9} - 4\frac{5}{9}$ **30.** $7\frac{17}{20} - 3\frac{13}{20}$

For more practice: **Skill Bank,** p. 422.

Subtracting from Whole Numbers

MODEL

Subtract.	Borrow 1 from 7.	
7	$7 = 6\frac{3}{3}$ ← Same	7
$-2\frac{1}{3}$	$-2\frac{1}{3} = 2\frac{1}{3}$ ← denominator	$-2\frac{1}{3}$
	$4\frac{2}{3}$	$4\frac{2}{3}$

Skill 37

 GET READY. Write as a mixed number.

1. $2 = 1\frac{?}{4}$ **2.** $5 = 4\frac{?}{7}$ **3.** $10 = 9\frac{?}{5}$ **4.** $12 = 11\frac{?}{10}$ **5.** $24 = 23\frac{?}{15}$

Write in lowest terms.

6. $4\frac{2}{4}$ **7.** $5\frac{6}{9}$ **8.** $12\frac{12}{36}$ **9.** $19\frac{9}{15}$ **10.** $27\frac{15}{20}$

 NOW USE SKILL 38. Write the answer in lowest terms.

11. 7
$-\frac{2}{5}$

12. 9
$-\frac{3}{8}$

13. 4
$-\frac{2}{3}$

14. 11
$-\frac{5}{9}$

15. 18
$-\frac{9}{10}$

16. 3
$-\frac{13}{16}$

17. 4
$-\frac{7}{15}$

18. 9
$-\frac{8}{19}$

19. 8
$-\frac{19}{20}$

20. 6
$-\frac{10}{13}$

21. 4
$-2\frac{4}{5}$

22. 9
$-3\frac{2}{3}$

23. 8
$-6\frac{7}{9}$

24. 9
$-7\frac{2}{5}$

25. 7
$-4\frac{5}{8}$

26. $6 - \frac{3}{8}$ **27.** $9 - \frac{5}{9}$ **28.** $11 - 8\frac{11}{12}$

146

For more practice: **Skill Bank**, p. 423.

Subtracting Mixed Numbers, with Borrowing

MODEL

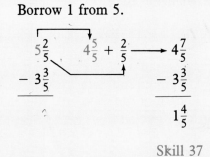

Subtract.

$5\frac{2}{5}$

$- 3\frac{3}{5}$

Borrow 1 from 5.

$5\frac{2}{5}$ $4\frac{5}{5} + \frac{2}{5} \longrightarrow 4\frac{7}{5}$

$- 3\frac{3}{5}$ $- 3\frac{3}{5}$

$1\frac{4}{5}$

Skill 37

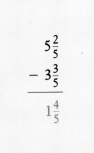

$5\frac{2}{5}$

$- 3\frac{3}{5}$

$1\frac{4}{5}$

GET READY. Borrow 1 from the whole number and write the equivalent mixed number.

1. $8\frac{1}{4}$ **2.** $6\frac{2}{3}$ **3.** $10\frac{3}{8}$ **4.** $12\frac{11}{15}$ **5.** $18\frac{5}{12}$

Write in lowest terms.

6. $3\frac{12}{16}$ **7.** $9\frac{9}{12}$ **8.** $12\frac{24}{60}$ **9.** $6\frac{80}{100}$ **10.** $11\frac{36}{144}$

NOW USE SKILL 39. Write the answer in lowest terms.

11. $8\frac{1}{4}$ **12.** $6\frac{2}{5}$ **13.** $7\frac{1}{8}$ **14.** $9\frac{1}{6}$ **15.** $5\frac{4}{7}$

   $- 5\frac{3}{4}$ $- 4\frac{3}{5}$ $- 4\frac{3}{8}$ $- 3\frac{5}{6}$ $- 3\frac{5}{7}$

16. $17\frac{2}{7}$ **17.** $18\frac{3}{5}$ **18.** $12\frac{13}{15}$ **19.** $37\frac{9}{14}$ **20.** $46\frac{26}{31}$

   $- 9\frac{5}{7}$ $- 6\frac{4}{5}$ $- 8\frac{14}{15}$ $- 25\frac{13}{14}$ $- 20\frac{27}{31}$

21. $40\frac{13}{21}$ **22.** $73\frac{9}{56}$ **23.** $80\frac{29}{37}$ **24.** $59\frac{3}{104}$ **25.** $30\frac{2}{99}$

   $- 24\frac{19}{21}$ $- 42\frac{31}{56}$ $- 67\frac{30}{37}$ $- 25\frac{37}{104}$ $- 16\frac{7}{99}$

For more practice: **Skill Bank,** p. 423.

Perimeter

The **perimeter** of a figure is the distance around it. For a polygon, the perimeter is the sum of the lengths of the sides.

MODEL

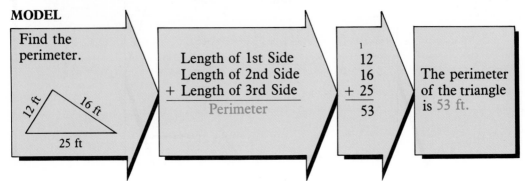

Find the perimeter.	Length of 1st Side Length of 2nd Side + Length of 3rd Side ——————— Perimeter

$$\begin{array}{r} \overset{1}{12} \\ 16 \\ +\ 25 \\ \hline 53 \end{array}$$

The perimeter of the triangle is 53 ft.

▶ **GET READY.**

1. $\begin{array}{r} 2.3 \\ +\ 1.6 \\ \hline \end{array}$

2. $\begin{array}{r} 19.5 \\ +\ 13.8 \\ \hline \end{array}$

3. $\begin{array}{r} 3.26 \\ +\ 5.88 \\ \hline \end{array}$

4. $\begin{array}{r} 7\frac{1}{4} \\ 4\frac{5}{8} \\ +\ 2\frac{3}{4} \\ \hline \end{array}$

5. $\begin{array}{r} 3\frac{3}{8} \\ 3\frac{3}{8} \\ +\ 3\frac{3}{8} \\ \hline \end{array}$

▶ **NOW USE APPLICATION A17. Find the perimeter.**

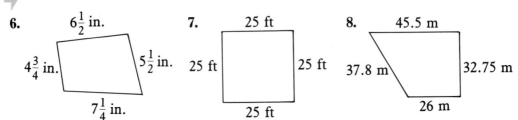

6. $6\frac{1}{2}$ in. $4\frac{3}{4}$ in. $5\frac{1}{2}$ in. $7\frac{1}{4}$ in.

7. 25 ft   25 ft   25 ft   25 ft

8. 45.5 m   37.8 m   32.75 m   26 m

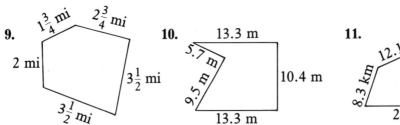

9. $1\frac{3}{4}$ mi $2\frac{3}{4}$ mi   2 mi $3\frac{1}{2}$ mi $3\frac{1}{2}$ mi

10. 13.3 m   5.7 m   9.5 m   10.4 m   13.3 m

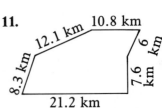

11. 8.3 km   12.1 km   10.8 km   6 km   7.6 km   21.2 km

If all the sides have the same length, you can find the perimeter by multiplying.

MODEL

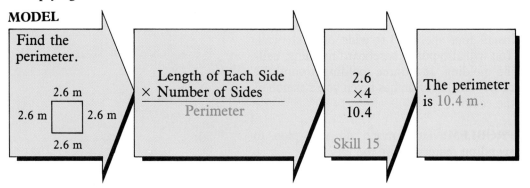

Find the perimeter.

2.6 m
2.6 m ☐ 2.6 m
2.6 m

Length of Each Side
× Number of Sides
—————————————
Perimeter

2.6
×4
————
10.4

Skill 15

The perimeter is 10.4 m.

NOW USE APPLICATION A17. Complete the table.

	12.	13.	14.	15.	16.	17.
Number of equal sides	9	4	3	7	5	6
Length of one side	14.2 m	20 ft	7.25 in.	4.6 km	35 in.	?
Perimeter	?	?	?	?	?	33 in.

18. Serge wants to put reflective tape around the edges of a stop sign. The sign has 8 sides, each 7.5 in. long. How much tape does he need?

19. Maria wants to make a picture frame 9 in. wide and 12 in. high. How long a piece of molding should she buy to make the four frame pieces?

20. Utah is shaped approximately as shown in the diagram. The distances are all in miles. What is the total length of Utah's borders?

21. Each of the 5 sides of the Pentagon is 921 ft long. If you were to walk all the way around the edge of the building, how far would you go?

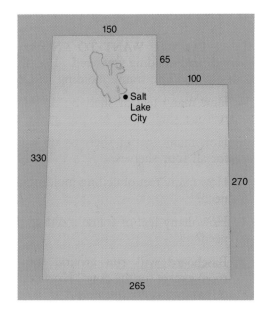

150
65
100
• Salt Lake City
330
270
265

For more practice: **Application Bank,** p. 455.

Hammersmith Carpentry Co.

You are doing the carpentry work to remodel the library's reserve book room. You will install four shelves 1 ft wide on each wall. You will also put in baseboard molding, ceiling molding, and frame molding around the door. The librarian has given you a sketch of the room.

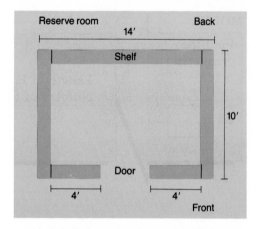

PROBLEM: How much do you plan to spend on materials for the job?

Solution Plan: Find the length of shelving and molding needed. Find the cost of each item. Add to find the total cost.

WHAT DO YOU KNOW? You know the number of shelves, the measurements of the room, and the width of the shelves.

1. For a single level of shelving, how many feet of shelving are needed on the **a.** left wall? **b.** right wall? **c.** back wall? **d.** front wall?

2. How many shelves will be on each wall?

3. What is the distance around the edge of the ceiling?

WHAT DO YOU WANT TO KNOW? You want to find how much shelving and molding you need. The librarian gives you a sketch of the door frame. The frame molding will be 6 in. wide.

4. How many feet of shelving do you need for one shelf level for the whole room?

5. How many feet of shelving do you need for all four shelves?

6. How many feet of ceiling molding do you need?

7. How many feet of frame molding do you need?

8. Baseboard will run around the entire room except for the door. How many feet of baseboard do you need?

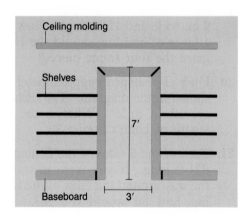

NOW SOLVE THE PROBLEM. Use the prices in the purchase order below to estimate the cost of materials. Round the lengths needed to the nearest ten feet; round the cost per foot to the nearest dollar. Multiply to find the estimated cost.

9. Shelving

10. Ceiling molding

11. Frame moding

12. Baseboard

13. Add to find the estimated total cost.

Figure the actual costs and fill out your purchase order for the materials.

	Quantity	Description	Price	Amount
	HAMMERSMITH CARPENTRY CO. 201 Shamrock Road Cheyenne, Wyoming 82001		**PURCHASE ORDER** Date: 7/22 Purchase Order No. 2462	
14.	? ft	Library shelf, 1 ft wide	$6.75/ft	?
15.	? ft	Baseboard	$1.20/ft	?
16.	? ft	Ceiling molding	$1.85/ft	?
17.	? ft	Frame molding	$2.30/ft	?
18.			TOTAL	?

Quiz

1. $\dfrac{15}{16}$ **2.** $13\dfrac{9}{10}$ **3.** 7 **4.** $10\dfrac{5}{12}$

     $-\dfrac{13}{16}$ $-10\dfrac{3}{10}$ $-3\dfrac{3}{8}$ $-3\dfrac{7}{12}$

Application A17

5. Find the perimeter of a hexagon (6 sides) with each side 3.25 cm long.

Problem Solving

6. You are building a solar heating system. Each collector panel is 3 feet by 9 feet. Tubes are arranged in each panel as shown. How many feet of tubing will you need for each panel? At $3 per foot, what is the total cost of the tubing for one panel?

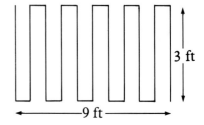

CONSUMER NOTE

        Your car should be checked and serviced regularly for safety, dependability, and fuel efficiency. The chart shows the recommended maintenance schedule for your car.

MAINTENANCE SCHEDULE	for cars driven under normal conditions.
Check fluid levels	Every 12 months or 7500 miles (whichever comes first)
Change oil, filter	Every 12 months or 7500 miles
Rotate radial tires	At 7500 miles, then every 15,000 miles
Check exhaust system	Every 12 months or 7500 miles
Mechanic's safety check	Every 12 months or 7500 miles
Tune-up	Every 30,000 miles

You expect to drive your new car about 15,000 miles a year. How many times should the following services be done during the first year?

1. check fluid levels   **2.** change oil, filter   **3.** mechanic's safety check

4. rotate radial tires   **5.** tune-up   **6.** check exhaust system

Weekly Time Card

Arnie Levesque is a cutter in a garment factory. He uses special tools to cut out pieces of fabric according to a pattern. Stitchers then sew the pieces together to make the garment.

Arnie keeps a time card at the factory. It shows the time he starts and stops working every day. On Monday, Arnie worked from 6:55 A.M. to 11:35 A.M. and from 12:25 P.M. to 3:30 P.M. How many hours did he work altogether?

Find the elapsed times and add. (page 142)

$$
\begin{array}{ll}
11{:}35 \text{ A.M.} \to & 10{:}95 \text{ A.M.} \\
-\ 6{:}55 \text{ A.M.} & -\ 6{:}55 \text{ A.M.} \\
\hline
& 4{:}40
\end{array}
\qquad
\begin{array}{ll}
3{:}30 \text{ P.M.} \to & 15{:}30 \text{ P.M.} \\
-\ 12{:}25 \text{ P.M.} & -\ 12{:}25 \text{ P.M.} \\
\hline
& 3{:}05
\end{array}
\qquad
\begin{array}{r}
4{:}40 \\
+\ 3{:}05 \\
\hline
7{:}45
\end{array}
$$

He worked a total of 7 hours 45 minutes.

Find the number of hours Arnie worked each day and the total number of hours for the week.

M & J Fashions Weekly Time Card			NAME: _Levesque, a._ Week of: _2/3_		
DATE	IN	OUT	IN	OUT	HOURS
2/3	6:55	11:35	12:25	3:30	7:45
2/4	6:50	12:05	12:35	4:03	?
2/5	7:05	1:15	2:00	5:40	?
2/6	6:50	11:50	12:25	3:35	?
2/7	7:00	12:00	12:30	4:20	?
				TOTAL FOR WEEK	?

Subtracting Fractions, Different Denominators

MODEL

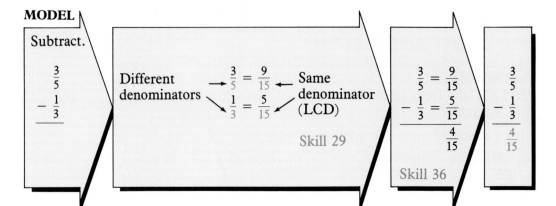

Subtract.

$\frac{3}{5}$
$-\frac{1}{3}$

Different denominators $\rightarrow \frac{3}{5} = \frac{9}{15}$ $\leftarrow$ Same denominator (LCD)

$\frac{1}{3} = \frac{5}{15}$

Skill 29

$\frac{3}{5} = \frac{9}{15}$
$-\frac{1}{3} = \frac{5}{15}$
$\frac{4}{15}$

Skill 36

$\frac{3}{5}$
$-\frac{1}{3}$
$\frac{4}{15}$

 GET READY. Write the LCD.

1. $\frac{1}{2}, \frac{1}{5}$ **2.** $\frac{3}{4}, \frac{7}{9}$ **3.** $\frac{5}{12}, \frac{4}{9}$ **4.** $\frac{1}{6}, \frac{8}{15}$

Write as fractions with the LCD.

5. $\frac{3}{4}, \frac{2}{3}$ **6.** $\frac{11}{12}, \frac{13}{20}$ **7.** $\frac{3}{8}, \frac{5}{6}$ **8.** $\frac{1}{2}, \frac{5}{7}$

 NOW USE SKILL 40. Write the answer in lowest terms.

9. $\frac{1}{2}$ $-\frac{1}{4}$ **10.** $\frac{5}{6}$ $-\frac{3}{8}$ **11.** $\frac{1}{6}$ $-\frac{1}{9}$ **12.** $\frac{1}{5}$ $-\frac{1}{7}$ **13.** $\frac{3}{4}$ $-\frac{1}{6}$

14. $\frac{1}{3}$ $-\frac{1}{8}$ **15.** $\frac{7}{8}$ $-\frac{1}{6}$ **16.** $\frac{7}{8}$ $-\frac{5}{6}$ **17.** $\frac{2}{3}$ $-\frac{3}{8}$ **18.** $\frac{5}{9}$ $-\frac{1}{6}$

19. $\frac{4}{5}$ $-\frac{5}{12}$ **20.** $\frac{3}{4}$ $-\frac{9}{14}$ **21.** $\frac{5}{6}$ $-\frac{13}{16}$ **22.** $\frac{11}{12}$ $-\frac{7}{18}$ **23.** $\frac{13}{21}$ $-\frac{3}{14}$

For more practice: **Skill Bank,** page 424.

Subtracting Mixed Numbers, Different Denominators

MODEL

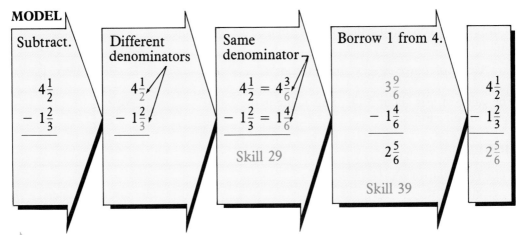

Subtract.	Different denominators	Same denominator	Borrow 1 from 4.	
$4\frac{1}{2}$	$4\frac{1}{2}$	$4\frac{1}{2} = 4\frac{3}{6}$	$3\frac{9}{6}$	$4\frac{1}{2}$
$-1\frac{2}{3}$	$-1\frac{2}{3}$	$-1\frac{2}{3} = 1\frac{4}{6}$	$-1\frac{4}{6}$	$-1\frac{2}{3}$
		Skill 29	$2\frac{5}{6}$	$2\frac{5}{6}$
			Skill 39	

 GET READY. Write as mixed numbers with the LCD.

1. $7\frac{3}{4}, 6\frac{1}{8}$ **2.** $5\frac{5}{6}, 4\frac{4}{5}$ **3.** $4\frac{5}{12}, 1\frac{1}{4}$ **4.** $2\frac{4}{9}, 3\frac{5}{6}$

 NOW USE SKILL 41. Write the answer in lowest terms.

5. $\begin{array}{r} 6\frac{6}{7} \\ -3\frac{4}{5} \\ \hline \end{array}$ **6.** $\begin{array}{r} 9\frac{1}{6} \\ -1\frac{3}{8} \\ \hline \end{array}$ **7.** $\begin{array}{r} 7\frac{2}{3} \\ -2\frac{5}{8} \\ \hline \end{array}$ **8.** $\begin{array}{r} 6\frac{5}{6} \\ -3\frac{5}{9} \\ \hline \end{array}$ **9.** $\begin{array}{r} 5\frac{2}{3} \\ -1\frac{1}{7} \\ \hline \end{array}$

10. $\begin{array}{r} 18\frac{5}{8} \\ -7\frac{7}{12} \\ \hline \end{array}$ **11.** $\begin{array}{r} 24\frac{5}{6} \\ -6\frac{19}{21} \\ \hline \end{array}$ **12.** $\begin{array}{r} 33\frac{13}{14} \\ -8\frac{1}{6} \\ \hline \end{array}$ **13.** $\begin{array}{r} 15\frac{2}{3} \\ -8\frac{5}{16} \\ \hline \end{array}$ **14.** $\begin{array}{r} 21\frac{4}{9} \\ -4\frac{7}{12} \\ \hline \end{array}$

15. $\begin{array}{r} 22\frac{19}{30} \\ -13\frac{11}{15} \\ \hline \end{array}$ **16.** $\begin{array}{r} 43\frac{16}{21} \\ -27\frac{8}{15} \\ \hline \end{array}$ **17.** $\begin{array}{r} 54\frac{17}{24} \\ -35\frac{9}{16} \\ \hline \end{array}$ **18.** $\begin{array}{r} 66\frac{9}{14} \\ -29\frac{17}{21} \\ \hline \end{array}$ **19.** $\begin{array}{r} 81\frac{16}{21} \\ -52\frac{17}{24} \\ \hline \end{array}$

20. $8\frac{5}{6} - 3\frac{2}{7}$ **21.** $14\frac{1}{6} - 8\frac{13}{16}$ **22.** $22\frac{9}{16} - 18\frac{11}{22}$

For more practice: **Skill Bank,** page 424.

Circumference

The distance around a circle is called the **circumference** of the circle. You can use the diameter of a circle to find the circumference.

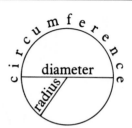

$$\text{Circumference} = \pi \times \text{diameter, or } C = \pi \times d$$

The value of π is about 3.14. This value is the same for all circles.

MODEL

| The diameter of a circle is 8 m. Find the circumference. | $C = \pi \times d$ $C = 3.14 \times 8 = 25.12$ | The circumference is 25.12 m. |

Because the diameter of a circle is twice the radius, the circumference can be expressed as follows:

$$\text{Circumference} = 2 \times \pi \times \text{radius, or } C = 2 \times \pi \times r$$

MODEL

| The radius of a circle is 4 m. Find the circumference. | $C = 2 \times \pi \times r$ $C = 2 \times 3.14 \times 4 = 25.12$ | The circumference is 25.12 m. |

 GET READY. Round the answer to the nearest tenth.

1. 3.14	2. 3.14	3. 3.14	4. 3.14
× 2	× 5	× 7	× 8.2

Complete the table.

	5.	6.	7.	8.	9.	10.	11.
Radius	7 cm	?	15 m	2.5 ft	0.75 yd	?	1500 ft
Diameter	?	25 m	?	?	?	185 m	?

NOW USE APPLICATION A18. Find the circumference of the circle. Round answers to the nearest tenth.

12. $d = 3$ in. **13.** $d = 6$ ft **14.** $d = 10$ yd **15.** $d = 13$ m

16. $d = 4.5$ in. **17.** $r = 12.3$ cm **18.** $r = 9.08$ m **19.** $r = 25.8$ yd

20. A pipe has a diameter of 3 in. If you place 4 strips of tape with no overlapping around the pipe, how many inches of tape would you need? Round answer to the nearest tenth of an inch.

21. Earth has a radius of about 4,000 miles. A satellite is in circular orbit around Earth 350 miles above the equator. How far does the satellite travel in one complete orbit?

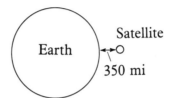

The figures below are made of semicircles (half-circles) and straight lines. Find the perimeter of the figure. Round answers to the nearest tenth. The dashed lines are not part of the perimeter.

22.

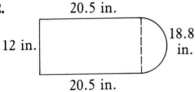

20.5 in.

12 in.

18.8 in.

20.5 in.

23.

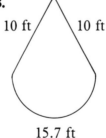

10 ft   10 ft

15.7 ft

24.

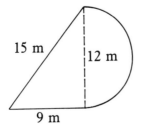

15 m

12 m

9 m

25.

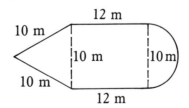

12 m

10 m

10 m

10 m

10 m

12 m

For more practice: **Application Bank,** p. 455.

Summer Garden

You are planning your summer garden. Tomatoes, strawberries, and cabbage are among the things you will plant. During the winter you read articles about some gardening methods you want to use this year. You plan to grow tomatoes in circular cages 2 ft across made from wire fencing. You will place a round cardboard "collar," 4 in. across, around each cabbage seedling to protect it from insects. Strawberries will be grown in a pyramid arrangement to save space.

PROBLEM: What materials will you need if you use these gardening methods?

Solution Plan: Find out how much wire or cardboard you need for each tomato cage or cabbage collar. Multiply to find the total amount of wire and cardboard needed. Find the amount of edging needed for each level of the strawberry pyramid.

WHAT DO YOU KNOW? You know the sizes of the tomato cages and cabbage collars.

1. What is the diameter of each tomato cage?

2. What is the diameter of each cabbage collar?

WHAT DO YOU WANT TO KNOW? You want to know how many tomato and cabbage plants you will have. Drawing sketches of the tomato and cabbage areas will help you to find out.

3. Your tomato patch is 4 ft by 10 ft. How many cages can fit in this space?

4. You will plant two 12 ft rows of cabbage, allowing 2 ft of row for each seedling. How many cabbages will you plant?

NOW SOLVE THE PROBLEM. Find the materials needed to carry out your plans.

5. What length of fence do you need for a tomato cage?

6. How many feet of fence do you need to make all the tomato cages?

7. Will a 50 ft roll of wire fence be enough for the tomato cages?

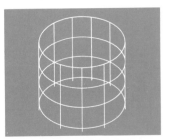

8. How long a piece of cardboard do you need for each cabbage collar (allow 1 in. extra for overlap)?

9. How many inches of cardboard do you need for all the cabbages?

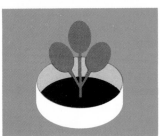

You draw a diagram of the strawberry pyramid. Each level is a circle, with metal edging 8 in. high around it.

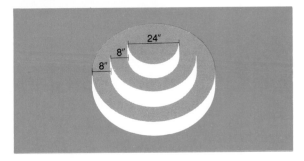

10. How much metal edging do you need for the whole strawberry pyramid?

11. Will a 35 ft roll of edging be enough for the strawberries?

Quiz

1. $\dfrac{2}{7}$
   $-\dfrac{1}{6}$

2. $\dfrac{7}{15}$
   $-\dfrac{5}{12}$

3. $7\dfrac{1}{5}$
   $-6\dfrac{1}{3}$

4. $12\dfrac{3}{8}$
   $-9\dfrac{3}{4}$

Application A18 **Find the circumference of the circle.**

5. $d = 13$ m

6. $r = 0.51$ ft

Problem Solving

7. Years ago farmers used an ox harnessed to a wheel to grind grain. A common arrangement is shown at the right. If the ox turned the wheel 2 complete turns every minute, how far did it walk in an hour?

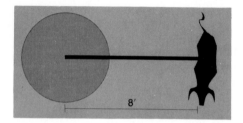

8'

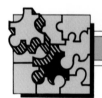 **EXTRA!**

Sports Winnings

1. Beth Daniel won $50,000 in the World Series of Golf. She played a total of 80 holes. How much did she earn per hole?

2. Johnny Rutherford won $1,502,425 at the Indianapolis 500 Mile Auto Race. How much did he earn each mile?

3. Genuine Risk won $250,550 in the Kentucky Derby. The race lasted 122 seconds. How much did Genuine Risk win per second? Round to the nearest cent.

Skills Review

Skill 33 (page 135)

1. $1\frac{1}{4}$ **2.** $5\frac{5}{6}$ **3.** $7\frac{4}{9}$ **4.** $6\frac{5}{7}$ **5.** $8\frac{5}{8}$ **6.** $20\frac{7}{12}$

$+\,3\frac{3}{4}$ $+\,3\frac{1}{6}$ $+\,2\frac{2}{9}$ $+\,4\frac{4}{7}$ $+\,5\frac{7}{8}$ $+\,38\frac{5}{12}$

Skill 35 (page 137)

7. $1\frac{1}{2}$ **8.** $4\frac{2}{7}$ **9.** $9\frac{2}{5}$ **10.** $31\frac{5}{6}$ **11.** $18\frac{4}{9}$ **12.** $13\frac{5}{8}$

$+\,3\frac{3}{7}$ $+\,6\frac{4}{5}$ $+\,4\frac{5}{8}$ $+\,16\frac{3}{8}$ $+\,24\frac{2}{3}$ $+\,28\frac{4}{9}$

Skill 38 (page 146)

13. 5 **14.** 14 **15.** 3 **16.** 4 **17.** 19 **18.** 43

$-\,\frac{1}{2}$ $-\,\frac{5}{6}$ $-\,\frac{7}{9}$ $-\,2\frac{11}{15}$ $-\,6\frac{23}{25}$ $-\,23\frac{17}{18}$

Skill 39 (page 147)

19. $6\frac{1}{3}$ **20.** $7\frac{5}{6}$ **21.** $9\frac{5}{12}$ **22.** $19\frac{13}{15}$ **23.** $31\frac{1}{6}$ **24.** $46\frac{7}{10}$

$-\,2\frac{2}{3}$ $-\,4\frac{1}{6}$ $-\,4\frac{7}{12}$ $-\,10\frac{14}{15}$ $-\,7\frac{5}{6}$ $-\,38\frac{9}{10}$

Skill 40 (page 154)

25. $\frac{8}{9}$ **26.** $\frac{13}{15}$ **27.** $\frac{5}{7}$ **28.** $\frac{7}{12}$ **29.** $\frac{5}{9}$ **30.** $\frac{5}{6}$

$-\,\frac{2}{5}$ $-\,\frac{2}{3}$ $-\,\frac{3}{8}$ $-\,\frac{3}{8}$ $-\,\frac{2}{5}$ $-\,\frac{3}{10}$

Skill 41 (page 155)

31. $7\frac{1}{4}$ **32.** $6\frac{2}{3}$ **33.** $10\frac{1}{5}$ **34.** $22\frac{1}{6}$ **35.** $18\frac{1}{4}$ **36.** $43\frac{7}{16}$

$-\,5\frac{1}{2}$ $-\,3\frac{3}{4}$ $-\,3\frac{5}{8}$ $-\,17\frac{2}{3}$ $-\,7\frac{9}{16}$ $-\,32\frac{3}{8}$

Applications Review

Application A15 **Complete the statement.** (page 138)

1. $11^2 = ?$ 2. $16^2 = ?$ 3. $\sqrt{144} = ?$ 4. $\sqrt{256} = ?$

5. You have 4 rows of cinder blocks with 4 blocks in each row. How many cinder blocks do you have?

Application A16 **The lengths of the sides of a triangle are given. Is it a right triangle?** (page 139)

6. 8 in., 10 in., 14 in.

7. 14 ft, 48 ft, 50 ft

8. 12 m, 16 m, 20 m

9. 10 cm, 14 cm, 17 cm

10. One end of a wire is attached to the top of a flagpole 15 ft tall. The other end of the wire is placed on the ground 8 feet from the base of the pole. The wire is taut, with no slack. How long is the wire?

Application A17 **Find the perimeter of the figure.** (pages 148–149)

11.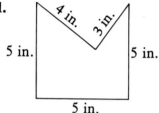

12. A square with each side 9 ft long.

13. On a baseball diamond, the distance between bases is 90 ft. What is the total distance around the diamond?

Application A18 **Complete the table. Round answers to the nearest tenth.** (pages 156–157)

	14.	15.	16.	17.
Radius	10 ft	?	16 yd	?
Diameter	?	13 in.	?	0.7 cm
Circumference	?	?	?	?

18. Jupiter, the largest planet in our solar system, has a diameter of about 86,000 miles. What is its circumference? Round to the nearest thousand.

Unit Test

1. $\dfrac{2}{5}$
$+\dfrac{4}{5}$

2. $\dfrac{7}{8}$
$+\dfrac{3}{8}$

Skill 33

3. $6\dfrac{5}{6}$
$+4\dfrac{1}{6}$

4. $6\dfrac{2}{3}$
$+7\dfrac{2}{3}$

Skill 34

5. $\dfrac{1}{9}$
$+\dfrac{7}{8}$

6. $\dfrac{1}{4}$
$+\dfrac{5}{12}$

Skill 35

7. $3\dfrac{3}{10}$
$+3\dfrac{7}{15}$

8. $4\dfrac{8}{15}$
$+12\dfrac{2}{9}$

Skill 36

9. $\dfrac{7}{10}$
$-\dfrac{3}{10}$

Skill 37

10. $19\dfrac{7}{16}$
$-10\dfrac{3}{16}$

Skill 38

11. 9
$-3\dfrac{5}{7}$

Skill 39

12. $17\dfrac{3}{20}$
$-3\dfrac{7}{20}$

Skill 40

13. $\dfrac{16}{25}$
$-\dfrac{2}{5}$

14. $\dfrac{6}{7}$
$-\dfrac{3}{8}$

Skill 41

15. $10\dfrac{5}{12}$
$-7\dfrac{7}{8}$

16. $5\dfrac{11}{16}$
$-2\dfrac{11}{12}$

Application A15

17. $21^2 = ?$

18. $\sqrt{900} = ?$

Application A16

19. The lengths of the sides of a triangle are 4 cm, 7.5 cm, and 8.5 cm. Is it a right triangle?

Application A17

20. Find the perimeter of a triangle with two sides 4 ft long and one side 5 ft long.

Application A18 **Find the circumference.**

21. $d = 8$ in.

22. $r = 23$ yd

163

6

Multiply and Divide Fractions

Area and Volume

Problem Solving

Unit Preview

 You will use these skills and applications in this unit. Which do you already know? Work each problem.

Skill 42

1. $\frac{3}{8} \times \frac{4}{9}$

Skill 43

2. $\frac{9}{10} \times 12$

Skill 44

3. $\frac{5}{6} \times 1\frac{5}{7}$

Skill 45

4. $3\frac{3}{5} \times 1\frac{7}{8}$

Skill 46 Find the reciprocal.

5. $\frac{5}{8}$

6. $2\frac{1}{2}$

Skill 47

7. $\frac{3}{4} \div \frac{5}{6}$

8. $\frac{5}{9} \div \frac{10}{3}$

Skill 48

9. $\frac{8}{9} \div 16$

Skill 49

10. $2\frac{2}{3} \div \frac{4}{9}$

Skill 50

11. $2\frac{4}{5} \div 7$

Skill 51

12. $3\frac{1}{3} \div 7\frac{1}{2}$

Application A19

13. Find the area of a square with sides 6 in. long.

Application A20

14. Find the area of the triangle.

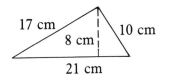

17 cm 8 cm 10 cm 21 cm

Application A21

15. Find the area of a circle whose diameter is 10 ft.

Application A22

16. A window is made up of a semicircle and a rectangle. Find the area, in square feet, of the glass.

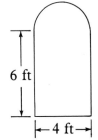

6 ft

|← 4 ft →|

Application A23

17. A crate is 5 ft long, 3 ft wide, and 2 ft high. Find the volume.

Application A24

18. A can is 4 in. tall. The diameter of its base is 2 in. Find the volume.

 Check your answers. If you had difficulty with any skill or application, be sure to study the corresponding lesson in this unit.

165

Multiplying Fractions

MODEL

Multiply.

$$\frac{1}{2} \times \frac{3}{4}$$

Multiply numerators

$$\frac{1}{2} \times \frac{3}{4} = \frac{1 \times 3}{2 \times 4} = \frac{3}{8}$$

Multiply denominators

$$\frac{1}{2} \times \frac{3}{4} = \frac{3}{8}$$

 GET READY.

1. $\frac{2}{5} \times \frac{3}{8} = \frac{? \times 3}{5 \times 8}$

2. $\frac{7}{8} \times \frac{1}{4} = \frac{7 \times ?}{8 \times 4}$

3. $\frac{8}{9} \times \frac{4}{5} = \frac{8 \times 4}{? \times 5}$

4. $\frac{1}{16} \times \frac{6}{7} = \frac{? \times 6}{16 \times 7}$

5. $\frac{6}{7} \times \frac{4}{15} = \frac{6 \times 4}{? \times 15}$

6. $\frac{7}{3} \times \frac{2}{5} = \frac{7 \times 2}{3 \times ?}$

 NOW USE SKILL 42. Write the answer in lowest terms.

7. $\frac{1}{3} \times \frac{1}{2}$

8. $\frac{1}{4} \times \frac{1}{5}$

9. $\frac{2}{3} \times \frac{4}{5}$

10. $\frac{5}{6} \times \frac{3}{4}$

11. $\frac{5}{9} \times \frac{4}{7}$

12. $\frac{5}{8} \times \frac{7}{10}$

13. $\frac{4}{5} \times \frac{1}{3}$

14. $\frac{5}{8} \times \frac{2}{5}$

15. $\frac{6}{11} \times \frac{4}{5}$

16. $\frac{9}{10} \times \frac{5}{6}$

17. $\frac{5}{6} \times \frac{3}{8}$

18. $\frac{2}{3} \times \frac{4}{9}$

19. $\frac{1}{10} \times \frac{3}{4}$

20. $\frac{2}{3} \times \frac{1}{12}$

21. $\frac{1}{6} \times \frac{11}{12}$

22. $\frac{3}{5} \times \frac{4}{15}$

23. $\frac{3}{5} \times \frac{3}{10}$

24. $\frac{4}{7} \times \frac{11}{14}$

25. $\frac{4}{5} \times \frac{7}{12}$

26. $\frac{11}{18} \times \frac{2}{5}$

27. $\frac{13}{20} \times \frac{12}{13}$

28. $\frac{10}{17} \times \frac{17}{30}$

29. $\frac{5}{6} \times \frac{14}{25}$

30. $\frac{17}{20} \times \frac{8}{9}$

31. $\frac{9}{22} \times \frac{11}{18}$

32. $\frac{8}{25} \times \frac{15}{16}$

33. $\frac{55}{56} \times \frac{8}{11}$

34. $\frac{32}{45} \times \frac{3}{4}$

For more practice: **Skill Bank,** page 425.

Multiplying Whole Numbers and Fractions

MODEL

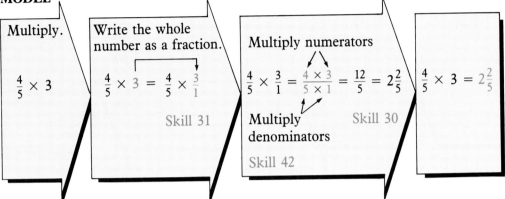

Multiply.

$\frac{4}{5} \times 3$

Write the whole number as a fraction.

$\frac{4}{5} \times 3 = \frac{4}{5} \times \frac{3}{1}$

Skill 31

Multiply numerators

$\frac{4}{5} \times \frac{3}{1} = \frac{4 \times 3}{5 \times 1} = \frac{12}{5} = 2\frac{2}{5}$

Multiply denominators

Skill 42

Skill 30

$\frac{4}{5} \times 3 = 2\frac{2}{5}$

 GET READY.

1. $2 = \frac{?}{1}$ **2.** $7 = \frac{?}{1}$ **3.** $8 = \frac{8}{?}$ **4.** $10 = \frac{?}{1}$ **5.** $12 = \frac{?}{1}$

Write as a whole number or a mixed number in lowest terms.

6. $\frac{9}{4}$ **7.** $\frac{8}{6}$ **8.** $\frac{6}{2}$ **9.** $\frac{9}{8}$ **10.** $\frac{7}{7}$

NOW USE SKILL 43. Write the answer in lowest terms.

11. $6 \times \frac{1}{4}$ **12.** $\frac{1}{5} \times 8$ **13.** $5 \times \frac{1}{7}$ **14.** $4 \times \frac{1}{8}$

15. $\frac{2}{5} \times 4$ **16.** $6 \times \frac{5}{8}$ **17.** $\frac{2}{3} \times 3$ **18.** $\frac{5}{6} \times 7$

19. $15 \times \frac{2}{3}$ **20.** $\frac{3}{4} \times 16$ **21.** $10 \times \frac{4}{9}$ **22.** $\frac{4}{5} \times 11$

23. $\frac{3}{5} \times 12$ **24.** $18 \times \frac{2}{3}$ **25.** $\frac{5}{6} \times 5$ **26.** $\frac{7}{8} \times 8$

27. $\frac{1}{16} \times 8$ **28.** $3 \times \frac{13}{15}$ **29.** $5 \times \frac{9}{10}$ **30.** $6 \times \frac{4}{17}$

31. $\frac{5}{3} \times 12$ **32.** $\frac{3}{2} \times 8$ **33.** $\frac{7}{4} \times 2$ **34.** $\frac{9}{5} \times 3$

For more practice: **Skill Bank,** page 425.

Multiplying Mixed Numbers and Fractions

MODEL

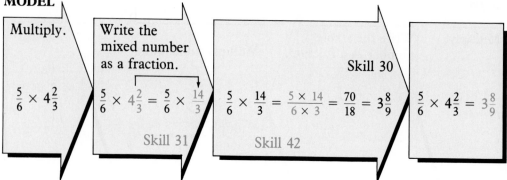

Multiply. | Write the mixed number as a fraction. | | Skill 30

$$\frac{5}{6} \times 4\frac{2}{3}$$

$$\frac{5}{6} \times 4\frac{2}{3} = \frac{5}{6} \times \frac{14}{3}$$

$$\frac{5}{6} \times \frac{14}{3} = \frac{5 \times 14}{6 \times 3} = \frac{70}{18} = 3\frac{8}{9}$$

$$\frac{5}{6} \times 4\frac{2}{3} = 3\frac{8}{9}$$

Skill 31   Skill 42

 GET READY. Write as a fraction.

1. $1\frac{1}{4}$ 2. $3\frac{1}{5}$ 3. $6\frac{2}{3}$ 4. $7\frac{5}{6}$ 5. $9\frac{1}{6}$

6. $2\frac{7}{10}$ 7. $1\frac{1}{16}$ 8. $4\frac{3}{11}$ 9. $2\frac{4}{15}$ 10. $3\frac{7}{12}$

Write as a whole number or mixed number in lowest terms.

11. $\frac{8}{5}$ 12. $\frac{9}{9}$ 13. $\frac{24}{9}$ 14. $\frac{92}{12}$ 15. $\frac{60}{25}$

 NOW USE SKILL 44. Write the answer in lowest terms.

16. $\frac{1}{4} \times 3\frac{1}{2}$ 17. $4\frac{1}{6} \times \frac{1}{3}$ 18. $\frac{4}{7} \times 5\frac{1}{3}$ 19. $5\frac{3}{8} \times \frac{1}{4}$

20. $\frac{3}{4} \times 4\frac{3}{8}$ 21. $6\frac{3}{5} \times \frac{4}{9}$ 22. $\frac{5}{9} \times 3\frac{6}{7}$ 23. $\frac{3}{4} \times 1\frac{4}{9}$

24. $8\frac{2}{3} \times \frac{3}{4}$ 25. $\frac{4}{7} \times 4\frac{1}{5}$ 26. $2\frac{4}{11} \times \frac{5}{8}$ 27. $\frac{5}{6} \times 2\frac{2}{15}$

28. $\frac{5}{13} \times 7\frac{4}{5}$ 29. $2\frac{1}{11} \times \frac{1}{10}$ 30. $\frac{11}{15} \times 1\frac{2}{3}$ 31. $2\frac{4}{5} \times \frac{2}{7}$

32. $1\frac{1}{10} \times \frac{4}{5}$ 33. $2\frac{2}{3} \times \frac{5}{12}$ 34. $2\frac{4}{15} \times \frac{3}{10}$ 35. $\frac{3}{11} \times 1\frac{7}{15}$

36. $3\frac{3}{5} \times \frac{2}{3}$ 37. $4\frac{4}{5} \times \frac{7}{8}$ 38. $1\frac{2}{5} \times \frac{3}{14}$ 39. $\frac{1}{12} \times 3\frac{5}{6}$

For more practice: **Skill Bank,** page 426.

Multiplying Mixed Numbers

MODEL

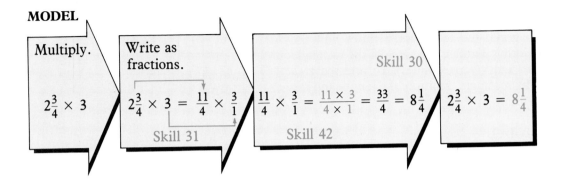

 GET READY. Write as a fraction.

1. $2\frac{1}{5}$ **2.** $3\frac{5}{6}$ **3.** $2\frac{7}{8}$ **4.** $3\frac{3}{5}$ **5.** $2\frac{3}{8}$

6. 16 **7.** $8\frac{9}{16}$ **8.** $1\frac{11}{18}$ **9.** 7 **10.** 15

 NOW USE SKILL 45. Write the answer in lowest terms.

11. $1\frac{5}{6} \times 2\frac{4}{5}$ **12.** $2\frac{3}{8} \times 3\frac{1}{2}$ **13.** $1\frac{3}{4} \times 1\frac{2}{3}$ **14.** $1\frac{5}{6} \times 1\frac{1}{2}$

15. $1\frac{3}{5} \times 3\frac{5}{6}$ **16.** $2\frac{3}{4} \times 1\frac{5}{7}$ **17.** $2\frac{5}{8} \times 3\frac{1}{5}$ **18.** $1\frac{5}{7} \times 2\frac{2}{9}$

19. $2\frac{2}{5} \times 1\frac{3}{4}$ **20.** $2\frac{3}{7} \times 2\frac{4}{5}$ **21.** $3\frac{3}{5} \times 2\frac{1}{7}$ **22.** $1\frac{5}{9} \times 3\frac{3}{7}$

23. $6 \times 5\frac{3}{7}$ **24.** $2\frac{1}{2} \times 4$ **25.** $5 \times 2\frac{6}{7}$ **26.** $3\frac{4}{7} \times 14$

27. $2\frac{5}{6} \times 3$ **28.** $4 \times 7\frac{5}{6}$ **29.** $3\frac{5}{8} \times 2$ **30.** $5 \times 4\frac{4}{5}$

31. $1\frac{9}{10} \times 5$ **32.** $3 \times 1\frac{5}{12}$ **33.** $2\frac{11}{15} \times 3$ **34.** $4 \times 1\frac{5}{16}$

35. $7\frac{7}{10} \times 4\frac{6}{11}$ **36.** $2\frac{4}{7} \times 7\frac{5}{9}$ **37.** $5\frac{1}{6} \times 4\frac{9}{10}$ **38.** $6\frac{2}{5} \times 1\frac{7}{8}$

For more practice: **Skill Bank,** page 426.

Area of a Rectangle

The amount of space enclosed by a figure is called the **area.** For the rectangle in the diagram, the length is 5 miles and the width is 3 miles. The area of the rectangle is 15 mi² (square miles). To find the area of a rectangle, multiply the length and the width.

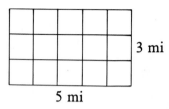

3 mi

5 mi

MODEL

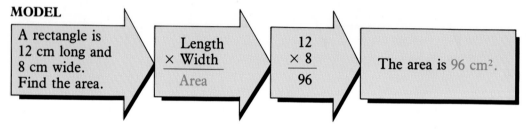

A rectangle is 12 cm long and 8 cm wide. Find the area.

$$\frac{\text{Length}}{\times \text{Width}}$$
Area

$$\frac{12}{\times 8}$$
$$96$$

The area is 96 cm².

 GET READY.

1. 18×28

2. 9.2×10.6

3. $17\frac{1}{2} \times 6\frac{3}{4}$

4. $39 \times 52\frac{1}{4}$

5. 8.9×4.13

6. $56\frac{3}{5} \times 14\frac{1}{3}$

 NOW USE APPLICATION A19. Find the area of the rectangle.

	7.	**8.**	**9.**	**10.**	**11.**	**12.**
Length	20 cm	$12\frac{1}{2}$ in.	4.2 mi	22 m	220 ft	$6\frac{1}{2}$ in.
Width	7 cm	8 in.	2.5 mi	9.4 m	88 ft	$\frac{3}{4}$ in.
Area	?	?	?	?	?	?

13. Which has the greater area, a square 12 cm on each side or a rectangle 10 cm wide and 14 cm long?

14. A rectangle is 2 feet wide and 4 feet long. Find its area in square inches.

15. Colorado is approximately a rectangle 365 miles long and 265 miles wide. What is the approximate area of Colorado?

For more practice: **Application Bank,** p. 456.

Area of a Triangle

The area of triangle ADE is half the area of rectangle $ABCD$. You can find the area of the triangle by taking half of the area of the rectangle.

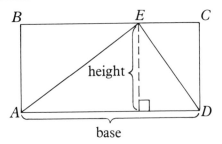

Area of triangle $= \frac{1}{2} \times$ area of rectangle

Area of triangle $= \frac{1}{2} \times$ base $\times$ height

$$A = \frac{1}{2} \times b \times h$$

MODEL

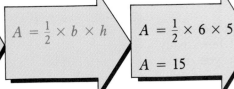

The base of a triangle measures 6 cm. The height measures 5 cm. Find the area.

$A = \frac{1}{2} \times b \times h$

$A = \frac{1}{2} \times 6 \times 5$

$A = 15$

The area is 15 cm².

 NOW USE APPLICATION A20. Complete the table.

	1.	2.	3.	4.	5.	6.
Base	12 m	6 cm	$3\frac{1}{2}$ ft	9 in.	50 yd	24 in.
Height	8 m	14 cm	4 ft	11 in.	120 yd	$3\frac{5}{8}$ in.
Area	?	?	?	?	?	?

Find the area of the triangle.

7.

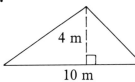

4 m

10 m

8.

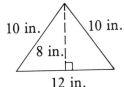

10 in. 10 in.

8 in.

12 in.

9.

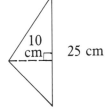

10 cm

25 cm

10. Which has the greater area, a triangle with a base of 12 cm and a height of 6 cm, or a triangle with a base of 10 cm and a height of 8 cm?

For more practice: **Application Bank**, page 456.

Home Energy Conservation

You are planning some home improvements to save on your heating expenses. You are going to insulate your attic, weatherstrip the outside doors, install insulated window shades in the bedrooms, and seal around bedroom windows with rope caulk. You work with a floor plan of your house.

PROBLEM: What materials do you need for the project?

Solution Plan: Find the area of the attic, and find how many rolls of insulation you need. Find the amount of fabric, weatherstrip, or caulk for one window or door. Multiply to find the total amount.

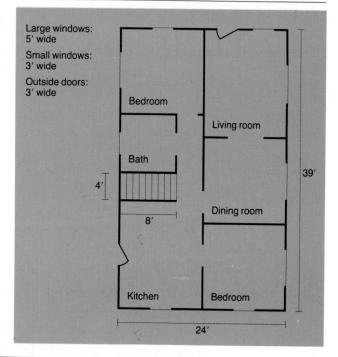

Large windows: 5' wide

Small windows: 3' wide

Outside doors: 3' wide

WHAT DO YOU KNOW? Use the floor plan.

1. How many small windows are being caulked and insulated?

2. How many large windows are being caulked and insulated?

3. How many outside doors are there?

4. What are the overall dimensions of the house?

5. What are the dimensions of the space used for the attic stairs?

WHAT DO YOU WANT TO KNOW? You want to know how much fabric and caulk you need.

You need two layers of insulated fabric for each window. Each window is 5 ft tall. Each outside door is 7 ft tall.

6. How many square feet of fabric do you need for each
   a. small window?   **b.** large window?

7. How many feet of rope caulk do you need around the edge of a
   a. small window?   **b.** large window?

8. How many feet of weatherstrip do you need to go around the edge of each outside door? (Do not include the bottom edge of the door.)

WHAT MORE DO YOU WANT? You want to know the floor space being insulated in the attic.

9. What is the total area of the attic?

10. You will not insulate the opening for the attic stairs. How many square feet is this?

11. How many square feet of attic insulation do you need?

NOW SOLVE THE PROBLEM.
Answer the question.

A roll of insulation covers 88 ft².

12. Estimate the number of rolls needed for the attic.

13. Figure the actual number of rolls needed.

14. How many feet of rope caulk do you need for the windows?

15. How many feet of weatherstrip do you need for the doors?

16. How many square feet of fabric do you need for the window shades?

Quiz

Skill 42   Skill 43   Skill 44   Skill 45

1. $\frac{6}{7} \times \frac{2}{3}$ **2.** $9 \times \frac{8}{15}$ **3.** $\frac{3}{4} \times 2\frac{2}{5}$ **4.** $13\frac{1}{2} \times 3$

Application A19

5. A rectangle is 5 in. wide and 20 in. long. Find its area.

Application A20

6. A triangle has a base of 10 m and a height of 12 m. Find its area.

Problem Solving

7. You are leasing two large offices for your company. The floor plan is shown at the right. How many square feet of office space do you have? What is the area of the larger office?

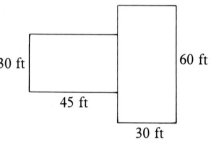

30 ft   45 ft   60 ft   30 ft

EXTRA!

You can make multiplying fractions easier when the numerator and the denominator have common factors. Divide numerator and denominator by the common factor before multiplying.

$$\frac{5}{6} \times \frac{7}{10} = \frac{\overset{1}{\cancel{5}} \times 7}{6 \times \cancel{10}_{2}} = \frac{1 \times 7}{6 \times 2} = \frac{7}{12}$$

Divide numerator and denominator by 5.

$$\frac{9}{10} \times \frac{5}{12} = \frac{9 \times \overset{1}{\cancel{5}}}{\cancel{10}_{2} \times 12} = \frac{\overset{3}{\cancel{9}} \times 1}{2 \times \cancel{12}_{4}} = \frac{3 \times 1}{2 \times 4} = \frac{3}{8}$$

Divide numerator and denominator by 5.

Divide numerator and denominator by 3.

Multiply.

1. $\frac{3}{4} \times \frac{5}{6}$ **2.** $\frac{7}{12} \times \frac{8}{9}$ **3.** $\frac{24}{5} \times \frac{7}{18}$ **4.** $\frac{16}{33} \times \frac{55}{24}$

5. $\frac{9}{16} \times \frac{8}{15}$ **6.** $\frac{27}{50} \times \frac{10}{9}$ **7.** $\frac{18}{35} \times \frac{14}{27}$ **8.** $\frac{21}{100} \times \frac{25}{24}$

174

Accountant

Leonard Kreidel is an accountant. One of his jobs is to prepare a *balance sheet* every quarter to report his company's financial status.

A **balance sheet** is a statement of the company's *assets* and *liabilities*. **Assets** are money owed by customers (accounts receivable), the company's property, and cash. **Liabilities** are the company's debts, such as bills to be paid (accounts payable), mortages, loans, and so on. **Owner's equity** is the value of assets owned outright, after liabilities are subtracted.

The total assets and the total of liabilities and owner's equity are shown by double underlines. These two amounts must be equal.

SPORT CITY, INC.
Balance Sheet
June 30

Assets		Liabilities	
Cash	$ 14,600	Accounts payable	$ 33,000
Accounts receivable	98,400	Mortgage	180,000
Inventory	126,000	Stock	200,000
Building, equipment	204,000		
Land	20,000	Total liabilities	$413,000
		Owner's Equity	$ 50,000
		Total liabilities	
Total assets	$463,000	and owner's equity	$463,000

Mercadante Enterprises had these assets and liabilities on December 31. Prepare a balance sheet. Be sure to include owner's equity.

Assets		Liabilities	
Cash	$ 562,500	Accounts payable	$2,065,000
Accounts receivable	3,750,000	Notes	3,375,000
Investments	23,500	Taxes owed	131,250
Inventory	3,450,000	Stock	1,857,000
Property	4,875,000	Mortgage	4,032,750

175

Skill 46

Finding Reciprocals

MODEL

Find the reciprocal.	Write as fractions.	**Reciprocals** are numbers whose product is 1.	Reciprocal
$\frac{5}{8}$	$\frac{5}{8}$	$\frac{5}{8} \times \frac{8}{5} = 1$	$\frac{5}{8} \longrightarrow \frac{8}{5}$
7	$7 = \frac{7}{1}$	$\frac{7}{1} \times \frac{1}{7} = 1$	$7 \longrightarrow \frac{1}{7}$
$2\frac{3}{4}$	$2\frac{3}{4} = \frac{11}{4}$	$\frac{11}{4} \times \frac{4}{11} = 1$	$2\frac{3}{4} \longrightarrow \frac{4}{11}$
	Skill 31	Skill 42	

 GET READY. Complete the statement.

1. $\frac{1}{2} \times \frac{2}{1} = ?$ **2.** $\frac{3}{8} \times \frac{8}{3} = ?$ **3.** $\frac{10}{7} \times \frac{7}{10} = ?$ **4.** $\frac{16}{5} \times \frac{5}{16} = ?$

5. $\frac{3}{4} \times ? = 1$ **6.** $\frac{4}{5} \times ? = 1$ **7.** $\frac{7}{8} \times ? = 1$ **8.** $\frac{9}{5} \times ? = 1$

9. $\frac{9}{10} \times ? = 1$ **10.** $\frac{7}{16} \times \frac{16}{7} = ?$ **11.** $\frac{18}{7} \times ? = 1$ **12.** $\frac{23}{15} \times \frac{15}{23} = ?$

NOW USE SKILL 46. Find the reciprocal.

13. $\frac{1}{3}$ **14.** $\frac{1}{4}$ **15.** $\frac{3}{5}$ **16.** $\frac{5}{9}$ **17.** $\frac{8}{5}$

18. $\frac{9}{10}$ **19.** $\frac{15}{16}$ **20.** $\frac{7}{13}$ **21.** $\frac{16}{19}$ **22.** $\frac{22}{9}$

23. 6 **24.** 7 **25.** 4 **26.** 3 **27.** 8

28. 81 **29.** 203 **30.** 778 **31.** 34 **32.** 1

33. $1\frac{1}{2}$ **34.** $2\frac{1}{4}$ **35.** $3\frac{4}{5}$ **36.** $1\frac{5}{6}$ **37.** $2\frac{7}{9}$

For more practice: **Skill Bank,** page 427.

Dividing Fractions

MODEL

Divide.

$$\frac{2}{3} \div \frac{3}{4}$$

Multiply by the reciprocal.

$$\frac{2}{3} \div \frac{3}{4} = \frac{2}{3} \times \frac{4}{3} = \frac{8}{9}$$

Reciprocal

Skill 46

$$\frac{2}{3} \div \frac{3}{4} = \frac{8}{9}$$

GET READY. Find the reciprocal.

1. $\frac{4}{7}$ 2. $\frac{9}{10}$ 3. $\frac{7}{11}$ 4. $\frac{15}{23}$ 5. $\frac{8}{31}$ 6. $\frac{11}{43}$

Complete the statement.

7. $\frac{1}{4} \div \frac{3}{4} = \frac{1}{4} \times ?$ 8. $\frac{3}{11} \div \frac{5}{8} = \frac{3}{11} \times ?$ 9. $\frac{2}{3} \div \frac{1}{6} = \frac{2}{3} \times ?$

10. $\frac{1}{10} \div \frac{3}{4} = ? \times \frac{4}{3}$ 11. $\frac{5}{16} \div \frac{7}{12} = \frac{5}{16} \times ?$ 12. $\frac{15}{19} \div \frac{9}{10} = ? \times \frac{10}{9}$

NOW USE SKILL 47. Write the answer in lowest terms.

13. $\frac{1}{2} \div \frac{3}{4}$ 14. $\frac{2}{3} \div \frac{4}{5}$ 15. $\frac{7}{8} \div \frac{1}{6}$ 16. $\frac{8}{9} \div \frac{1}{3}$ 17. $\frac{2}{3} \div \frac{1}{5}$

18. $\frac{3}{4} \div \frac{3}{5}$ 19. $\frac{4}{7} \div \frac{5}{6}$ 20. $\frac{7}{8} \div \frac{5}{9}$ 21. $\frac{3}{4} \div \frac{1}{4}$ 22. $\frac{1}{5} \div \frac{4}{5}$

23. $\frac{7}{9} \div \frac{3}{7}$ 24. $\frac{4}{7} \div \frac{4}{5}$ 25. $\frac{5}{2} \div \frac{1}{2}$ 26. $\frac{2}{3} \div \frac{7}{6}$ 27. $\frac{6}{7} \div \frac{2}{9}$

28. $\frac{8}{9} \div \frac{4}{7}$ 29. $\frac{3}{7} \div \frac{12}{5}$ 30. $\frac{2}{3} \div \frac{17}{6}$ 31. $\frac{3}{8} \div \frac{10}{3}$ 32. $\frac{5}{6} \div \frac{15}{8}$

33. $\frac{2}{7} \div \frac{1}{10}$ 34. $\frac{1}{12} \div \frac{3}{5}$ 35. $\frac{5}{6} \div \frac{11}{15}$ 36. $\frac{1}{20} \div \frac{5}{9}$ 37. $\frac{7}{8} \div \frac{21}{20}$

38. $\frac{15}{16} \div \frac{3}{8}$ 39. $\frac{5}{12} \div \frac{5}{4}$ 40. $\frac{21}{25} \div \frac{7}{15}$ 41. $\frac{12}{25} \div \frac{2}{5}$ 42. $\frac{16}{25} \div \frac{4}{5}$

For more practice: **Skill Bank,** page 427.

Dividing Whole Numbers and Fractions

MODEL

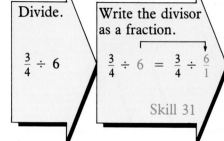

Divide.	Write the divisor as a fraction.	Multiply by the reciprocal.	
$\frac{3}{4} \div 6$	$\frac{3}{4} \div 6 = \frac{3}{4} \div \frac{6}{1}$	$\frac{3}{4} \div \frac{6}{1} = \frac{3}{4} \times \frac{1}{6} = \frac{3}{24} = \frac{1}{8}$	$\frac{3}{4} \div 6 = \frac{1}{8}$
	Skill 31	Skill 47	

▶ **GET READY. Find the reciprocal.**

1. 4   **2.** 8   **3.** 11   **4.** 25   **5.** 16   **6.** 32

7. $\frac{4}{5}$ **8.** $\frac{3}{7}$ **9.** $\frac{10}{11}$ **10.** $2\frac{1}{5}$ **11.** $4\frac{3}{11}$ **12.** $1\frac{14}{15}$

▶ **NOW USE SKILL 48. Write the answer in lowest terms.**

13. $2 \div \frac{1}{4}$ **14.** $8 \div \frac{1}{2}$ **15.** $3 \div \frac{2}{3}$ **16.** $5 \div \frac{3}{4}$ **17.** $4 \div \frac{5}{6}$

18. $4 \div \frac{3}{8}$ **19.** $4 \div \frac{4}{5}$ **20.** $2 \div \frac{2}{3}$ **21.** $3 \div \frac{1}{10}$ **22.** $5 \div \frac{5}{12}$

23. $8 \div \frac{9}{10}$ **24.** $9 \div \frac{21}{10}$ **25.** $\frac{1}{8} \div 2$ **26.** $\frac{7}{9} \div 3$ **27.** $\frac{3}{4} \div 2$

28. $\frac{4}{5} \div 3$ **29.** $\frac{5}{8} \div 4$ **30.** $\frac{7}{9} \div 7$ **31.** $\frac{5}{3} \div 10$ **32.** $\frac{7}{6} \div 1$

33. $\frac{1}{10} \div 2$ **34.** $\frac{9}{16} \div 3$ **35.** $\frac{7}{15} \div 7$ **36.** $\frac{15}{11} \div 6$ **37.** $18 \div \frac{12}{5}$

38. $\frac{18}{25} \div 9$ **39.** $14 \div \frac{7}{3}$ **40.** $196 \div \frac{14}{13}$ **41.** $100 \div \frac{25}{7}$ **42.** $31 \div \frac{6}{5}$

43. $270 \div \frac{1}{3}$ **44.** $6 \div \frac{1}{48}$ **45.** $1 \div \frac{20}{9}$ **46.** $\frac{13}{40} \div \frac{13}{40}$ **47.** $\frac{15}{8} \div 1$

For more practice: **Skill Bank,** page 427.

Dividing Mixed Numbers and Fractions

MODEL

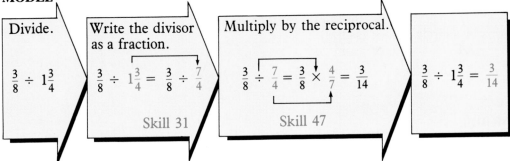

Divide.

$$\frac{3}{8} \div 1\frac{3}{4}$$

Write the divisor as a fraction.

$$\frac{3}{8} \div 1\frac{3}{4} = \frac{3}{8} \div \frac{7}{4}$$

Skill 31

Multiply by the reciprocal.

$$\frac{3}{8} \div \frac{7}{4} = \frac{3}{8} \times \frac{4}{7} = \frac{3}{14}$$

Skill 47

$$\frac{3}{8} \div 1\frac{3}{4} = \frac{3}{14}$$

 GET READY. Write as a fraction.

1. $1\frac{1}{4}$ **2.** $6\frac{4}{5}$ **3.** $2\frac{6}{7}$ **4.** $1\frac{3}{8}$ **5.** $2\frac{2}{3}$

6. $3\frac{5}{8}$ **7.** $2\frac{9}{10}$ **8.** $1\frac{5}{12}$ **9.** $2\frac{5}{16}$ **10.** $4\frac{7}{20}$

 NOW USE SKILL 49. Write the answer in lowest terms.

11. $2\frac{1}{4} \div \frac{1}{8}$ **12.** $3\frac{1}{3} \div \frac{1}{4}$ **13.** $1\frac{1}{6} \div \frac{2}{3}$ **14.** $3\frac{3}{4} \div \frac{7}{8}$

15. $2\frac{2}{3} \div \frac{5}{6}$ **16.** $3\frac{5}{9} \div \frac{4}{3}$ **17.** $2\frac{1}{4} \div \frac{1}{16}$ **18.** $7\frac{1}{3} \div \frac{8}{9}$

19. $3\frac{3}{5} \div \frac{7}{10}$ **20.** $2\frac{13}{16} \div \frac{5}{8}$ **21.** $\frac{1}{3} \div 4\frac{1}{4}$ **22.** $\frac{4}{9} \div 6\frac{2}{7}$

23. $\frac{5}{11} \div 2\frac{8}{11}$ **24.** $\frac{3}{4} \div 1\frac{5}{16}$ **25.** $\frac{4}{15} \div 2\frac{6}{15}$ **26.** $\frac{5}{16} \div 1\frac{5}{24}$

27. $1\frac{4}{5} \div \frac{32}{75}$ **28.** $8\frac{5}{6} \div \frac{1}{6}$ **29.** $\frac{22}{17} \div 1\frac{4}{51}$ **30.** $4\frac{2}{7} \div \frac{5}{9}$

31. $8\frac{7}{9} \div \frac{2}{3}$ **32.** $\frac{4}{3} \div 3\frac{5}{6}$ **33.** $2\frac{7}{8} \div \frac{3}{4}$ **34.** $4\frac{1}{7} \div \frac{3}{7}$

For more practice: **Skill Bank,** page 428.

Area of a Circle

To find the area of a circle, multiply π (pi) by the square of the radius.

Area of a circle $= \pi \times$ (radius)^2
$A \qquad = \pi \times \quad r^2$

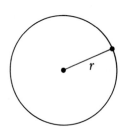

MODEL

A circle has a radius of 5 cm. Find the area. $\Rightarrow$ $A = \pi \times r^2$ $\Rightarrow$ $A = 3.14 \times 5^2$
$A = 3.14 \times 25$
$A = 78.5$ $\Rightarrow$ The area of the circle is 78.5 cm^2.

 GET READY. Complete the statement.

1. $6^2 = ?$ **2.** $17^2 = ?$ **3.** $\left(4\frac{1}{2}\right)^2 = ?$ **4.** $(2.5)^2 = ?$

 NOW USE APPLICATION A21. Complete the table. Round to the nearest tenth.

	5.	6.	7.	8.	9.	10.
Radius	10 cm	22 ft	3.5 m	15 mi	6.2 cm	8.3 km
Area	?	?	?	?	?	?

11. A circle has a diameter of 4 m. Find the area.

12. A circle has a diameter of 30 in. Find the area.

13. Which has the larger area, a circle with a radius of 4 cm or a square 7 cm on each side?

14. A lighthouse is visible for 12 miles in all directions. Over how many square miles can the lighthouse be seen?

For more practice: **Application Bank,** page 456.

Area of Irregular Figures

Simple figures, such as rectangles, triangles, and circles, can be combined to form other figures. To find the area of the figure shown, find the areas of a rectangle and a triangle.

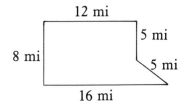

Area of rectangle	Area of triangle
$A = 1 \times w$	$A = \frac{1}{2} \times b \times h$
$A = 8 \times 12$	$A = \frac{1}{2} \times 3 \times 4$
$A = 96$	$A = 6$
Application A19	Application A20

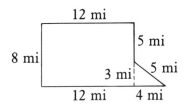

The area of the entire figure is 96 + 6, or 102, square miles.

GET READY. Find the area.

1. Rectangle
length: 9 in.
width: 6 in.

2. Circle
radius: 15 cm

3. Triangle
base: 7 ft
height: 24 ft

NOW USE APPLICATION A22. Find the area of the figure.

4.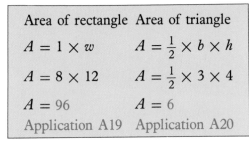
10 in.
7 in.
14 in.

5. 4 cm
8 cm
4 cm
8 cm

6. 9 ft
12 ft
15 ft

7. 3 yd   3 yd
4 yd
10 yd   10 yd
4 yd
3 yd   3 yd

8.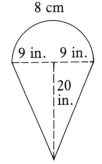
9 in.   9 in.
20 in.

9.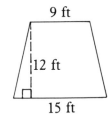
8 mi
8 mi   8 mi
8 mi

For more practice: **Application Bank,** page 457.

Agricultural Experiment

You are a soybean farmer. You decide to use a revolving sprinkler system on one field as an experiment. You want to know how much you can expect to increase your income from this field. Each sprinkler is a boom 200 ft long that waters a circular area.

PROBLEM: How much more income do you expect if you use the revolving sprinkler system?

Solution Plan: Find out how much you would expect to make without the sprinkler system. Compare this with the amount you expect to make using the sprinklers.

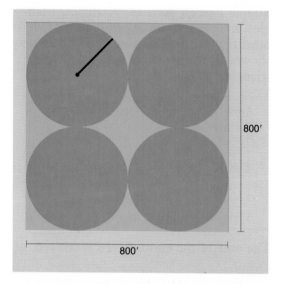

WHAT DO YOU KNOW? You know the dimensions of the field. Use the diagram.

1. What is the shape of the field?

2. What are the length and width of the field?

3. How many circular areas will be watered?

4. What is the radius of each circular area?

WHAT DO YOU WANT TO KNOW? You want to know the acreage.

One acre is equivalent to 43,560 square feet.

5. What is the area, in square feet, of the entire field?

6. Round 43,560 to the nearest ten thousand to estimate the area of the entire field in acres.

7. Find the actual area of the entire field to the nearest tenth of an acre.

$$\begin{array}{r} \text{Area in acres} \\ 43,560\overline{)\text{Area in sq ft}} \\ \textit{Skill 20} \end{array}$$

8. Find the area in square feet covered by the revolving sprinklers.

9. Round 43,560 and the area from problem 8 to the nearest ten thousand to estimate the watered area in acres.

10. Find the actual acreage of the watered section to the nearest tenth.

WHAT MORE DO YOU WANT? You want to know the expected crop yield.

11. Using your old method, you can produce 32 bushels per acre on the entire square field. How many bushels would you produce?

12. Using the sprinkler system, you expect a yield of 50 bushels per acre on the watered area. How many bushels would you produce altogether?

NOW SOLVE THE PROBLEM.
You expect to sell your soybeans for $7.20 a bushel.

13. Using your old method, how much would you receive for the crop grown on the entire square field?

14. How much do you expect to receive for the crop grown using the sprinkler system?

15. How much more income do you expect, using the sprinkler system?

Quiz

Skill 46 **Find the reciprocal.**

Skill 47

1. $\dfrac{3}{8}$ 2. 5   3. $2\dfrac{2}{5}$ 4. $\dfrac{8}{15} \div \dfrac{3}{5}$ 5. $\dfrac{5}{8} \div \dfrac{1}{2}$

Skill 48

Skill 49

6. $4 \div \dfrac{4}{5}$ 7. $\dfrac{3}{4} \div 6$ 8. $4\dfrac{1}{2} \div \dfrac{6}{7}$ 9. $\dfrac{1}{2} \div 1\dfrac{1}{4}$

Application A21

10. Find the area of a circle with radius 8 cm.

Application A22

11. Find the area of the figure.

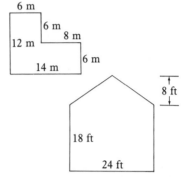

Problem Solving

12. You are painting the end of your house. It has the given dimensions. A gallon of paint will cover 250 square feet. This paint comes in gallon cans only. How many cans do you need?

CONSUMER NOTE

When you rent an apartment, you may have many housing expenses in addition to the rent. For example, the rent for a one-bedroom apartment in Cornell Gardens is $325 a month. Tenants pay their own telephone and utility bills. There is also a monthly parking fee of $25 per car.

Find each family's total housing cost.

1. Rent: $425.00
   Telephone: 12.00
   Electricity: 32.50
   Parking: 50.00

2. Rent: $275.00
   Telephone: 18.50
   Electricity: 16.00

3. Rent: $230.00
   Telephone: 34.75
   Electricity: 28.00
   Gas: 19.25
   Parking: 25.00

4. Rent: $450.00
   Telephone: 19.25
   Electricity: 35.00
   Gas: 16.75
   Parking: 50.00

Reading Time Zone Maps

One morning Greg Olsen, a New Yorker, called his sister in Oregon. She wasn't as pleased as he had expected. It was 4:00 A.M. in Oregon!

New York and Oregon are in different time zones. Starting in the west, each time zone is one hour earlier than the zone immediately east of it. The time difference between New York and Oregon is 3 hours.

The map below shows the time zones throughout the United States.

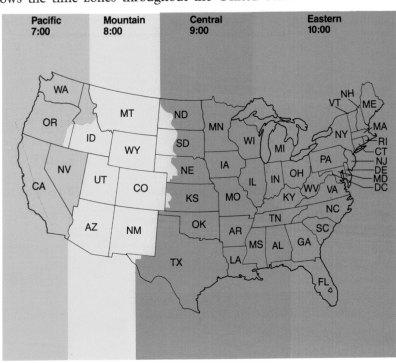

Name the time zone for the given state.

1. Iowa   2. California   3. Maine   4. Alaska

5. Idaho   6. Hawaii   7. Texas   8. Louisiana

It is now 9:00 A.M. in North Carolina. What is the time in the following states?

9. Arkansas   10. Nevada   11. Montana   12. Colorado

13. West Virginia   14. Vermont   15. Mississippi   16. Washington

Dividing Mixed and Whole Numbers

MODEL

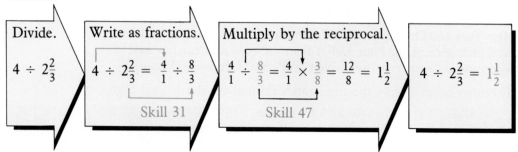

Divide.

$4 \div 2\frac{2}{3}$

Write as fractions.

$4 \div 2\frac{2}{3} = \frac{4}{1} \div \frac{8}{3}$

Skill 31

Multiply by the reciprocal.

$\frac{4}{1} \div \frac{8}{3} = \frac{4}{1} \times \frac{3}{8} = \frac{12}{8} = 1\frac{1}{2}$

Skill 47

$4 \div 2\frac{2}{3} = 1\frac{1}{2}$

 GET READY. Complete the statement.

1. $1\frac{5}{6} = \frac{?}{6}$

2. $8\frac{3}{4} = \frac{?}{4}$

3. $6\frac{5}{8} = \frac{53}{?}$

4. $1\frac{5}{9} = \frac{14}{?}$

5. $2 = \frac{?}{1}$

6. $1 = \frac{?}{1}$

7. $9 = \frac{9}{?}$

8. $12 = \frac{12}{?}$

Write as a whole number or mixed number in lowest terms.

9. $\frac{8}{4}$

10. $\frac{4}{3}$

11. $\frac{15}{3}$

12. $\frac{36}{10}$

13. $\frac{24}{18}$

14. $4\frac{6}{9}$

15. $3\frac{4}{6}$

16. $\frac{14}{8}$

17. $\frac{92}{16}$

18. $\frac{225}{15}$

 NOW USE SKILL 50. Write the answer in lowest terms.

19. $4\frac{1}{3} \div 1$

20. $3\frac{1}{5} \div 4$

21. $6\frac{3}{4} \div 6$

22. $5\frac{7}{8} \div 8$

23. $2\frac{2}{5} \div 4$

24. $1\frac{4}{7} \div 5$

25. $1\frac{5}{9} \div 12$

26. $3\frac{1}{10} \div 2$

27. $1\frac{7}{18} \div 10$

28. $1\frac{4}{15} \div 16$

29. $17 \div 4\frac{1}{4}$

30. $5 \div 3\frac{2}{5}$

31. $8 \div 5\frac{5}{6}$

32. $2 \div 3\frac{1}{2}$

33. $5 \div 4\frac{4}{15}$

34. $13 \div 1\frac{6}{7}$

35. $15 \div 3\frac{8}{9}$

36. $7\frac{4}{5} \div 21$

37. $18\frac{5}{9} \div 5$

38. $36 \div 7\frac{1}{5}$

39. $3\frac{3}{7} \div 3$

40. $6\frac{5}{12} \div 3$

41. $8 \div 6\frac{2}{3}$

42. $5 \div 6\frac{4}{21}$

For more practice: **Skill Bank**, page 428.

Dividing Mixed Numbers

MODEL

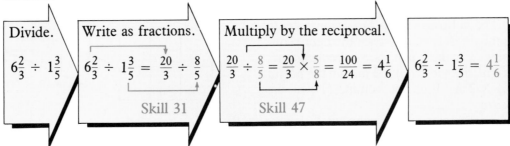

Divide.

$6\frac{2}{3} \div 1\frac{3}{5}$

Write as fractions.

$6\frac{2}{3} \div 1\frac{3}{5} = \frac{20}{3} \div \frac{8}{5}$

Skill 31

Multiply by the reciprocal.

$\frac{20}{3} \div \frac{8}{5} = \frac{20}{3} \times \frac{5}{8} = \frac{100}{24} = 4\frac{1}{6}$

Skill 47

$6\frac{2}{3} \div 1\frac{3}{5} = 4\frac{1}{6}$

GET READY. Complete the statement.

1. $2\frac{4}{5} = \frac{?}{5}$

2. $1\frac{5}{8} = \frac{?}{8}$

3. $4\frac{3}{7} = \frac{?}{7}$

4. $3\frac{2}{9} = \frac{?}{9}$

5. $3\frac{1}{10} = \frac{?}{10}$

6. $2\frac{3}{16} = \frac{35}{?}$

7. $2\frac{5}{24} = \frac{53}{?}$

8. $1\frac{13}{16} = \frac{?}{16}$

9. $21\frac{5}{6} = \frac{131}{?}$

10. $18\frac{3}{8} = \frac{?}{8}$

Find the reciprocal.

11. $8\frac{3}{4}$

12. $5\frac{7}{6}$

13. $3\frac{2}{3}$

14. $3\frac{8}{9}$

15. $6\frac{4}{7}$

NOW USE SKILL 51. Write the answer in lowest terms.

16. $1\frac{1}{4} \div 1\frac{1}{5}$

17. $2\frac{1}{3} \div 3\frac{1}{6}$

18. $2\frac{4}{5} \div 2\frac{5}{8}$

19. $1\frac{1}{2} \div 1\frac{1}{2}$

20. $1\frac{3}{5} \div 1\frac{2}{3}$

21. $3\frac{2}{3} \div 1\frac{2}{9}$

22. $6\frac{4}{5} \div 3\frac{1}{3}$

23. $5\frac{2}{3} \div 2\frac{5}{6}$

24. $2\frac{2}{5} \div 1\frac{1}{2}$

25. $1\frac{2}{7} \div 1\frac{1}{5}$

26. $2\frac{1}{7} \div 2\frac{1}{2}$

27. $6\frac{1}{3} \div 2\frac{3}{8}$

28. $9\frac{3}{8} \div 2\frac{1}{2}$

29. $4\frac{3}{8} \div 1\frac{3}{4}$

30. $13\frac{1}{5} \div 1\frac{1}{5}$

31. $12\frac{3}{5} \div 6\frac{3}{5}$

32. $14\frac{3}{4} \div 4\frac{1}{2}$

33. $18\frac{3}{4} \div 1\frac{7}{8}$

34. $13\frac{3}{5} \div 2\frac{2}{5}$

35. $9\frac{4}{5} \div 1\frac{13}{15}$

36. $11\frac{3}{5} \div 6\frac{2}{3}$

37. $10\frac{1}{2} \div 4\frac{3}{5}$

38. $12\frac{3}{4} \div 2\frac{1}{8}$

39. $12\frac{1}{3} \div 12\frac{1}{2}$

40. $12\frac{4}{7} \div 2\frac{1}{4}$

41. $20\frac{5}{8} \div 5\frac{11}{16}$

42. $13\frac{4}{5} \div 6\frac{1}{2}$

43. $26\frac{4}{9} \div 2\frac{6}{7}$

For more practice: **Skill Bank,** page 428.

Volume of a Prism

A **rectangular prism** has the shape of a box. Each face of the box is a rectangle.

The **volume** of a 3-dimensional figure is the measure of the space the figure contains. For the prism shown, the length (l) is 4 m, the width (w) is 3 m, and the height (h) is 2 m. Find the volume (V) of the prism.

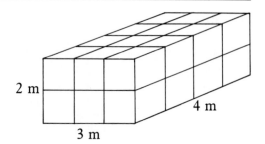

2 m

3 m

4 m

MODEL

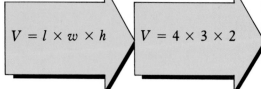 $V = l \times w \times h$ → $V = 4 \times 3 \times 2$ → $V = 24$ →

The volume is 24 m³.
↑
(cubic meters)

 GET READY. Multiply.

1. $8 \times 6 \times 2$ 2. $12 \times 5 \times 17$ 3. $22 \times 10 \times 13$

4. $14 \times 3.6 \times 20$ 5. $7.3 \times 5 \times 2$ 6. $9 \times 4 \times 3\frac{3}{4}$

 NOW USE APPLICATION A23. Find the volume of the prism.

	7.	8.	9.	10.	11.	12.
Length	9 cm	2.5 m	6 ft	12 in.	16 cm	3.6 m
Width	6 cm	1 m	3 ft	12 in.	7.2 cm	1 m
Height	3 m	2 m	2 ft	12 in.	25 cm	10 m

13. One carton is 15 in. long, 12 in. wide, and 8 in. high. Another carton is 12 in. long, 12 in. wide, and 9 in. high. Which has the greater volume?

14. In a cube, the length, width, and height are the same. What is the volume of a cube that is 3 ft by 3 ft by 3 ft?

15. A swimming pool is 30 ft long, 15 ft wide, and 4 ft deep. One cubic foot is equivalent to 7.5 gallons. How many gallons of water are needed to fill the pool?

For more practice: **Application Bank,** page 457.

Volume of a Cylinder

A **cylinder** has the shape of a can. The base is a circle.

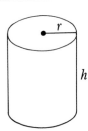

To find the volume of a cylinder, multiply the area of the base by the height. Since the base is a circle, the area of the base is $\pi \times r^2$.

Volume of a cylinder $=$ (area of base) $\times$ height
$$= \pi \times r^2 \times h$$

The radius of the base of a cylinder is 6 cm, and its height is 10 cm. Find the volume.

MODEL

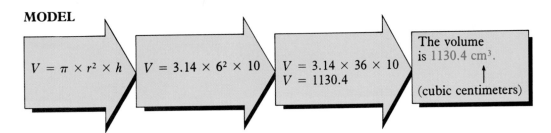

$V = \pi \times r^2 \times h$ → $V = 3.14 \times 6^2 \times 10$ → $V = 3.14 \times 36 \times 10$ → The volume is 1130.4 cm³.
$V = 1130.4$
(cubic centimeters)

 GET READY. Multiply.

1. $3.14 \times 5^2 \times 12$ **2.** $3.14 \times 10^2 \times 7$ **3.** $3.14 \times 8^2 \times 2.5$

 NOW USE APPLICATION A24. Find the volume of the cylinder.

	4.	5.	6.	7.	8.
Radius of base	25 ft	3 cm	3 cm	10 m	18 in.
Height	14 ft	6 cm	12 cm	6.2 m	30 in.

9. One can is 6 in. high, with a base of radius 4 in. Another can is 5 in. high, with a base of radius 5 in. Which has the greater volume?

10. A round cistern is 4 ft across and 5 ft deep. One cubic foot is equivalent to 7.5 gallons. How many gallons of water will the cistern hold?

For more practice: **Application Bank,** page 457.

Tank Farm

You are a home heating oil distributor. You have a tank farm with 6 tanks, each 50 ft high and 100 ft across.

PROBLEM: How long should you go before you have your tanks refilled?

Solution Plan: Set a reserve margin. Use your expected average monthly oil consumption to find out when you reach your reserve margin.

WHAT DO YOU KNOW? You know how many tanks you have and their dimensions.

1. How many tanks do you have?

2. What is the shape of each tank?

3. How tall is each tank?

4. What is the radius of the base of a tank?

WHAT DO YOU WANT TO KNOW? You want to know your total storage capacity.

5. Find the capacity in cubic feet of one tank.

6. What is the capacity in cubic feet of all the tanks?

7. There are 7.5 gallons per cubic foot. Find your total storage capacity in gallons.

WHAT MORE DO YOU WANT?
You want to find the expected average monthly oil consumption.

8. You have 15,000 customers. They use an average of 250 gallons a month each. How many gallons of oil per month do your customers use altogether?

NOW SOLVE THE PROBLEM.
You set your reserve margin at half your capacity. You will refill your tanks when you have only half your capacity left in storage.

9. You begin the season with all tanks full. To the nearest million gallons, how much oil is this?

10. You will refill your tanks when you have reached your reserve margin. To the nearest million gallons, how much oil will you have sold by that time?

11. To the nearest million gallons, how much oil do your customers use per month altogether?

12. Use the numbers from Exercises 10 and 11 to estimate, to the nearest tenth, how many months it will take before you will refill your tanks.

191

Skill 50

1. $2\frac{2}{3} \div 4$

2. $3 \div 2\frac{1}{4}$

Skill 51

3. $1\frac{7}{8} \div 1\frac{1}{4}$

4. $3\frac{1}{5} \div 2\frac{2}{3}$

Application A23

5. A package of foam insulation is $2\frac{1}{2}$ ft wide, 4 ft long, and 2 ft high. Will it hold a cubic yard (27 ft³) of insulation?

Application A24

6. A round silo is 18 ft across and 30 ft tall. How many cubic feet of silage will it hold?

Problem Solving

7. A reservoir is 50 feet long, 50 feet wide, and 15 feet deep. It can be filled at a rate of 100 gallons per minute. How long does it take to fill the reservoir? (1 ft³ = 7.5 gal)

COMPUTER BITS

Components of a Computer System

Twice a year a public utility sends water bills to its 2500 customers. To compute the bills, each customer's water meter is read to determine the number of cubic feet of water used. Then a formula is used to calculate the fee. The meter readings are an example of *raw data*. The process of changing the raw data into a more usable form (the typed bill) is called *data processing*. Computers are used for *electronic data processing* (EDP).

The utility company recently installed a computer system to use in the billing process. The meter readings are entered into the computer through an *input* device called a CRT terminal. The computer is *programmed* to process the raw data into a usable form. All data processing, such as calculating the bills and alphabetizing by customer name, is performed by the *central processing unit* (CPU). The bills are then printed on special forms by an *output* device called a line printer.

The basic data processing cycle changes raw data into usable information.

INPUT $\longrightarrow$ **PROCESS** $\longrightarrow$ **OUTPUT**

Skills Review

Skill 42 (page 166)

1. $\frac{1}{3} \times \frac{1}{2}$

2. $\frac{1}{4} \times \frac{3}{5}$

3. $\frac{4}{7} \times \frac{1}{8}$

4. $\frac{5}{6} \times \frac{3}{7}$

5. $\frac{3}{4} \times \frac{5}{16}$

6. $\frac{2}{7} \times \frac{4}{5}$

7. $\frac{6}{11} \times \frac{3}{4}$

8. $\frac{2}{3} \times \frac{5}{6}$

9. $\frac{2}{3} \times \frac{11}{12}$

10. $\frac{10}{21} \times \frac{12}{25}$

Skill 45 (page 169)

11. $1\frac{1}{6} \times 2\frac{1}{4}$

12. $3\frac{2}{5} \times 2\frac{1}{4}$

13. $8\frac{1}{3} \times 1\frac{3}{5}$

14. $4\frac{3}{4} \times 6\frac{2}{3}$

15. $2\frac{1}{10} \times 1\frac{1}{4}$

16. $4\frac{3}{8} \times 2\frac{4}{5}$

17. $3\frac{1}{5} \times 2\frac{5}{6}$

18. $7 \times 4\frac{2}{3}$

Skill 46 **Find the reciprocal.** (page 176)

19. 5

20. 7

21. $\frac{1}{6}$

22. $\frac{5}{9}$

23. $1\frac{3}{4}$

Skill 48 (page 178)

24. $6 \div \frac{1}{8}$

25. $9 \div \frac{2}{3}$

26. $6 \div \frac{5}{4}$

27. $3 \div \frac{1}{10}$

28. $\frac{9}{10} \div 6$

29. $\frac{11}{16} \div 3$

30. $\frac{12}{5} \div 2$

31. $12 \div \frac{4}{5}$

32. $\frac{8}{15} \div 4$

33. $\frac{5}{9} \div 10$

Skill 49 (page 179)

34. $6\frac{3}{5} \div \frac{3}{10}$

35. $9\frac{5}{7} \div \frac{5}{7}$

36. $1\frac{7}{8} \div \frac{5}{6}$

37. $\frac{1}{5} \div 3\frac{1}{4}$

38. $\frac{1}{16} \div 1\frac{1}{2}$

39. $\frac{5}{12} \div 1\frac{2}{3}$

40. $\frac{2}{5} \div 3\frac{2}{5}$

41. $\frac{4}{7} \div 8\frac{4}{5}$

Skill 51 (page 187)

42. $7\frac{1}{2} \div 3\frac{1}{8}$

43. $5\frac{1}{3} \div 1\frac{3}{5}$

44. $2\frac{2}{3} \div 1\frac{3}{4}$

45. $5\frac{3}{4} \div 4\frac{2}{3}$

46. $2\frac{7}{9} \div 1\frac{5}{6}$

47. $9\frac{3}{5} \div 1\frac{5}{7}$

48. $4\frac{3}{4} \div 2\frac{5}{6}$

49. $9\frac{7}{9} \div 2\frac{3}{4}$

Applications Review

Application A19 **Find the area of the rectangle.** (page 170)

1. length: $6\frac{1}{2}$ in.   width: $4\frac{3}{4}$ in.
2. length: 7 m   width: 9 m
3. length: 14 yd   width: 12 yd
4. length: 6.7 ft   width: 3.2 ft

Application A20 **Find the area of the triangle.** (page 171)

5.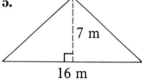
7 m
16 m

6.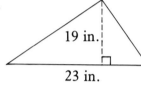
19 in.
23 in.

7.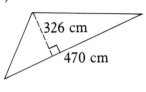
326 cm
470 cm

Application A21 **Find the area of the circle.** (page 180)

8. radius: 4 in.
9. diameter: 3 mi
10. radius: 9.5 ft
11. diameter: 68 m

Application A22 **Find the area of the figure.** (page 181)

12.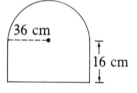
36 cm
16 cm

13.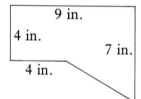
9 in.
4 in.
7 in.
4 in.

14.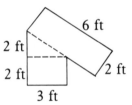
6 ft
2 ft
2 ft
2 ft
3 ft

Application A23 **Find the volume of the prism.** (page 188)

15. length: 15 in.   width: 12 in.   height: 19 in.
16. length: 85 cm   width: 78 cm   height: 34 cm
17. length: 21 in.   width: 36 in.   height: $20\frac{1}{2}$ in.

Application A24 **Which cylinder has the greater volume?** (page 189)

18. a.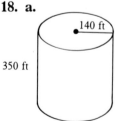
140 ft
350 ft

b.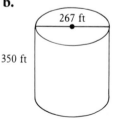
267 ft
350 ft

19. a.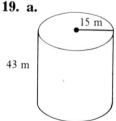
15 m
43 m

b.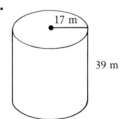
17 m
39 m

Unit Test

Skill 42

1. $\frac{5}{9} \times \frac{3}{7}$ **2.** $\frac{8}{11} \times \frac{7}{9}$

Skill 43

3. $\frac{14}{15} \times 5$ **4.** $12 \times \frac{25}{36}$

Skill 44

5. $\frac{6}{7} \times 2\frac{3}{4}$ **6.** $4\frac{1}{5} \times \frac{6}{7}$

Skill 45

7. $3\frac{2}{3} \times 1\frac{7}{8}$ **8.** $5\frac{1}{2} \times 6\frac{2}{3}$

Skill 46 **Find the reciprocal.**

9. $\frac{7}{4}$ **10.** 3

Skill 47

11. $\frac{4}{5} \div \frac{6}{7}$ **12.** $\frac{7}{10} \div \frac{9}{2}$

Skill 48

13. $\frac{9}{10} \div 18$ **14.** $12 \div \frac{4}{5}$

Skill 49

15. $4\frac{4}{5} \div \frac{3}{10}$ **16.** $7\frac{7}{8} \div \frac{9}{14}$

Skill 50

17. $2\frac{5}{8} \div 7$ **18.** $4\frac{4}{7} \div 8$

Skill 51

19. $3\frac{5}{6} \div 5\frac{3}{4}$ **20.** $8\frac{2}{3} \div 2\frac{4}{5}$

Application A19 **Find the area.**

21.

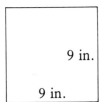

9 in.

9 in.

Application A20 **Find the area.**

22.

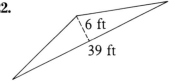

6 ft

39 ft

Application A21 **Find the area.**

23.

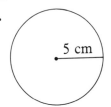

5 cm

Application A22 **Find the area.**

24.

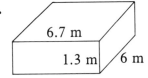

6 ft

4 ft

9 ft

10 ft

Application A23 **Find the volume.**

25.

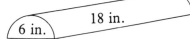

6.7 m

1.3 m 6 m

Application A24 **Find the volume.**

26.

18 in.

6 in.

195

Skills

Skill 24 **Write in lowest terms.** (page 104)

1. $\frac{6}{12}$
2. $\frac{14}{54}$
3. $2\frac{45}{60}$
4. $5\frac{63}{81}$

Skill 25 **Write in higher terms.** (page 105)

5. $\frac{2}{5} = \frac{?}{15}$
6. $\frac{13}{25} = \frac{?}{100}$
7. $4\frac{3}{8} = 4\frac{?}{56}$
8. $7\frac{6}{7} = 7\frac{?}{21}$

Skill 29 **Write as fractions with the LCD.** (page 115)

9. $\frac{1}{3}, \frac{1}{2}$
10. $\frac{6}{11}, \frac{7}{9}$
11. $\frac{13}{20}, \frac{3}{4}, \frac{2}{5}$
12. $\frac{5}{16}, \frac{21}{32}, \frac{3}{8}$

Skill 30 **Write as a whole number or mixed number in lowest terms.** (page 122)

13. $\frac{9}{5}$
14. $\frac{28}{7}$
15. $\frac{39}{4}$
16. $\frac{76}{2}$
17. $\frac{108}{4}$

Skill 31 **Write as a fraction.** (page 123)

18. 6
19. $3\frac{7}{8}$
20. $18\frac{4}{9}$
21. $27\frac{3}{5}$
22. $53\frac{2}{5}$

Skill 33 (page 135)

23. $2\frac{4}{5}$
   $+ 3\frac{2}{5}$

24. $5\frac{7}{9}$
   $+ 7\frac{4}{9}$

25. $8\frac{8}{11}$
   $+ 6\frac{4}{11}$

26. $9\frac{6}{7}$
   $+ 4\frac{5}{7}$

Skill 35 (page 137)

27. $3\frac{5}{7}$
   $+ 6\frac{3}{5}$

28. $2\frac{5}{6}$
   $+ 9\frac{1}{2}$

29. $85\frac{2}{3}$
   $+ 16\frac{1}{9}$

30. $21\frac{4}{5}$
   $+ 59\frac{3}{8}$

Skill 38 (page 146)

31. 7
   $- \frac{3}{8}$

32. 6
   $- \frac{7}{9}$

33. 12
   $- \frac{9}{10}$

34. 10
   $- \frac{13}{15}$

Skill 39 (page 147)

35. $9\frac{5}{12}$
    $-\ 3\frac{7}{12}$

36. $11\frac{3}{10}$
    $-\ 6\frac{9}{10}$

37. $26\frac{1}{6}$
    $-\ 17\frac{5}{6}$

38. $78\frac{5}{9}$
    $-\ 64\frac{7}{9}$

Skill 40 (page 154)

39. $\frac{7}{9}$
    $-\ \frac{1}{5}$

40. $\frac{11}{12}$
    $-\ \frac{11}{36}$

41. $\frac{5}{6}$
    $-\ \frac{3}{10}$

42. $\frac{6}{7}$
    $-\ \frac{9}{11}$

Skill 41 (page 155)

43. $8\frac{1}{6}$
    $-\ 6\frac{5}{8}$

44. $13\frac{5}{8}$
    $-\ 9\frac{3}{4}$

45. $29\frac{5}{9}$
    $-\ 15\frac{7}{12}$

46. $35\frac{5}{6}$
    $-\ 27\frac{6}{7}$

Skill 42 (page 166)

47. $\frac{1}{6} \times \frac{2}{3}$

48. $\frac{3}{7} \times \frac{4}{5}$

49. $\frac{6}{7} \times \frac{2}{3}$

50. $\frac{5}{8} \times \frac{7}{10}$

Skill 45 (page 169)

51. $2\frac{3}{8} \times 4$

52. $2\frac{2}{9} \times 5\frac{2}{5}$

53. $6\frac{2}{7} \times 3$

54. $11\frac{5}{7} \times 1\frac{3}{4}$

Skill 47 (page 177)

55. $\frac{5}{9} \div \frac{5}{12}$

56. $\frac{6}{5} \div \frac{1}{4}$

57. $\frac{5}{8} \div \frac{1}{6}$

58. $\frac{1}{3} \div \frac{8}{9}$

Skill 51 (page 187)

59. $2\frac{1}{5} \div 3\frac{1}{5}$

60. $4\frac{2}{7} \div 1\frac{1}{5}$

61. $6\frac{2}{3} \div 3\frac{1}{3}$

62. $8\frac{3}{4} \div 1\frac{3}{4}$

Cumulative Review: Units 4–6

Applications

Application A11 (page 106)

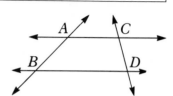

1. Name a pair of parallel lines.

2. Name the lines that intersect at point B.

3. List the corners of the polygon.

Application A12 (page 107)

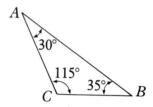

4. Is ∠*A* acute, right, or obtuse?

5. Is ∠*C* acute, right, or obtuse?

Application A13 What is the measure of ∠*A*? (pages 116–117)

6. 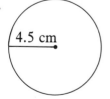   7.   8.   **A** 9.

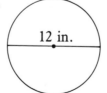   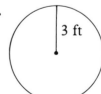

Application A14 Name the type of quadrilateral. (pages 124–125)

10. *DABG*   11. *DEFG*   12. *ABFE*

Application A15 (page 138)

13. $\sqrt{49} = ?$ 14. $\sqrt{81} = ?$ 15. $\sqrt{144} = ?$

Application A16 Find the length of the segment. (page 139)

16. $\overline{VZ}$ 17. $\overline{VX}$

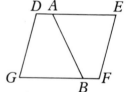

Application A17 Find the perimeter of the triangle.
(pages 148–149)

18. *VZY*   19. *VXY*   20. *VZX*

Application A18 Find the circumference of the circle. (pages 156–157)

21.   22.   23.

Application A19 **Find the area of the rectangle.** (page 170)

24.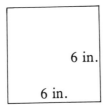

6 in.

6 in.

25.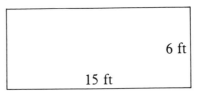

6 ft

15 ft

Application A20 **Find the area of the triangle.** (page 171)

26.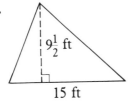

$9\frac{1}{2}$ ft

15 ft

27. base: 10 cm height: 23 cm

28. base: $8\frac{1}{2}$ in. height: 14 in.

Application A21 **Find the area of the circle.** (page 180)

29. radius: 14 in.

30. diameter: 5 m

Application A23 **Find the volume of the prism.** (page 188)

31.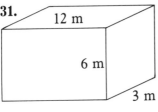

12 m

6 m

3 m

32.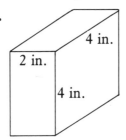

4 in.

2 in.

4 in.

33. length: 16 cm
width: 5.4 cm
height: 7.8 cm

34. A concrete slab 18 yards wide, 22 yards long, and a yard thick is being made. How many cubic yards of concrete are needed?

Application A24 **Find the volume of the cylinder.** (page 189)

35.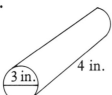

4 in.

3 in.

36.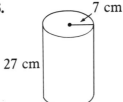

7 cm

27 cm

37.

4 ft 3 ft

7

Rates, Ratios, Proportions

Probability and Statistics

Problem Solving

Unit Preview

 You will use these skills and applications in this unit. Which do you already know? Work each problem.

Skill 52 **Write the ratio as a fraction.**   **Skill 53** **Are the ratios equal?**

1. 3 out of 7 jets   **2.** 17 out of 30 teachers   **3.** $\frac{2}{3}, \frac{4}{6}$ **4.** $\frac{8}{12}, \frac{4}{8}$

Skill 54 **Solve the proportion.**

5. $\frac{5}{10} = \frac{8}{a}$ **6.** $\frac{12}{15} = \frac{d}{20}$ **7.** $\frac{n}{40} = \frac{5}{16}$

Skill 55 **Write the equal rates as a proportion.**

8. 12 stops in 2 hours $= t$ stops in 3 hours

9. 310 persons in 4 square miles $= c$ persons in 3 square miles

Skill 56 **Write the equal rates as a proportion. Solve the proportion.**

10. 150 cars on 2 parking levels $= x$ cars on 3 parking levels

Skill 57 **Find the unit rate.**

11. 30 errors on 15 pages   **12.** 12 touchdowns in 10 games

Application A25

13. You have 3 red and 9 black pencils in a drawer. You take a pencil from the drawer without looking. Find the probability that you will take a black pencil.

Application A26

14. A study shows 2 out of every 10 consumers prefer Brand X. Of 3000 consumers, how many would you expect to prefer Brand X?

Application A27   **Application A28**

15. Find the mean: 33, 15, 23, 25.   **16.** Find the median: 13, 25, 99, 7, 41, 53, 2.

Check your answers. If you had difficulty with any skill or application, be sure to study the corresponding lesson in this unit.   201

Writing Ratios

MODEL

Write the ratio using a colon (:) and as a fraction.

19 wins and 25 losses

The ratio of wins to losses is written

19:25

$\frac{19}{25}$

(read "19 to 25")

GET READY. Write in lowest terms.

1. $\frac{9}{81}$ **2.** $\frac{14}{49}$ **3.** $\frac{36}{72}$ **4.** $\frac{65}{90}$ **5.** $\frac{42}{91}$ **6.** $\frac{142}{200}$

NOW USE SKILL 52. Write the ratio using a colon.

7. 46 miles out of 190 miles

8. 11 winners out of 46 players

9. 28 of the 73 guests

10. 6 volunteers from a class of 28

Write the ratio as a fraction.

11. 2 out of 3 doctors

12. 9 out of 10 times

13. 7 apples in every dozen

14. 10 mL of salt to 45 mL of water

Write the ratio as a fraction in lowest terms.

15. 2 hours every day

16. 2 copies in every 6 copies

17. 95¢ out of a dollar

18. 6 days out of 9 days

19. 16 out of 56 samples

20. 8 oranges in every dozen

21. 32 of the 96 years

22. 14 of the 35 invitations

23. There are 37 cars in the parking lot. Of the total 6 are red, 9 are blue, and 4 are brown. Use a colon to write the following ratios.

    a. blue cars to red cars

    b. red cars to brown cars

    c. red cars to the total number

    d. blue cars to the total number

    e. brown cars to the total number

For more practice: **Skill Bank,** page 429.

Determining Equal Ratios

MODELS

Are the ratios equal?

$$\frac{5}{9}, \frac{25}{45}$$

Cross multiply.

$$\frac{5}{9} \diagdown \frac{25}{45}$$

$$5 \times 45 \qquad 9 \times 25$$

Are the products equal?

$$5 \times 45 \overset{?}{=} 9 \times 25$$

$$225 \overset{?}{=} 225 \quad \text{Yes}$$

The ratios are equal.

$$\frac{5}{9} = \frac{25}{45}$$

Are the ratios equal?

$$\frac{4}{7}, \frac{32}{63}$$

Cross multiply.

$$\frac{4}{7} \diagdown \frac{32}{63}$$

$$4 \times 63 \qquad 7 \times 32$$

Are the products equal?

$$4 \times 63 \overset{?}{=} 7 \times 32$$

$$252 \overset{?}{=} 224 \quad \text{No}$$

The ratios are not equal.

GET READY. Complete the statement.

1. $\frac{2}{3}, \frac{6}{9}$; $2 \times ? = 3 \times 6$

2. $\frac{1}{4}, \frac{3}{12}$; $1 \times 12 = ? \times 3$

3. $\frac{3}{5}, \frac{6}{10}$; $? \times 10 = 5 \times 6$

4. $\frac{2}{3}, \frac{10}{15}$; $2 \times 15 = 3 \times ?$

5. $\frac{4}{5}, \frac{40}{50}$; $4 \times ? = 5 \times 40$

6. $\frac{12}{24}, \frac{3}{6}$; $12 \times 6 = ? \times 3$

NOW USE SKILL 53. Are the ratios equal?

7. $\frac{1}{2}, \frac{3}{6}$

8. $\frac{3}{5}, \frac{4}{7}$

9. $\frac{1}{3}, \frac{3}{9}$

10. $\frac{3}{8}, \frac{6}{9}$

11. $\frac{2}{3}, \frac{4}{6}$

12. $\frac{1}{5}, \frac{4}{8}$

13. $\frac{1}{4}, \frac{3}{8}$

14. $\frac{3}{4}, \frac{6}{8}$

15. $\frac{2}{4}, \frac{6}{12}$

16. $\frac{5}{6}, \frac{9}{12}$

17. $\frac{4}{9}, \frac{4}{12}$

18. $\frac{3}{8}, \frac{6}{16}$

19. $\frac{1}{4}, \frac{5}{16}$

20. $\frac{4}{8}, \frac{11}{20}$

21. $\frac{10}{25}, \frac{2}{5}$

22. $\frac{18}{21}, \frac{6}{7}$

23. $\frac{4}{20}, \frac{14}{70}$

24. $\frac{15}{42}, \frac{9}{26}$

25. $\frac{9}{16}, \frac{12}{22}$

26. $\frac{14}{32}, \frac{6}{15}$

For more practice: **Skill Bank,** page 429.

Solving Proportions

A **proportion** is a statement that two ratios are equal.

MODEL

Solve the proportion.

$$\frac{2}{3} = \frac{13}{m}$$

Cross multiply.

$$\frac{2}{3} \diagup\!\!\!\!\diagdown \frac{13}{m}$$

$$2 \times m = 3 \times 13$$

$$2 \times m = 39$$

$$2 \times m = 39$$

Divide $m = 39 \div 2$

$$m = 19\frac{1}{2}$$

Solution:

$$m = 19\frac{1}{2}$$

⮞ **GET READY. Complete the statement.**

1. $\frac{2}{3} = \frac{4}{6}$; $2 \times ? = 3 \times 4$

2. $\frac{6}{9} = \frac{12}{18}$; $6 \times 18 = ? \times 12$

3. $\frac{8}{12} = \frac{2}{3}$; $8 \times 3 = 12 \times ?$

4. $\frac{5}{10} = \frac{10}{20}$; $? \times 20 = 10 \times 10$

5. $\frac{27}{36} = \frac{3}{4}$; $27 \times 4 = 36 ? 3$

6. $\frac{1}{6} = \frac{5}{30}$; $1 \times ? = 6 \times 5$

7. $\frac{2}{8} = \frac{1}{4}$; $2 \times 4 = 8 \times ?$

8. $\frac{6}{30} = \frac{1}{5}$; $6 ? 5 = 30 \times 1$

Are the ratios equal? Write yes or no.

9. $\frac{3}{4}, \frac{7}{8}$

10. $\frac{18}{36}, \frac{1}{2}$

11. $\frac{4}{8}, \frac{12}{16}$

12. $\frac{7}{28}, \frac{2}{4}$

13. $\frac{11}{121}, \frac{5}{55}$

14. $\frac{6}{8}, \frac{4}{6}$

15. $\frac{4}{10}, \frac{18}{45}$

16. $\frac{16}{24}, \frac{10}{15}$

17. $\frac{3}{20}, \frac{6}{40}$

18. $\frac{4}{5}, \frac{6}{7}$

19. $\frac{1}{2}, \frac{11}{24}$

20. $\frac{5}{6}, \frac{15}{18}$

21. $\frac{3}{5}, \frac{5}{8}$

22. $\frac{4}{7}, \frac{36}{63}$

23. $\frac{13}{15}, \frac{26}{30}$

Write two equal ratios.

24. $\frac{1}{8}$

25. $\frac{3}{5}$

26. $\frac{1}{2}$

27. $\frac{1}{4}$

28. $\frac{7}{11}$

29. $\frac{4}{5}$

30. $\frac{2}{3}$

31. $\frac{5}{8}$

32. $\frac{9}{10}$

33. $\frac{3}{7}$

34. $\frac{7}{9}$

35. $\frac{6}{7}$

Complete the statement.

36. $2 \times m = 26$
$m = 26 \div ?$

37. $5 \times p = 160$
$p = 160 \div ?$

38. $3 \times w = 51$
$w = 51 \div ?$

39. $t \times 40 = 280$
$t = 280 \div ?$

40. $n \times 59 = 118$
$n = 118 \div ?$

41. $a \times 81 = 243$
$a = 243 \div ?$

NOW USE SKILL 54. Solve the proportion.

42. $\frac{2}{3} = \frac{6}{a}$

43. $\frac{3}{4} = \frac{6}{n}$

44. $\frac{3}{2} = \frac{9}{y}$

45. $\frac{1}{2} = \frac{3}{b}$

46. $\frac{5}{4} = \frac{c}{7}$

47. $\frac{2}{7} = \frac{14}{c}$

48. $\frac{5}{9} = \frac{d}{7}$

49. $\frac{5}{6} = \frac{m}{8}$

50. $\frac{3}{n} = \frac{9}{15}$

51. $\frac{2}{a} = \frac{8}{12}$

52. $\frac{5}{3} = \frac{y}{21}$

53. $\frac{3}{4} = \frac{d}{35}$

54. $\frac{3}{n} = \frac{24}{80}$

55. $\frac{y}{8} = \frac{9}{25}$

56. $\frac{c}{7} = \frac{8}{9}$

57. $\frac{n}{8} = \frac{7}{36}$

58. $\frac{2}{7} = \frac{10}{m}$

59. $\frac{3}{4} = \frac{z}{7}$

60. $\frac{2}{19} = \frac{4}{a}$

61. $\frac{1}{4} = \frac{13}{b}$

62. $\frac{4}{7} = \frac{y}{22}$

63. $\frac{2}{5} = \frac{z}{43}$

64. $\frac{7}{8} = \frac{m}{13}$

65. $\frac{2}{c} = \frac{8}{20}$

66. $\frac{d}{9} = \frac{9}{15}$

67. $\frac{4}{d} = \frac{20}{75}$

68. $\frac{9}{b} = \frac{27}{33}$

69. $\frac{7}{c} = \frac{42}{72}$

70. $\frac{8}{7} = \frac{s}{63}$

71. $\frac{12}{x} = \frac{3}{1}$

72. $\frac{4}{b} = \frac{8}{18}$

73. $\frac{m}{6} = \frac{10}{30}$

74. $\frac{t}{4} = \frac{5}{1}$

75. $\frac{42}{b} = \frac{4}{1}$

76. $\frac{39}{x} = \frac{13}{7}$

77. $\frac{m}{12} = \frac{9}{5}$

78. $\frac{64}{v} = \frac{8}{7}$

79. $\frac{x}{34} = \frac{3}{2}$

80. $\frac{9}{24} = \frac{s}{7}$

81. $\frac{u}{36} = \frac{27}{9}$

82. $\frac{t}{9} = \frac{56}{72}$

83. $\frac{46}{u} = \frac{2}{5}$

84. $\frac{7}{9} = \frac{m}{72}$

85. $\frac{12}{36} = \frac{6}{p}$

86. $\frac{7}{11} = \frac{24}{b}$

87. $\frac{c}{9} = \frac{96}{72}$

88. $\frac{8}{9} = \frac{d}{108}$

89. $\frac{11}{13} = \frac{121}{v}$

For more practice: **Skill Bank,** page 429.

Probability

Probability is a measure of chance. Suppose you put 4 green marbles and 2 red marbles in a jar, shake them up, and take one out without looking. There is an equal chance of getting any one of the marbles. Since there are 6 marbles altogether and 2 are red, the probability of getting a red marble is $\frac{2}{6}$, or $\frac{1}{3}$.

The probability of getting a green marble is $\frac{4}{6}$, or $\frac{2}{3}$.

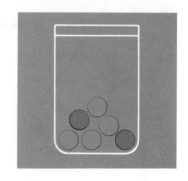

$$\text{Probability of an event} = \frac{\text{number of favorable results}}{\text{total number of results}}$$

In the example above, there are no yellow marbles in the jar. The probability of getting a yellow marble is $\frac{0}{6}$, or 0. It is *impossible* to get a yellow marble.

The probability that you get either a red or a green marble is $\frac{6}{6}$, or 1. You are *certain* to get a red or a green marble.

MODEL

| There are 8 girls and 6 boys in a club. One name is chosen at random. What is the probability that a girl's name is chosen? | $\dfrac{\text{favorable results}}{\text{total results}} = \dfrac{8}{14}$ | The probability of choosing a girl's name is $\dfrac{8}{14}$, or $\dfrac{4}{7}$. |

 GET READY. Write in lowest terms.

1. $\frac{3}{6}$ 2. $\frac{10}{14}$ 3. $\frac{6}{6}$ 4. $\frac{0}{8}$ 5. $\frac{25}{100}$

6. $\frac{6}{8}$ 7. $\frac{0}{15}$ 8. $\frac{10}{50}$ 9. $\frac{13}{52}$ 10. $\frac{4}{52}$

 NOW USE APPLICATION A25.

11. A bag contains 3 red marbles and 2 white marbles. You take one out without looking. What is the probability of getting
    a. a red marble?   **b.** a white marble?

12. You spin the spinner shown at the right. Find the probability of the following events.

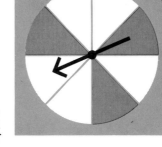

   a. Spinner lands on white.

   b. Spinner lands on red.

   c. Spinner lands on green.

13. You have a numbered cube. It has six faces, each with one of the numbers from 1 through 6. You drop the cube. Find the probability of the following events.

   a. You get a 2.
                            b. You get an even number.

   c. You get an odd number.
             d. You get a number less than 5.

   e. You get a number less than 7.
       f. You get a number greater than 6.

   g. You get an odd number less than 5.
      h. You get a number greater than 2.

14. You shuffle a standard deck of 52 playing cards and draw one card. Find the probability of the following events.

   a. You get a heart.

   b. You get a black card.

   c. You get an ace.

   d. You get a face card (jack, queen, or king).

   e. You get a card that is not a diamond.

15. You flip a coin 50 times. How many times would you expect to get a head?

16. To enter a lottery, you must choose a number from 0 to 9999. One of these numbers is picked as the winning number. What is the probability that your number is picked as the winner?

17. Who has a better chance of winning a drawing: Earl with 5 tickets out of 1000 tickets, or Sally with 1000 tickets out of 1,000,000 tickets?

18. A spinner has 18 orange sectors and 12 blue sectors. All sectors are the same size. What is the chance the spinner will stop on blue?

For more practice: **Application Bank,** page 458.

Bull Market

You are an analyst for a stockbroker. You advise customers on what stocks to buy or sell. You use the *price-earnings ratio,* or P/E, and the *dividend yield ratio* as guidelines. You recommend buying a stock when its P/E is low. You also recommend buying a stock when its dividend yield ratio is high.

You are presently considering five stocks. The business section of the newspaper has some information useful to you.

PROBLEM: Choose two stocks that you will advise customers to buy.

Solution Plan: Find the P/E and dividend yield ratio for each stock. Use these ratios to decide which stocks to recommend.

	Dividend	Price (in $)
American Metals	$1.80	$22\frac{1}{4}$
General Finance	1.60	$16\frac{7}{8}$
National Oil	3.60	$68\frac{3}{4}$
Precision Computer	4.80	45
Resthaven Hotels	.80	$26\frac{3}{8}$

WHAT DO YOU WANT TO KNOW? You want to know the dividend yield ratio. The dividend yield ratio indicates how good a return you get on your investment. To compute the ratio, you round the price to the nearest dollar and divide it into the dividend.

You figure that the dividend yield ratio for American Metals is 0.08, to the nearest hundredth.

$$\frac{\text{Dividend}}{\text{Price}} \frac{\text{yield ratio}}{)\text{Dividend}} \qquad 22\overline{)1.80}^{\,0.081}$$

Find the dividend yield ratio to the nearest hundredth.

1. American Metals
2. General Finance
3. National Oil
4. Precision Computer
5. Resthaven Hotels

WHAT MORE DO YOU WANT? You need to know the P/E for each stock.

Earnings per share can be found by reading a company's annual report. To compute the P/E, you round the price to the nearest dollar and divide by the earnings per share. A low P/E indicates that a stock is worth more than its present price.

Earnings per Share Source: Annual Reports	
American Metals	$3.22
General Finance	1.04
National Oil	2.46
Precision Computer	3.47
Resthaven Hotels	1.22

You figure that the P/E for American Metals is 7, to the nearest whole number.

$$\overset{\text{P/E}}{\text{Earnings}\overline{)\text{Price}}} \qquad 3.22\overline{)22}^{\;6.8}$$
per share

Find the P/E to the nearest whole number.

6. American Metals

7. General Finance

8. National Oil

9. Precision Computer

10. Resthaven Hotels

NOW SOLVE THE PROBLEM.

11. Which stock has the highest dividend yield ratio?

12. Which stock has the lowest dividend yield ratio?

13. Which stock has the highest P/E?

14. Which stock has the lowest P/E?

15. Which two stocks do you advise your customers to buy? Why?

Skill 52 Write the ratio as a fraction.

1. 2 out of 5 books   **2.** 17 out of 100 mice   **3.** 10¢ of every dollar

Skill 53 Are the ratios equal?   **Skill 54 Solve the proportion.**

4. $\frac{12}{20}, \frac{18}{30}$ **5.** $\frac{5}{12}, \frac{8}{19}$ **6.** $\frac{20}{c} = \frac{8}{14}$ **7.** $\frac{5}{8} = \frac{9}{n}$

Application A25

8. You shuffle a standard deck of playing cards and draw one card. What is the probability that you get a king?

Problem Solving

9. Find the P/E for each stock, to the nearest whole number.

	Price (in $)	Earnings per share
Soco, Inc.	$30\frac{1}{4}$	$2.20
A & M Co.	75	3.15

CALCULATOR DISPLAYS

Proportions

Each spot-welder on the Major Motors assembly line processes 7 cars every 10 minutes. How many cars does each welder process in 8 hours? (Remember that 8 hours = 480 minutes.)

KEY-IN	DISPLAY	
7	7	$\frac{7}{10} = \frac{c}{480}$
⊠	7	
480	480	$7 \times 480 = 10 \times c$
⩵	3360 ⟶	$3360 = 10 \times c$
÷	3360	
10	10	$3360 \div 10 = c$
⩵	336 ⟶	$336 = c$

Each welder processes 336 cars in 8 hours.

Use a calculator to solve the proportion.

1. $\frac{9}{36} = \frac{x}{108}$ **2.** $\frac{n}{19} = \frac{15}{57}$ **3.** $\frac{7}{h} = \frac{35}{22}$ **4.** $\frac{8}{15} = \frac{36}{t}$ **5.** $\frac{8}{a} = \frac{32}{63}$

Scale Drawing

Below is a scale drawing of the floor plan of a house. It is used to show the actual measurements of the house on a reduced scale.

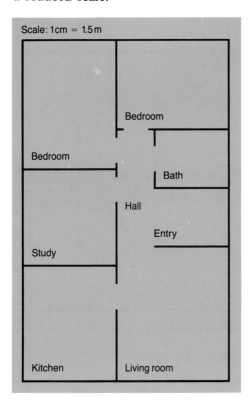

Scale: 1 cm = 1.5 m

Bedroom

Bedroom

Bath

Hall

Entry

Study

Kitchen

Living room

Use a ruler to find the scale length of the house. What is the actual length?

$$\frac{\text{scale length (cm)}}{\text{actual length (m)}} \quad \frac{1}{1.5} = \frac{9}{l} \longleftarrow \text{The scale length is 9 cm.}$$

$$1 \times l = 1.5 \times 9$$
$$l = 13.5$$

The actual length of the house is 13.5 m.

Complete the table.

Room	Length		Width	
	Scale	Actual	Scale	Actual
kitchen	3 cm	?	2.5 cm	?
study	?	?	?	?
bathroom	?	?	?	?
living room	?	?	?	?
entry	?	?	?	?

Writing Rates as Fractions

MODEL

Write the rate as a fraction.

85 pages in 30 minutes

$\dfrac{85}{30}$

 NOW USE SKILL 55. Write the rate as a fraction.

1. 100 meters in 20 seconds

2. 3 cups in 2 hours

3. 3 liters for 120 kilometers

4. 3 take-offs in 5 minutes

5. $36 for 2 skirts

6. 50 points in 4 games

7. 630 people in 3 planes

8. 3 holidays in 2 months

9. 120 guests in 10 days

10. 85 responses to 140 invitations

11. 39 shows in 4 weeks

12. 200 gallons for 800 people

13. 70 dogs for 3 exhibitions

14. 12 commercials in 60 minutes

15. 110 cars in 7 lots

16. 4 fouls in 3 games

17. 4 weeks' supplies for $262

18. 3 trips in 7 weeks

19. 380 seats in 40 rows

20. 17 houses in 10 miles

21. 42 birds for 7 cages

22. 80 places for 120 applicants

23. $36 for 8 tickets

24. 2 tablespoons for 3 packages

25. 7 toll booths in 430 miles

26. 86 telephones in 30 offices

27. 5 airplanes on 3 runways

28. 5 touchdowns for 30 points

29. 4 days in 2 months

30. 6 pounds in 7 weeks

31. 12 letters in 5 days

32. 25 floors in 10 seconds

33. 4 cars for 15 people

34. 1987 people at 9 performances

35. 93 concerts in 17 weeks

36. 8 visits in 7 weeks

Writing Equal Rates

MODEL

Write the equal rates as a proportion.

7 people in 2 cars = x people in 6 cars

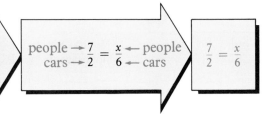

$$\text{people} \rightarrow \frac{7}{2} = \frac{x}{6} \leftarrow \text{people}$$
$$\text{cars} \rightarrow \quad \quad \leftarrow \text{cars}$$

$$\frac{7}{2} = \frac{x}{6}$$

 GET READY. Complete the proportion.

37. 680 kilometers in 10 hours = x kilometers in 6 hours $\quad \frac{680}{10} = \frac{x}{?}$

38. 10 gallons in 3 hours = b gallons in 6 hours $\quad \frac{?}{3} = \frac{b}{6}$

39. 1500 tons in 4 grain elevators = 3000 tons in t grain elevators $\quad \frac{1500}{4} = \frac{3000}{?}$

40. 10 kilograms for \$7.00 = s kilograms for \$140.00 $\quad \frac{10}{?} = \frac{s}{140}$

 NOW USE SKILL 55. Write the equal rates as a proportion.

41. 825 km traveled in 30 days = n km traveled in 6 days

42. 680 people sitting in 8 rows = q people sitting in 13 rows

43. \$46.00 for 3 shirts = \$138.00 for m shirts

44. 4 kilograms for 80¢ = 1 kilogram for r cents

45. 3 games in 2 hours = 9 games in w hours

46. 7 trains in 11 minutes = h trains in 55 minutes

47. 36 units on 2 floors = p units on 10 floors

48. \$150 for 3 meters = \$900 for y meters

49. 8 bottles for 4 days = z bottles for 15 days

50. 15 books in 2 semesters = 45 books in n semesters

51. 243 cars parked in 7 lots = e cars parked in 10 lots

52. 9 buses in 15 minutes = 27 buses in v minutes

53. 768 students in 2 schools = m students in 11 schools

54. 5 movies in 3 months = 20 movies in x months

For more practice: **Skill Bank,** page 430.

Using Equal Rates in Proportions

MODEL

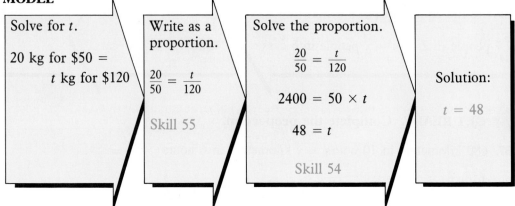

Solve for t.

20 kg for \$50 =
t kg for \$120

Write as a proportion.

$$\frac{20}{50} = \frac{t}{120}$$

Skill 55

Solve the proportion.

$$\frac{20}{50} = \frac{t}{120}$$

$$2400 = 50 \times t$$

$$48 = t$$

Skill 54

Solution:

$$t = 48$$

 GET READY. Complete the proportion.

1. 425 kilometers in 8 hours = s kilometers in 16 hours $\quad \frac{425}{8} = \frac{?}{16}$

2. 76 trains in 2 days = 190 trains in h days $\quad \frac{76}{2} = \frac{190}{?}$

3. t liters in 40 seconds = 300 liters in 50 seconds $\quad \frac{t}{40} = \frac{300}{?}$

4. 24 grams for x dollars = 120 grams for \$75 $\quad \frac{?}{x} = \frac{120}{75}$

Complete the statement.

5. $\frac{18}{3} = \frac{n}{1}$; $18 \times 1 = 3 \times$?

6. $\frac{32}{4} = \frac{4}{s}$; $32 \times$? $= 4 \times 4$

7. $\frac{b}{21} = \frac{5}{1}$; $b \times 1 =$? $\times 5$

8. $\frac{48}{12} = \frac{4}{m}$; ? $\times m = 12 \times 4$

9. $\frac{3}{8} = \frac{12}{b}$; $3 \times$? $= 8 \times 12$

10. $\frac{2}{7} = \frac{12}{t}$; $2 \times t =$? $\times 12$

11. $\frac{n}{5} = \frac{8}{10}$; $n \times 10 = 5 \times$?

12. $\frac{a}{12} = \frac{6}{18}$; ? $\times 18 = 12 \times 6$

214

Solve the proportion.

13. $\frac{3}{10} = \frac{x}{80}$ **14.** $\frac{4}{21} = \frac{16}{s}$ **15.** $\frac{a}{17} = \frac{10}{85}$ **16.** $\frac{6}{m} = \frac{36}{66}$

17. $\frac{6}{24} = \frac{s}{73}$ **18.** $\frac{3}{40} = \frac{12}{g}$ **19.** $\frac{36}{n} = \frac{72}{3}$ **20.** $\frac{12}{6} = \frac{n}{18}$

21. $\frac{7}{t} = \frac{24}{36}$ **22.** $\frac{5}{16} = \frac{4}{m}$ **23.** $\frac{u}{7} = \frac{15}{8}$ **24.** $\frac{16}{25} = \frac{x}{100}$

NOW USE SKILL 56. Write the equal rates as a proportion. Solve the proportion.

25. 350 kilometers on 25 liters = x kilometers on 4 liters

26. 275 miles in 1 day = 825 miles in s days

27. 30 deliveries in 14 hours = 75 deliveries in t hours

28. 8 commercials in 5 minutes = w commercials in 45 minutes

29. $38.40 earned for 8 hours = t dollars earned for 40 hours

30. 2 loaves in 3 days = 16 loaves in z days

31. 36 passengers in 1 bus = 180 passengers in n buses

32. 9 wins in 18 games = x wins in 20 games

33. $475 for one month's rent = b dollars for 12 months' rent

34. 64 pages read in 20 minutes = 150 pages read in t minutes

35. 4 teaspoons for 8 cups = k teaspoons for 10 cups

36. 120 people in 2 rows = m people in 20 rows

37. 1 kilometer in 15 minutes = p kilometers in 45 minutes

38. 1440 filters made in 6 hours = c filters made in 5 hours

39. 8 games in 4 hours = r games in 10 hours

40. 3 radios fixed in 5 hours = h radios fixed in 20 hours

41. $560 saved in 2 months = t dollars saved in 24 months.

42. 40 words typed per minute = j words typed in 18 minutes

43. 18 responses in 3 days = 138 responses in s days

44. 37 problems per page = w problems in 164 pages

45. 246 letters in 82 mailboxes = q letters in 1 mailbox

For more practice: **Skill Bank,** page 430.

215

Sampling and Prediction

In **sampling,** information is obtained from a small group of people. This small group is called the **sample.** The information about the sample is used to make predictions about a larger **population.**

The city library received a grant to buy some new books. The librarian took a poll to get an idea of what types of books to order. The library has a total membership of 7564. A sample of 175 people gave the responses shown.

Type of Book	Favorable Votes
Novels	53
Mysteries	75
Classics	21
Instructional	26
Total in Sample	175

The formula below can be used to make predictions based on sampling.

$$\frac{\text{Number favorable in sample}}{\text{Total in sample}} = \frac{\text{Number favorable in population}}{\text{Total in population}}$$

MODEL

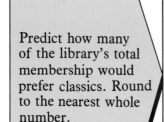

Predict how many of the library's total membership would prefer classics. Round to the nearest whole number.

Use the formula.

$$\frac{21}{175} = \frac{n}{7564}$$

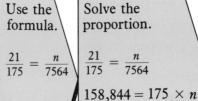

Solve the proportion.

$$\frac{21}{175} = \frac{n}{7564}$$

$$158,844 = 175 \times n$$

$$907.68 = n$$

Skill 54

You predict that 908 members would prefer the classics.

 GET READY.

1.	846 × 7	2.	370 × 34	3.	515 × 82	4.	792 × 27	5.	1243 × 63

Solve the proportion. Round to the nearest whole number.

6. $\frac{5}{8} = \frac{n}{32}$

7. $\frac{23}{46} = \frac{n}{13}$

8. $\frac{7}{20} = \frac{n}{376}$

9. $\frac{56}{70} = \frac{n}{800}$

NOW USE APPLICATION A26. Round to the nearest whole number.

10. The local hardware store plans to extend evening hours on Wednesdays. To get an idea of customer preferences, the store took a poll of 185 customers.

Extend Hours to:	Favorable Votes
8:30 P.M.	61
9:30 P.M.	85
10:30 P.M.	39

The store serves an estimated population of 12,000 people. How many customers do you predict would prefer the store to be open until 8:30 P.M.? until 9:30 P.M.? until 10:30 P.M.?

11. Mifton Soaps is developing a new product. They asked 70 people to test the fragrances Mifton is considering for a new bath soap. The table shows the results of the poll.

Fragrance	Favorable Votes
A	22
B	18
C	30

The estimated market size for this product is 258,000 people. How many people do you predict would prefer fragrance A? fragrance B? fragrance C?

12. John Abbott is running for office in this year's Student Government election. According to the by-laws, nominees for president must receive more than 350 votes on the first ballot before they are listed as candidates. This year there are 5 nominees. A poll of 78 students was conducted to predict the winners of the first ballot.

The student population is 2450. According to the poll, how many votes is John Abbott expected to receive on the first ballot?

How many nominees are expected to receive enough votes on the first ballot to become candidates?

Nominee	Votes
Miriam Abbas	11
John Abbott	16
Steven Chen	12
Elaine Manning	18
Carol White	21

For more practice: **Application Bank,** page 458.

A & S Bootique

Alvarez and Suda decide to open a shoe store. They form a partnership to own their new business. As part of their partnership agreement, they arrange a plan stating how they will divide their profits.

Alvarez, manager of the store, will get a salary of $10,000 from the store's profits. Suda will get a salary of $5000. In addition, they will divide the remaining profit between themselves in the same ratio as their original investments in the business. When they form the partnership, Alvarez invests $25,000 and Suda invests $75,000.

PROBLEM: How much does each partner earn?

Solution Plan: Find out how much profit there is, and figure each partner's share of the profit.

WHAT DO YOU KNOW? Answer the questions.

1. How much did Alvarez invest in the business?

2. How much did Suda invest in the business?

3. What is their total investment in the business?

4. What is each partner's salary?

WHAT DO YOU WANT TO KNOW? You want to know how much money the partners divide. Use the worksheet.

5. How much profit did the store earn?

6. How much profit is left after the partners receive their salaries? This amount is called the Remainder in the partnership agreement.

A&S Bootique Worksheet For year ended Dec. 31		
Expenses		
Inventory	$20,000	00
Wages	22,000	00
Rent	10,500	00
Other	14,000	00
Income	118,000	00

NOW SOLVE THE PROBLEM. The Remainder is divided according to the partners' investments in the business. The ratio of each partner's share of the Remainder to the total Remainder is equal to the ratio of each partner's investment to the total investment. Complete the earnings statement.

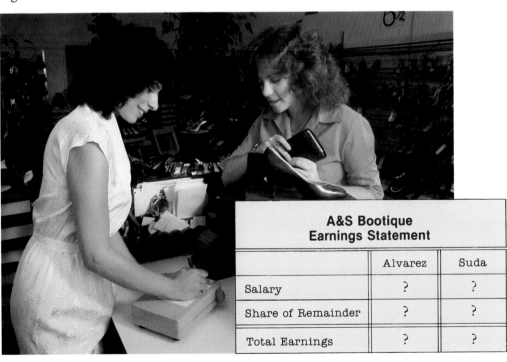

A&S Bootique Earnings Statement		
	Alvarez	Suda
Salary	?	?
Share of Remainder	?	?
Total Earnings	?	?

Skill 55 **Write the equal rates as a proportion.**

1. 15,000 books in 2 libraries = c books in 4 libraries

2. 32 suits on 4 racks = 48 suits on n racks

Skill 56 **Write the equal rates as a proportion. Solve the proportion.**

3. 100 cards in 5 boxes = x cards in 4 boxes

4. 330 words in 15 lines = x words in 6 lines

Application A26

5. A poll shows 2 out of 3 people approve of the mayor's performance. How many voters out of 20,000 do you predict would approve?

Problem Solving

6. You invest $3000 in your business. Your partner invests $7000. You are dividing a profit of $12,000 in the same ratio as the investments. What is your share of the profit?

CONSUMER NOTE

The Consumer Price Index (CPI) is the current cost of goods and services that would have cost $100 in 1967. Given the current CPI, you can use the formula below to compare the cost of a product or service in 1967 with the current cost.

$$1967 \text{ cost} = \frac{100 \times \text{Current cost}}{\text{Current CPI}}$$

The Daniels paid $68.77 for groceries this week. The current CPI is 294.6. How much would the groceries have cost in 1967?

$$1967 \text{ cost} = \frac{100 \times \$68.77}{294.6} = \$23.34$$

Suppose the current CPI is 302.3. To the nearest cent, find the cost in 1967.

1. $68,450 house

2. $5700 annual rent

3. $89.93 grocery bill

4. $6475 car

5. $15.87 meal

6. $8806 pickup truck

Computer Careers

Society places great importance on computers. Despite computers' increasing ability to operate without help from humans, people remain the most important part of a computer system.

Installation and maintenance of a computer's physical components, or *hardware*, is performed by *field engineers*. Field engineers usually have a background in electronics, and education in a technical discipline.

Computer operators look after the physical operation of the system. Operators receive specialized training for the equipment they are using. They mount magnetic tapes and disks, load punched cards into the card reader, and distribute the printed output to the proper departments.

Data entry clerks put data into a form that the machine can use, such as information key-punched onto computer cards. Often, data entry is done by entering information directly into the computer. The data entry clerk sits at a CRT terminal and types information into a program which is being executed by the computer. The information is then either processed immediately or stored for future processing.

Systems analysts and *applications programmers* work with computers. Systems analysts use computers to solve problems. They use techniques such as mathematical modeling, operations research, and sampling to analyze business situations and determine the best solution to the problem. Then they prepare instructions for the computer programmer to follow. Applications programmers are responsible for writing the computer programs and making sure they work. The programmer also provides a written description of each program and its use. Both systems analysts and applications programmers usually have education beyond high school.

Activity

Make a list of computer-related job openings found in the help wanted section of a newspaper. Include in your list education requirements, years of experience needed, and salary range.

221

Finding Unit Rates

MODEL

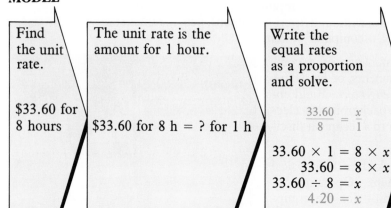

Find the unit rate.

$33.60 for 8 hours

The unit rate is the amount for 1 hour.

$33.60 for 8 h = ? for 1 h

Write the equal rates as a proportion and solve.

$$\frac{33.60}{8} = \frac{x}{1}$$

$33.60 \times 1 = 8 \times x$
$33.60 = 8 \times x$
$33.60 \div 8 = x$
$4.20 = x$

$33.60 for 8 hours is the same as $4.20 for 1 hour.

The unit rate is $4.20 per hour.

 GET READY. Solve the proportion.

1. $\frac{40}{8} = \frac{s}{1}$ **2.** $\frac{96}{8} = \frac{m}{1}$ **3.** $\frac{50}{5} = \frac{x}{1}$ **4.** $\frac{80}{20} = \frac{a}{1}$

5. $\frac{108}{12} = \frac{n}{1}$ **6.** $\frac{2720}{20} = \frac{v}{1}$ **7.** $\frac{54.40}{8} = \frac{m}{1}$ **8.** $\frac{144}{9} = \frac{b}{1}$

 NOW USE SKILL 57. Find the unit rate.

9. 175 words in 5 minutes

10. $600 for 48 dinners

11. 680 kilometers in 10 hours

12. 768 students in 4 classes

13. 3750 miles in 12 minutes

14. 53,410 volumes in 7 libraries

15. 819 jelly beans in 3 jars

16. 780 employees on 5 floors

17. 34 windows in 17 hours

18. $1260 saved in 12 months

19. 43 repairs in 5 days

20. 75 calls in 2 hours

21. 360 people in 6 rooms

22. 48 hits in 20 games

23. 920 miles in 2 days

24. 136 patients in 16 hours

Using Unit Rates

MODEL

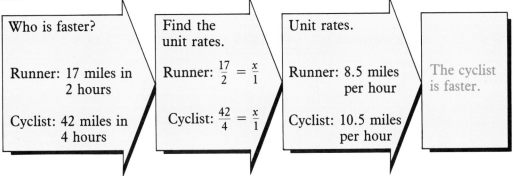

Who is faster?

Runner: 17 miles in 2 hours

Cyclist: 42 miles in 4 hours

Find the unit rates.

Runner: $\frac{17}{2} = \frac{x}{1}$

Cyclist: $\frac{42}{4} = \frac{x}{1}$

Unit rates.

Runner: 8.5 miles per hour

Cyclist: 10.5 miles per hour

The cyclist is faster.

Compare the unit rates to answer the question.

25. Car A: 186 miles on 6 gallons.
Car B: 224 miles on 8 gallons.
Which car gets better mileage?

26. Bus A: 700 kilometers on 190 liters.
Bus B: 180 kilometers on 45.6 liters.
Which bus gets better mileage?

27. Train: 572 miles in 11 hours.
Bus: 704 miles in 16 hours.
Which is faster?

28. Plane A: 2080 miles in 8 hours.
Plane B: 2590 miles in 6 hours.
Which is faster?

29. Ruth: 2120 bolts in 8 hours.
Joseph: 3240 bolts in 12 hours.
Who works faster?

30. Alfred: 105 sales in 5 days.
Wanda: 161 sales in 7 days.
Who had the higher average
daily sales?

31. Cashews: 850 calories in 5 ounces.
Pecans: 2460 calories in 12 ounces.
Which has more calories per ounce?

32. Charlie: $73.60 in 16 hours.
George: $192.00 in 40 hours.
Who makes the higher hourly wage?

33. Anna: $1280 in 8 weeks.
Marie: $620 in 4 weeks.
Who saved the most per week?

34. Flour A: $2.17 for 5 pounds.
Flour B: $1.65 for 3 pounds.
Which is the better buy?

35. Cereal A: $1.36 for 8 ounces.
Cereal B: $1.70 for 10 ounces.
Which is the better buy?

36. Notebook A: $1.20 for 80 pages.
Notebook B: $1.45 for 120 pages.
Which is the better buy?

37. Copper: 89 grams for 10 cm^3.
Iron: 68 grams for 9 cm^3.
Which is more dense?
(Density is mass per unit volume.)

38. Water: 103 grams for 103 cm^3.
Aluminum: 104 grams for 36 cm^3.
Which is more dense?
(Density is mass per unit volume.)

For more practice: **Skill Bank,** page 430.

Mean

The **mean** of a set of numbers is the *average* of the numbers. To find the mean, add the numbers, then divide by the number of items in the set.

$$\text{Mean} = \frac{\text{Sum}}{\text{Number of Items}}$$

MODEL

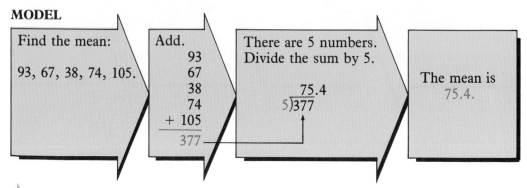

Find the mean:

93, 67, 38, 74, 105.

Add.
```
   93
   67
   38
   74
 + 105
  377
```

There are 5 numbers.
Divide the sum by 5.

```
     75.4
 5)377
```

The mean is
75.4.

GET READY. Find the sum of the numbers.

1. 29, 38, 67, 99   **2.** 43, 67, 52, 88, 107   **3.** 65, 90, 31, 27, 63

NOW USE APPLICATION A27. Find the mean.
Round to the nearest tenth when necessary.

4. 6, 7, 9, 11, 12, 12   **5.** 6, 7, 11, 12, 14

6. 16, 19, 20, 22, 23, 24   **7.** 16, 19, 20, 22, 23, 25

8. 95, 43, 67, 88, 67, 24   **9.** 87, 81, 75, 76, 92, 81

10. 916, 102, 168, 897, 207, 396, 303   **11.** 105, 196, 713, 512, 309, 109

12. Angela bought 3 books at the school fair. She paid $1.35 for one, $2.10 for another, and $3.45 for the third. On the average, how much did Angela spend on each book?

13. Brenda scored 115, 132, 124, and 156 to win the junior bowling championship. What was her average score?

14. The attendance at 6 baseball games was 20,876; 31,138; 18,533; 25,653; 33,179; and 27,845. What was the mean attendance?

15. George drove 589 miles on 19 gallons of gasoline. What was the average number of miles per gallon?

For more practice: **Application Bank,** page 459.

Median, Mode, and Range

When a set of numbers is arranged in order, you can find the **median, mode,** and **range.**

Median: The middle number.
Mode: The most frequent number.
Range: The difference between the largest and the smallest numbers.

If there is an even number of items, the median is the mean of the two middle numbers. If no number occurs more than once, there is no mode.

MODEL

Find the median, mode, and range:	Arrange the numbers in order.	Median	Median:	13
13, 8, 9, 24, 24	8, 9, 13, 24, 24	8, 9, 13, 24, 24	Mode:	24
		Mode	Range:	16
		Range = 24 − 8 = 16		

 GET READY. Write in order from smallest to largest.

1. 28, 40, 36, 18, 24

2. 62, 55, 73, 41, 27, 30, 35

3. 4674, 4682, 4666

4. 1260, 1130, 1473, 1213, 1320

 NOW USE APPLICATION A28. Find the median and range.

5. 42, 43, 51, 58, 59

6. 63, 64, 64, 67, 81, 81, 86

7. 486, 397, 409, 486, 382

8. 121, 118, 136, 118, 136

9. 264, 127, 385, 469, 583

10. 600, 300, 100, 200, 500

11. 8000, 1000, 7000

12. 2462, 1370, 3382, 1265

13. $64, $86, $58, $86

14. $14, $26, $18, $23, $47, $68

15. $420, $380, $565, $225, $565

16. $4360, $3872, $2870, $1890, $5445

17. Five football players weighed 230 lb, 255 lb, 272 lb, 255 lb, and 267 lb. What are the median and mode of these weights?

18. Scores for a quiz are 88, 56, 76, 82, 93, 83, 88, 74, 67, 59, 93, 76, 80, 88, 94, 81, and 77. What is the median? the mode? the range?

For more practice: **Application Bank,** page 459.

Driving a Bargain

You bought a used car two years ago for $4200. Since then, you have driven it 17,500 miles. You kept a record of all your car expenses.

Car Expense Record			
1st Year		**2nd Year**	
Insurance	$360	License	$ 66
Tune-up	65	Repair brakes	165
License	66	Tune-up	60
Repair clutch	200	Insurance	400
Gas	480	Gas	390
		Repair muffler	60

PROBLEM: What was your average cost per mile to own and operate your car?

Solution Plan: Find the total of your car expenses. Figure the unit rate, in cents per mile, for your car.

WHAT DO YOU KNOW? You know your car expenses for the past two years.

1. What was the total cost for insurance?

2. What was the total cost for licensing?

3. What was your total repair and maintenance expense?

4. How much did you spend on gas altogether?

WHAT DO YOU WANT TO KNOW? You want to know the cost of *depreciation*. This is a "hidden" cost. **Depreciation** is the decrease in the value of your car because of age and wear and tear.

5. How much did you pay for the car?

6. At the end of two years you find that the resale value of your car is $3000. How much is the depreciation?

WHAT MORE DO YOU WANT? You want to know the total of all expenses for the two years.

Complete the expense summary.

	Item	Amount
7.	Insurance	?
8.	Repair, maintenance	?
9.	License	?
10.	Gas	?
11.	Depreciation	?

NOW SOLVE THE PROBLEM.

12. What was the total of your car expenses, including depreciation?

13. Find the cost, in cents per mile, for your car. Round to the nearest cent.

Quiz

Skill 57 **Find the unit rate.**

1. $19.95 for 5 gallons

2. $198.00 for 2 suits

3. Suarez: 1000 m in 150 sec   Parr: 1500 m in 230 sec   Who is faster?

Application A27

4. Find the mean:
48, 52, 39, 41, 55.

Application A28

5. Your baseball team played 5 extra-inning games last season. The games lasted 15, 10, 11, 12, and 14 innings. What was the median length of these games?

Problem Solving

6. You have the following automobile expenses for the year: Insurance, $950; repair, $250; license, $50; gas, $725; depreciation, $1500. You drove 13,200 miles. To the nearest cent, what was the cost per mile?

CONSUMER NOTE

Unit prices help you compare the costs of products sold in packages or containers of different sizes. If price is your concern, unit prices will help you find which item is the best buy.

$$\text{Unit price} = \frac{\text{Price}}{\text{Measure or Count}}$$

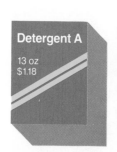 

Which is the better buy?

Detergent A
$$\frac{118¢}{13 \text{ oz}} = 9.08¢ \text{ per oz}$$

Detergent B
$$\frac{215¢}{20 \text{ oz}} = 10.75¢ \text{ per oz}$$

Detergent A is the better buy.

Find the unit prices. Which is the better buy?

1. Tomato paste
33¢ for 6 oz
53¢ for 10 oz

2. Paper napkins
$.65 for 50
$2.10 for 140

3. Salad oil
$1.35 for 25 oz
$1.82 for 35 oz

Skills Review

Skill 53 Are the ratios equal? (page 203)

1. $\frac{1}{2}, \frac{4}{8}$ 2. $\frac{3}{4}, \frac{4}{9}$ 3. $\frac{2}{3}, \frac{4}{7}$ 4. $\frac{7}{9}, \frac{4}{5}$ 5. $\frac{4}{5}, \frac{13}{15}$

6. $\frac{2}{3}, \frac{30}{45}$ 7. $\frac{8}{9}, \frac{23}{27}$ 8. $\frac{3}{8}, \frac{15}{40}$ 9. $\frac{4}{13}, \frac{20}{65}$ 10. $\frac{5}{13}, \frac{15}{39}$

Skill 54 Solve the proportion. (pages 204–205)

11. $\frac{15}{b} = \frac{5}{1}$ 12. $\frac{24}{6} = \frac{4}{t}$ 13. $\frac{m}{8} = \frac{15}{6}$ 14. $\frac{b}{3} = \frac{20}{15}$

15. $\frac{67}{4} = \frac{m}{5}$ 16. $\frac{4}{a} = \frac{18}{27}$ 17. $\frac{t}{6} = \frac{14}{35}$ 18. $\frac{c}{4} = \frac{7}{5}$

Skill 55 Write the rate as a fraction. (pages 212–213)

19. 6 kilometers walked in 40 minutes   20. 320 km driven on 30 liters

21. 240 people in 4 rows   22. 4 cars repaired for $376.20

23. 64 radios produced in 40 hours   24. $1984 earned in 16 weeks

Skill 56 Write as a proportion. Solve the proportion. (pages 214–215)

25. 1470 bottles in 3 hours = t bottles in 40 hours

26. $160 saved in 4 weeks = $800 saved in m weeks

27. w passengers in 5 planes = 696 passengers in 6 planes

Skill 57 Find the unit rate. (pages 222–223)

28. $48 for 30 transistors   29. 480 miles in 2 minutes

30. 96 calls in 4 hours   31. 195 windows in 3 buildings

Compare the unit rates to answer the question.

32. Wally: $2100 spent in 16 weeks.
Sarah: $1728 spent in 12 weeks.
Who had the greater weekly expenses?

33. Brand A: $17.76 for 8 ounces.
Brand B: $14.88 for 6 ounces.
Which is the better buy?

Application A25 **Find the probability.** (pages 206–207)

1. The spinner lands on blue.
2. The spinner lands on 6.
3. The spinner lands on red.
4. The spinner lands on 4.
5. The spinner lands on 1.
6. The spinner lands on 2.
7. The spinner lands on 5.

Application A26 **Round to the nearest whole number.** (pages 216–217)

8. The newsstand sells a wide variety of magazines. The manager recorded the types of magazines sold in one week.

The newsstand serves about 6000 customers. How many customers do you predict would prefer fashion magazines? sports magazines? weekly news magazines?

Type of Magazine	Number Sold
Professional	15
Fashion	76
Home	43
Sports	49
News Weeklies	80
Foreign	12
Total Sold	275

Application A27 **Find the mean.** (page 224)

9. 5, 4, 8, 2, 9, 5, 4, 7
10. 45, 89, 23, 65, 70, 99, 43, 85, 66
11. 101, 403, 356, 210, 367, 438
12. 1291, 3091, 1678, 2208, 2584, 3182

Application A28 (page 225)

13. Atlas Industries kept a record of the number of vacation days employees took each month.

Jan.	Feb.	Mar.	Apr.	May	June	July	Aug.	Sept.	Oct.	Nov.	Dec.
25	28	35	40	40	60	75	90	85	78	50	30

What is the median number of vacation days per month? What is the mode? What is the range?

Unit Test

Skill 52 **Write the ratio as a fraction.** Skill 53 **Are the ratios equal?**

1. 65 out of 70 children

2. 19 out of 55 applicants

3. $\frac{4}{5}, \frac{6}{7}$

4. $\frac{9}{12}, \frac{18}{24}$

Skill 54 **Solve the proportion.**

5. $\frac{4}{8} = \frac{6}{c}$ **6.** $\frac{4}{16} = \frac{5}{t}$ **7.** $\frac{15}{a} = \frac{2}{5}$ **8.** $\frac{n}{19} = \frac{5}{8}$

Skill 55 **Write the equal rates as a proportion.**

9. 525 miles in 5 hours $= t$ miles in 3 hours

Skill 56

Write the equal rates as a proportion. Solve the proportion.

10. $42 for 8 hours $= t$ dollars for 10 hours

11. 186 miles on 6 gallons $= 217$ miles on m gallons

Skill 57 **Find the unit rate.**

12. 96¢ for 3 apples

13. $50 for 4 shirts

Application A25

14. You have a numbered cube. It has six faces, each with one of the numbers from 1 through 6. You drop the cube. What is the probability that you get a number less than 4?

Application A26

15. In Green Village, 47 out of 100 voters polled favor a tax cut proposal. If 10,000 people vote, how many do you expect to vote for the proposal?

Application A27

16. Donald Said scored 22, 19, 23, 10, and 33 points in the last five basketball games. What was his mean score for these games?

Application A28

17. The scores on an exam were 79, 95, 83, 80, 60, 98, 83, 75, and 88. Find the median, mode and range of the scores.

8

Fractions, Decimals, Percents

Bar, Line, and Circle Graphs

Problem Solving

Unit Preview

 You will use these skills and applications in this unit.
Which do you already know? Work each problem.

Skill 58 Write as a decimal.
Round to the nearest hundredth.

1. $\frac{5}{6}$ 2. $\frac{19}{20}$

Skill 59 Write as a fraction in lowest terms.

3. 0.75 4. 0.84

Skill 60 Write as a decimal.
Round to the nearest hundredth.

5. $3\frac{5}{6}$ 6. $7\frac{3}{8}$

Skill 61 Write as a mixed number in lowest terms.

7. 2.6 8. 7.55

Skill 62 **Skill 63** **Skill 64** **Skill 65**

9. $\frac{3}{5} = \frac{?}{100}$ 10. $\frac{35}{100} = \underline{\quad?\quad}\%$ 11. $0.086 = \underline{\quad?\quad}\%$ 12. $\frac{3}{8} = \underline{\quad?\quad}\%$

Skill 66 Write as a decimal.

13. 42% 14. 3.8%

Skill 67 Write in lowest terms.

15. $60\% = \frac{?}{?}$ 16. $72\% = \frac{?}{?}$

Application A29

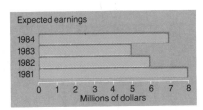

17. When do you expect earnings to increase over the preceding year?

Application A30

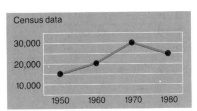

18. How much did the population decrease between 1970 and 1980?

Application A31

19. Which expense was the largest?

20. Which cost more, research or supplies?

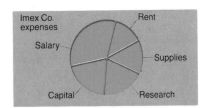

Check your answers. If you had difficulty with any skill or application, be sure to study the corresponding lesson in this unit.

233

Writing Fractions as Decimals

MODEL

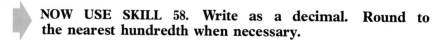

Write as a decimal. Round to the nearest hundredth.

$$\frac{4}{7}$$

Divide to thousandths.

$$\frac{0.571}{7)4.000}$$

Skill 20

Round to the nearest hundredth.

$$0.571 \to 0.57$$

Skill 2

To the nearest hundredth,

$$\frac{4}{7} = 0.57$$

GET READY. Round to the nearest hundredth.

1. 0.579   **2.** 0.332   **3.** 0.527   **4.** 0.834

5. 0.013   **6.** 0.937   **7.** 0.841   **8.** 0.724

NOW USE SKILL 58. Write as a decimal. Round to the nearest hundredth when necessary.

9. $\frac{1}{2}$ **10.** $\frac{1}{4}$ **11.** $\frac{3}{4}$ **12.** $\frac{1}{5}$ **13.** $\frac{2}{5}$ **14.** $\frac{3}{5}$

15. $\frac{4}{5}$ **16.** $\frac{1}{6}$ **17.** $\frac{5}{6}$ **18.** $\frac{1}{10}$ **19.** $\frac{3}{10}$ **20.** $\frac{7}{10}$

21. $\frac{2}{3}$ **22.** $\frac{1}{20}$ **23.** $\frac{3}{20}$ **24.** $\frac{7}{20}$ **25.** $\frac{13}{20}$ **26.** $\frac{1}{25}$

27. $\frac{2}{25}$ **28.** $\frac{3}{25}$ **29.** $\frac{7}{25}$ **30.** $\frac{1}{50}$ **31.** $\frac{1}{100}$ **32.** $\frac{37}{100}$

33. $\frac{1}{7}$ **34.** $\frac{4}{7}$ **35.** $\frac{4}{9}$ **36.** $\frac{7}{15}$ **37.** $\frac{3}{16}$ **38.** $\frac{4}{75}$

Write as a decimal. Round to the nearest thousandth when necessary.

39. $\frac{1}{8}$ **40.** $\frac{3}{8}$ **41.** $\frac{5}{8}$ **42.** $\frac{7}{8}$ **43.** $\frac{7}{30}$ **44.** $\frac{5}{12}$

45. $\frac{7}{40}$ **46.** $\frac{23}{40}$ **47.** $\frac{1}{16}$ **48.** $\frac{5}{16}$ **49.** $\frac{11}{18}$ **50.** $\frac{16}{75}$

For more practice: **Skill Bank,** page 431.

Writing Decimals as Fractions

MODEL

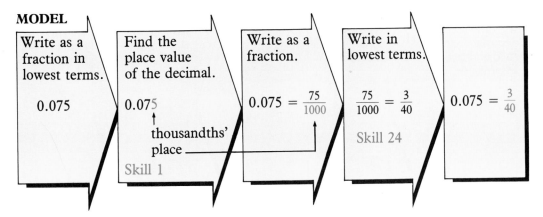

Write as a fraction in lowest terms.	Find the place value of the decimal.	Write as a fraction.	Write in lowest terms.	
0.075	0.075 ↑ thousandths' place — Skill 1	$0.075 = \frac{75}{1000}$	$\frac{75}{1000} = \frac{3}{40}$ Skill 24	$0.075 = \frac{3}{40}$

GET READY. Write as a fraction in lowest terms or as a whole number.

1. $\frac{2}{10}$ 2. $\frac{4}{10}$ 3. $\frac{5}{10}$ 4. $\frac{8}{10}$ 5. $\frac{10}{10}$ 6. $\frac{6}{10}$

7. $\frac{20}{100}$ 8. $\frac{30}{100}$ 9. $\frac{45}{100}$ 10. $\frac{68}{100}$ 11. $\frac{75}{100}$ 12. $\frac{100}{100}$

13. $\frac{84}{1000}$ 14. $\frac{34}{1000}$ 15. $\frac{96}{1000}$ 16. $\frac{218}{1000}$ 17. $\frac{325}{1000}$ 18. $\frac{648}{1000}$

NOW USE SKILL 59. Write as a fraction in lowest terms.

19. 0.1   20. 0.2   21. 0.5   22. 0.8   23. 0.9

24. 0.02   25. 0.08   26. 0.06   27. 0.05   28. 0.09

29. 0.25   30. 0.34   31. 0.48   32. 0.15   33. 0.75

34. 0.64   35. 0.86   36. 0.26   37. 0.24   38. 0.85

39. 0.125   40. 0.322   41. 0.275   42. 0.650   43. 0.875

44. 0.102   45. 0.225   46. 0.404   47. 0.375   48. 0.608

49. 0.025   50. 0.030   51. 0.058   52. 0.175   53. 0.104

54. 0.715   55. 0.648   56. 0.825   57. 0.144   58. 0.742

For more practice: **Skill Bank,** page 431.

Writing Mixed Numbers as Decimals

MODEL

Write as a decimal. Round to the nearest hundredth.	Rewrite as a fraction.	Divide to thousandths.	Round to the nearest hundredth.	To the nearest hundredth,
$3\frac{1}{8}$	$3\frac{1}{8} = \frac{25}{8}$	$\dfrac{3.125}{8)\overline{25.000}}$	$3.125 \to 3.13$	$3\frac{1}{8} = 3.13$
	Skill 31	Skill 20	Skill 2	

 GET READY. Write as a fraction.

1. $2\frac{1}{4}$ 2. $3\frac{5}{6}$ 3. $6\frac{3}{8}$ 4. $10\frac{4}{9}$ 5. $12\frac{3}{5}$

NOW USE SKILL 60. Write as a decimal. Round to the nearest hundredth when necessary.

6. $1\frac{1}{3}$ 7. $3\frac{1}{4}$ 8. $5\frac{1}{6}$ 9. $7\frac{1}{8}$ 10. $4\frac{1}{5}$

11. $1\frac{2}{9}$ 12. $4\frac{3}{8}$ 13. $5\frac{4}{5}$ 14. $2\frac{6}{7}$ 15. $3\frac{3}{5}$

16. $10\frac{1}{3}$ 17. $12\frac{3}{4}$ 18. $14\frac{1}{6}$ 19. $16\frac{5}{8}$ 20. $20\frac{1}{2}$

21. $13\frac{3}{8}$ 22. $16\frac{2}{3}$ 23. $24\frac{5}{6}$ 24. $35\frac{3}{7}$ 25. $28\frac{7}{9}$

26. $2\frac{3}{10}$ 27. $3\frac{4}{9}$ 28. $6\frac{5}{8}$ 29. $8\frac{8}{11}$ 30. $4\frac{9}{13}$

Write as a decimal. Round to the nearest thousandth when necessary.

31. $7\frac{5}{9}$ 32. $3\frac{1}{15}$ 33. $9\frac{6}{13}$ 34. $2\frac{8}{9}$ 35. $6\frac{7}{12}$

36. $3\frac{4}{7}$ 37. $5\frac{4}{9}$ 38. $8\frac{6}{11}$ 39. $7\frac{9}{40}$ 40. $4\frac{1}{3}$

For more practice: **Skill Bank,** page 431.

Writing Decimals as Mixed Numbers

MODEL

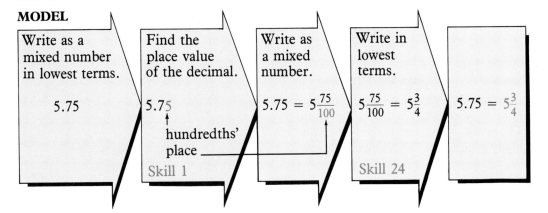

 GET READY. Write in lowest terms.

1. $8\frac{2}{8}$ **2.** $3\frac{3}{6}$ **3.** $4\frac{6}{9}$ **4.** $6\frac{2}{4}$ **5.** $9\frac{3}{9}$

6. $12\frac{2}{6}$ **7.** $16\frac{3}{9}$ **8.** $18\frac{6}{8}$ **9.** $20\frac{6}{9}$ **10.** $2\frac{2}{10}$

11. $5\frac{5}{20}$ **12.** $7\frac{24}{40}$ **13.** $8\frac{22}{50}$ **14.** $12\frac{4}{18}$ **15.** $15\frac{6}{24}$

Write as a fraction in lowest terms.

16. 0.24 **17.** 0.45 **18.** 0.36 **19.** 0.60 **20.** 0.85

 NOW USE SKILL 61. Write as a mixed number in lowest terms.

21. 6.2 **22.** 7.5 **23.** 7.4 **24.** 9.3 **25.** 7.20

26. 3.45 **27.** 1.62 **28.** 8.53 **29.** 4.02 **30.** 8.05

31. 7.04 **32.** 9.01 **33.** 40.25 **34.** 18.50 **35.** 26.16

36. 28.64 **37.** 4.600 **38.** 7.375 **39.** 8.400 **40.** 4.800

41. 5.040 **42.** 3.048 **43.** 9.625 **44.** 6.060 **45.** 8.175

46. 24.215 **47.** 42.428 **48.** 51.364 **49.** 67.328

50. 133.448 **51.** 174.256 **52.** 245.080 **53.** 456.825

For more practice: **Skill Bank,** page 431.

Bar Graphs

A *bar graph* is a way to present information. Bars (rectangles) and a scale are used to compare quantities. Each quantity is represented by the length of a bar. To read a bar graph, use the scale to find the length of the bar.

A group of high school seniors were asked what they planned to do after graduation. The results are shown in the bar graph. How many plan to go to technical school?

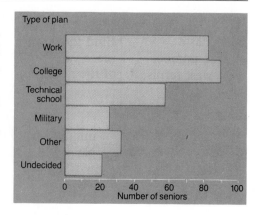

MODEL

Find the technical school bar. ➤ Estimate the corresponding value on the scale at the bottom of the graph. ➤ About 58 seniors plan to enter technical school.

 GET READY. Use the graph above to answer the question.

1. Which bar is longest?

2. Which bar is shortest?

 NOW USE APPLICATION A29.

The heights of a group of people were measured as part of a health survey. The results are shown in this bar graph.

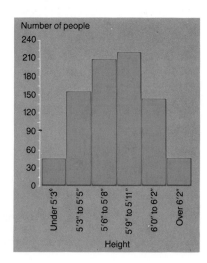

3. How many people measure under 5'3"?

4. Which height range has more people: under 5'3", or over 6'2"?

5. How many people measure between 5'9" and 5'11"?

6. Leslie is 5'8" tall. How many people in the survey are taller than Leslie?

7. Which height ranges have more than 150 people?

The Johnsons got their "electric use profile" in the January bill.

8. Which month shows the lowest use?

9. Which month shows the highest use?

10. What was the average daily use in September?

11. Estimate the total usage for September.

12. Which months show the same average use?

13. Which months averaged more than 20 kW · h per day?

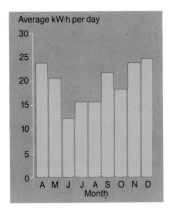

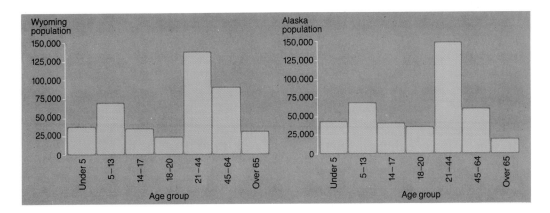

The two bar graphs above show the population of different age groups living in Wyoming and Alaska. These two states have about the same total population.

14. Which state has more people over 65?

15. Which state has more people under 21?

16. In which age group does Alaska have the most people?

17. In Wyoming, are more people 18–20 or 14–17 years old?

18. Which age groups have more than 75,000 people?

For more practice: **Application Bank**, page 460.

Estimated Gas Bill

You have received your gas bill. You notice that this is an estimated bill. This means your meter was not read this month. The gas company estimated your gas usage based on amounts used in past months. You suspect that the estimated amount is too high.

PROBLEM: What would your bill have been for the amount of gas you actually used?

Solution Plan: Find the amount of gas used. Use the gas company's rate table to figure the cost. Add any special charges.

WHAT DO YOU KNOW? You have information listed on your gas bill.

RESIDENTIAL RATE		SER	PERIOD COVERED		NO. MOS.	PRESENT READ	USE IN CCF	RATE	AMOUNT	
HUNDRED CU FEET CCF			FROM	TO						
PER MO	PRICE									
1	3.000	GE	4 | 10	5 | 12	1	2424	183	1	89	10
9	.448									
15	.332									
50	.279									
125	.252	SURCHARGE	NEXT SCHEDULED METER READING DATE			COST OF GAS ADJUSTMENT			TOTAL AMOUNT	
OVER	.241	$.27	JUN 11			$.1948			89	10

VALLEY GAS — ESTIMATED BILL

1. What was the estimated amount of gas used, in CCF?

2. How much is the gas adjustment charge?

3. There is a surcharge added to every bill to cover the gas company's home conservation services. How much is the surcharge?

4. How much is the total bill?

WHAT DO YOU WANT TO KNOW? You want to know how much gas you actually used.

5. What is the estimated present meter reading?

6. Find last month's meter reading.

7. You read your meter yourself to find the correct present reading. The reading is 2397. Find the amount of gas you actually used this month.

> Present reading
> − Last month's reading
> _____
> Amount used

8. Was the gas company's estimated use too high, or too low?

WHAT MORE DO YOU WANT? The gas company figures your bill in three parts. The surcharge is $.27.

9. The gas adjustment charge is a cost per unit (CCF) on all gas used. Find your gas adjustment charge on the actual amount of gas you used. Round to the nearest cent.

10. The basic rate structure is given in the table on your bill. The first unit costs $3.00. *Each* of the next 9 units costs $.448. Each of the next 15 units costs $.332, and so on. Figure the basic charge, using the rate table, for your actual gas usage. Round to the nearest cent.

NOW SOLVE THE PROBLEM.

11. Your total bill includes the basic charge, the gas adjustment charge, and the surcharge. Find the total bill for your actual usage.

12. If the estimated bill is too high, then the excess amount you pay will actually count towards paying for the gas used next month. How much is this excess amount?

Quiz

Skill 58 **Write as a decimal. Round to the nearest hundredth.**

1. $\frac{5}{9}$　　　　**2.** $\frac{13}{20}$

Skill 59
Write as a fraction in lowest terms.

3. 0.06　　　　**4.** 0.275

Skill 60 **Write as a decimal. Round to the nearest hundredth.**

5. $3\frac{3}{4}$　　　　**6.** $18\frac{2}{3}$

Skill 61 **Write as a mixed number in lowest terms.**

7. 3.6　　　　**8.** 9.325

Application A29

9. Which model had the highest sales?

10. How many cars were sold altogether?

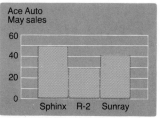

Problem Solving

11. You used 201 units of gas last month. The gas adjustment charge was $.1824 per unit. The total basic charge was $48.19. If the surcharge was $.32, what was your bill last month?

CALCULATOR DISPLAYS

Repeating Decimals

Often when one whole number is divided by another the digits in the answer form a pattern. Here is an example:

$$13\overline{)5.000000000000000} = 0.384615384615\ldots$$

The pattern is 384615 . . . . The dots show that the pattern keeps repeating.

Use your calculator to find the pattern.

1. $9\overline{)7}$　　　　**2.** $11\overline{)6}$

3. $99\overline{)23}$　　　　**4.** $37\overline{)23}$

Checkbook Register

A checkbook register lets you keep a record of all changes in your checking account. Record each check you write. Subtract the amount from the previous balance.

CHECK NO.	DATE	CHECK ISSUED TO OR DESCRIPTION OF DEPOSIT	AMOUNT OF CHECK		AMOUNT OF DEPOSIT		BALANCE	
							273	08
							- 26	55
714	4/23	Telephone Co.	26	55			246	53

Record each deposit. Add the amount to the previous balance.

CHECK NO.	DATE	CHECK ISSUED TO OR DESCRIPTION OF DEPOSIT	AMOUNT OF CHECK		AMOUNT OF DEPOSIT		BALANCE	
							273	08
							- 26	55
714	4/23	Telephone Co.	26	55			246	53
							+ 74	36
	4/25	Deposit			74	36	320	89

Find the balance after each check or deposit.

1.

AMOUNT OF CHECK		AMOUNT OF DEPOSIT		BALANCE	
				387	25
81	94			?	
5	32			?	
		50	00	?	

2.

AMOUNT OF CHECK		AMOUNT OF DEPOSIT		BALANCE	
				182	25
12	98			?	
		28	53	?	
		45	80	?	

3. Previous balance: $307.52
   Checks: $15.68
        $34.17
        $52.95
   Deposits: $100.00
        $41.75

4. Previous balance: $523.19
   Checks: $41.23
        $17.95
        $112.38
   Deposit: $75.95

243

Writing Fractions with Denominators of 100

MODEL

Write as a fraction with a denominator of 100.

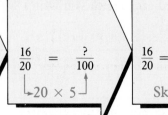

$$\frac{16}{20} = \frac{?}{100}$$

$$\frac{16}{20} = \frac{?}{100}$$

$$\overset{\llcorner}{20 \times 5}$$

$$\frac{16}{20} = \frac{16 \times 5}{20 \times 5}$$

Skill 25

$$\frac{16}{20} = \frac{80}{100}$$

GET READY. Complete the statement.

1. $5 \times \underline{} = 100$

2. $10 \times \underline{} = 100$

3. $25 \times \underline{} = 100$

4. $2 \times \underline{} = 100$

5. $\frac{5}{8} = \frac{?}{16}$

6. $\frac{4}{6} = \frac{?}{24}$

7. $\frac{7}{4} = \frac{?}{12}$

8. $\frac{5}{7} = \frac{20}{?}$

9. $\frac{6}{8} = \frac{?}{40}$

10. $\frac{5}{9} = \frac{25}{?}$

11. $\frac{2}{3} = \frac{?}{30}$

12. $\frac{6}{5} = \frac{?}{35}$

NOW USE SKILL 62. Write as a fraction with a denominator of 100.

13. $\frac{1}{4}$

14. $\frac{1}{2}$

15. $\frac{1}{5}$

16. $\frac{1}{10}$

17. $\frac{1}{20}$

18. $\frac{2}{5}$

19. $\frac{4}{5}$

20. $\frac{3}{4}$

21. $\frac{3}{10}$

22. $\frac{7}{10}$

23. $\frac{3}{20}$

24. $\frac{4}{25}$

25. $\frac{3}{50}$

26. $\frac{6}{25}$

27. $\frac{9}{50}$

28. $\frac{7}{20}$

29. $\frac{12}{25}$

30. $\frac{27}{20}$

31. $\frac{17}{25}$

32. $\frac{21}{25}$

33. $\frac{16}{25}$

34. $\frac{19}{20}$

35. $\frac{29}{50}$

36. $\frac{41}{50}$

37. $\frac{13}{25}$

38. $\frac{31}{10}$

39. $\frac{48}{25}$

40. $\frac{59}{20}$

41. $\frac{79}{50}$

42. $\frac{66}{25}$

43. $\frac{30}{10}$

44. $\frac{50}{50}$

45. $\frac{32}{25}$

46. $\frac{41}{20}$

47. $\frac{13}{10}$

For more practice: **Skill Bank**, page 432.

Writing Hundredths as Percents

MODEL

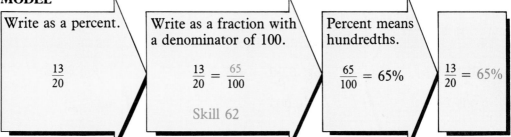

Write as a percent.

$$\frac{13}{20}$$

Write as a fraction with a denominator of 100.

$$\frac{13}{20} = \frac{65}{100}$$

Skill 62

Percent means hundredths.

$$\frac{65}{100} = 65\%$$

$$\frac{13}{20} = 65\%$$

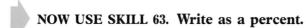 **GET READY.** Write as a fraction with a denominator of 100.

1. $\frac{1}{2}$ 2. $\frac{1}{4}$ 3. $\frac{1}{5}$ 4. $\frac{3}{5}$ 5. $\frac{1}{10}$

6. $\frac{3}{20}$ 7. $\frac{11}{20}$ 8. $\frac{27}{50}$ 9. $\frac{6}{5}$ 10. $\frac{12}{4}$

11. $\frac{30}{10}$ 12. $\frac{50}{25}$ 13. $\frac{37}{50}$ 14. $\frac{19}{20}$ 15. $\frac{21}{25}$

NOW USE SKILL 63. Write as a percent.

16. $\frac{5}{100}$ 17. $\frac{8}{100}$ 18. $\frac{9}{100}$ 19. $\frac{3}{100}$ 20. $\frac{7}{100}$

21. $\frac{10}{100}$ 22. $\frac{15}{100}$ 23. $\frac{12}{100}$ 24. $\frac{11}{100}$ 25. $\frac{19}{100}$

26. $\frac{32}{100}$ 27. $\frac{45}{100}$ 28. $\frac{34}{100}$ 29. $\frac{76}{100}$ 30. $\frac{55}{100}$

31. $\frac{2}{5}$ 32. $\frac{5}{10}$ 33. $\frac{4}{5}$ 34. $\frac{9}{20}$ 35. $\frac{9}{25}$

36. $\frac{13}{20}$ 37. $\frac{7}{10}$ 38. $\frac{1}{4}$ 39. $\frac{17}{20}$ 40. $\frac{3}{25}$

41. $\frac{3}{2}$ 42. $\frac{9}{10}$ 43. $\frac{15}{4}$ 44. $\frac{52}{25}$ 45. $\frac{11}{50}$

46. $\frac{3}{10}$ 47. $\frac{24}{20}$ 48. $\frac{13}{25}$ 49. $\frac{8}{5}$ 50. $\frac{34}{50}$

51. $\frac{3}{4}$ 52. $\frac{16}{25}$ 53. $\frac{7}{4}$ 54. $\frac{20}{20}$ 55. $\frac{49}{50}$

For more practice: **Skill Bank,** page 432.

Writing Decimals as Percents

MODEL

Write as a percent.	Move the decimal point two places to the right. Write a percent sign.	
0.4 = ?%	0.40 → 40%	0.4 = 40%
0.005 = ?%	0.005 → 0.5%	0.005 = 0.5%
1.326 = ?%	1.326 → 132.6%	1.326 = 132.6%

 GET READY. Write as a fraction with a denominator of 100.

1. 0.23   **2.** 0.06   **3.** 2.37   **4.** 0.7   **5.** 6

 NOW USE SKILL 64. Write as a percent.

6. 0.07	**7.** 0.03	**8.** 0.08	**9.** 0.09
10. 0.30	**11.** 0.17	**12.** 0.19	**13.** 0.33
14. 0.66	**15.** 0.83	**16.** 0.52	**17.** 0.49
18. 0.58	**19.** 0.64	**20.** 0.77	**21.** 0.90
22. 3.1	**23.** 4.7	**24.** 6.2	**25.** 7.9
26. 6.02	**27.** 5.08	**28.** 7.06	**29.** 9.09
30. 3.25	**31.** 8.49	**32.** 6.37	**33.** 7.00
34. 0.063	**35.** 0.045	**36.** 0.050	**37.** 0.023
38. 0.314	**39.** 0.425	**40.** 0.328	**41.** 0.663
42. 0.006	**43.** 0.008	**44.** 0.002	**45.** 0.009
46. 9.01	**47.** 6.294	**48.** 0.0893	**49.** 0.209
50. 8	**51.** 0.0029	**52.** 1.084	**53.** 2.354

For more practice: **Skill Bank,** page 432.

Writing Fractions as Percents

MODEL

Write as a percent.	Write as a decimal. Round to the nearest thousandth.	Move the decimal point two places to the right. Write a percent sign.	
$\frac{3}{4} = ?\%$	$\frac{3}{4} = 0.75$ Skill 58	$0.75 \rightarrow 75\%$	$\frac{3}{4} = 75\%$
$1\frac{5}{12} = ?\%$	$1\frac{5}{12} = 1.417$ Skill 60	$1.417 \rightarrow 141.7\%$ Skill 64	$1\frac{5}{12} = 141.7\%$

GET READY. Write as a percent.

1. 0.36　　　　**2.** 0.48　　　　**3.** 0.93　　　　**4.** 0.77

Write as a decimal. Round to the nearest thousandth when necessary.

5. $\frac{4}{5}$　　　　**6.** $\frac{9}{25}$　　　　**7.** $\frac{7}{12}$　　　　**8.** $\frac{5}{8}$

 NOW USE SKILL 65. Write as a percent.

9. $\frac{1}{8}$　　　　**10.** $\frac{1}{2}$　　　　**11.** $\frac{1}{5}$　　　　**12.** $\frac{1}{4}$

13. $\frac{4}{5}$　　　　**14.** $\frac{2}{3}$　　　　**15.** $\frac{7}{8}$　　　　**16.** $\frac{4}{9}$

17. $\frac{3}{10}$　　　　**18.** $\frac{5}{16}$　　　　**19.** $\frac{12}{25}$　　　　**20.** $\frac{9}{16}$

21. $\frac{7}{10}$　　　　**22.** $\frac{19}{25}$　　　　**23.** $\frac{11}{20}$　　　　**24.** $\frac{21}{50}$

25. $3\frac{1}{4}$　　　　**26.** $5\frac{1}{5}$　　　　**27.** $6\frac{1}{8}$　　　　**28.** $3\frac{1}{2}$

29. $2\frac{5}{8}$　　　　**30.** $3\frac{3}{5}$　　　　**31.** $8\frac{9}{10}$　　　　**32.** $2\frac{1}{3}$

33. $10\frac{2}{5}$　　　　**34.** $3\frac{7}{20}$　　　　**35.** $6\frac{3}{4}$　　　　**36.** $2\frac{41}{50}$

For more practice: **Skill Bank,** page 433.　　　　　　　　247

Line Graphs

A *line graph* is a way of presenting information. A line graph uses line segments to show how a quantity increases or decreases.

The Montaldos want to borrow money to expand their store. In order to give the bank an idea of the store's financial status, they compiled this graph. The graph shows the store's sales for the past nine years. How much were sales for the ninth year?

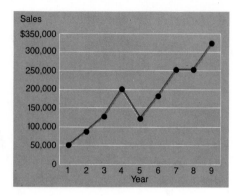

MODEL

Find the right year on the "Year" scale.

Estimate the corresponding number on the "Sales" scale.

Sales in the ninth year were about $325,000.

   NOW USE APPLICATION A30. Use the graph above to answer the question.

1. How much were sales for the third year?

2. How much were sales for the fourth year?

3. How much did sales increase between the third and fourth years?

4. Which year had the lowest sales?

5. In which year were sales $180,000?

6. In which year were sales lower than the year before?

7. In which year were sales the same as the year before?

8. What was the year when sales were more than $200,000 for the first time?

The graph shows the prime interest rate (the rate banks charge their best customers) for a six-month period.

9. What was the lowest interest rate in this period?

10. What was the highest interest rate?

11. Between which two months did the rate increase the most?

12. How much higher was the rate in February than in December?

13. During which months was the rate less than 19%?

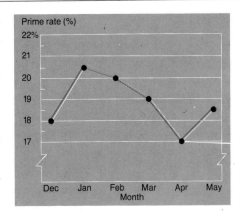

The graph shows the population of the town of Mineral, according to the census taken every ten years.

14. What was the population in 1970?

15. What was the first year the population was more than 12,000?

16. In which decade did the town's population increase the most?

17. Which census years showed the same population?

18. What is the increase in population since 1900?

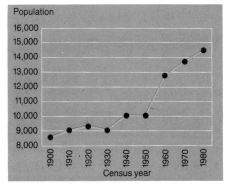

Jane and Joe are twins. The graphs show their heights in centimeters from ages 11 to 16, measured every year on their birthday.

19. How tall was Joe at 13?

20. How tall was Jane at 16?

21. Who was taller at 15?

22. When did Joe grow the most?

23. When did Jane grow the least?

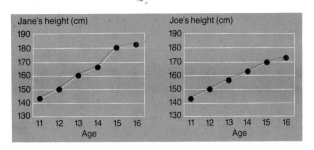

For more practice: **Application Bank,** page 460.

Protein Pro

You belong to a food cooperative. You have been asked to study human protein sources and requirements. You will report your findings at the next meeting.

PROBLEM: What are some desirable protein sources? What are the advantages of choosing foods containing complementary proteins?

Solution Plan: Find the information you need. Analyze it, and decide which protein sources seem to be desirable. Calculate how much more protein you can get by combining certain foods in your meals.

WHAT DO YOU WANT TO KNOW? You want to know some desirable protein sources.

You find a table relating protein and calorie content in certain foods. The table gives the number of calories you consume in order to get one gram of usable protein.

Food	Calories per Gram of Usable Protein	Food	Calories per Gram of Usable Protein
Chicken	7	Oatmeal	41
Corn	37	Peas	25
Egg	14	Potato	60
Haddock	5	Spaghetti noodles	50
Hamburger	15	Tuna	9
Kidney beans	39	Yogurt	27

1. List the five foods that have the fewest calories per gram of usable protein.

2. Some co-op members are concerned about cholesterol and other considerations. They have eliminated foods from animal sources, including eggs but not dairy products. List the three foods for these people that have the fewest calories per gram of usable protein.

250

The average person needs about 43 grams of usable protein a day. How many calories would you consume if you got all of your protein from the given food?

3. Haddock

4. Egg

5. Yogurt

6. Potatoes

WHAT MORE DO YOU WANT TO KNOW? You want to know about the advantages of foods that contain complementary proteins.

You research complementary proteins. You learn that the protein in one food may contain components missing in the protein of another food. These proteins are called *complementary*.

You work out an example to show this idea to the co-op members. The beef equivalence ratio (BER) is used to express the usable protein in a food in terms of the amount of beef containing the same amount of protein.

You find that $1\frac{1}{2}$ cups of beans contain 62 grams of protein. Beans have a BER of 2.87. You figure that the beans contain as much usable protein as 178 grams of beef, to the nearest gram.

$$\text{grams of protein} \times \text{BER} = \text{equivalent amount of beef}$$

$$62 \times 2.87 = 178$$

7. Rice: 4 cups contain 59 grams of protein.
BER is 3.32.
4 cups of rice contain as much usable protein as __?__ grams of beef.

8. The beans and rice contain as much usable protein, in total, as __?__ grams of beef.

9. Beans and rice eaten together:
a. $1\frac{1}{2}$ cups of beans contain __?__ grams of protein.
b. 4 cups of rice contain __?__ grams of protein.
c. The rice and beans contain a total of __?__ grams of protein together.
d. The BER of the bean-rice combination is 4.38. The beans and rice together contain as much usable protein as __?__ grams of beef.

10. Compare the answers to problem 8 and problem 9d. By eating the beans and rice together (problem 9d) you get the equivalent of __?__ more grams of beef than if you ate the beans and rice separately (problem 8).

251

Quiz

Skill 62

1. $\dfrac{16}{20} = \dfrac{?}{100}$

2. $\dfrac{9}{25} = \dfrac{?}{100}$

Skill 63

3. $\dfrac{67}{100} = \underline{\ ?\ }\%$

4. $\dfrac{138}{100} = \underline{\ ?\ }\%$

Skill 64

5. $0.82 = \underline{\ ?\ }\%$

6. $1.236 = \underline{\ ?\ }\%$

Skill 65

7. $\dfrac{5}{8} = \underline{\ ?\ }\%$

8. $\dfrac{4}{5} = \underline{\ ?\ }\%$

Application A30

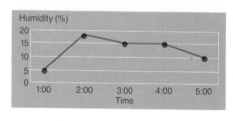

9. How much did the humidity drop between 4:00 and 5:00?

10. When was the humidity the highest?

Problem Solving

11. You prepare a dish for dinner that contains 8 grams of usable protein per serving. How many servings would you need to eat to get your 43 gram daily requirement of usable protein? Round to the nearest whole number.

EXTRA!

Media Magic uses a *pictograph* to report its annual magazine sales.

1. Which magazine had the highest sales? the lowest sales?

2. Media Magic sold 2,500,000 copies of *Fashion World*. Find the sales for **a.** *News Review* **b.** *Family Living* **c.** *Teen Times* **d.** *Sports Spots*

Laboratory Technician

Laboratory tests play an important role in the detection, diagnosis, and treatment of many diseases. Laboratory technicians get their training in 2-year programs after graduating from high school. They attend community or junior colleges, trade schools, or technical institutions.

A laboratory technician performs highly skilled tests and laboratory procedures. Technicians may work in several areas or specialize in one field.

Grace Warton is a laboratory technician for a pharmaceutical firm. She is testing a new antibiotic on a special strain of mice. After administering the antibiotic, Grace does a blood test every hour. She records the data. The line graph below shows the number of units of the antibiotic present in 1 mL of blood at hourly intervals.

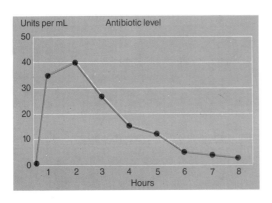

Find the antibiotic level at the given time.

1. 1 hour

2. 2 hours

3. 4 hours

4. 6 hours

5. Which reading showed the first drop in the level of the antibiotic?

6. At which reading did the level of the antibiotic become less than 10 units per mL?

Writing Percents as Decimals

MODEL

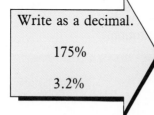

Write as a decimal.

175%

3.2%

Move the decimal point two places to the left. Remove the percent sign.

175% → 1.75

003.2% → 0.032

175% = 1.75

3.2% = 0.032

 GET READY. Write as a percent.

1. 0.05	**2.** 0.07	**3.** 0.09	**4.** 0.02
5. 0.17	**6.** 0.20	**7.** 0.65	**8.** 0.57
9. 0.258	**10.** 1.47	**11.** 2.38	**12.** 9.416

NOW USE SKILL 66. Write as a decimal.

13. 4%	**14.** 8%	**15.** 3%	**16.** 1%
17. 2%	**18.** 5%	**19.** 9%	**20.** 7%
21. 11%	**22.** 14%	**23.** 16%	**24.** 18%
25. 23%	**26.** 45%	**27.** 56%	**28.** 63%
29. 100%	**30.** 300%	**31.** 700%	**32.** 500%
33. 180%	**34.** 285%	**35.** 195%	**36.** 399%
37. 432%	**38.** 712%	**39.** 306%	**40.** 550%
41. 0.5%	**42.** 0.9%	**43.** 0.2%	**44.** 0.6%
45. 6.2%	**46.** 9.4%	**47.** 7.6%	**48.** 5.3%
49. 2.3%	**50.** 3.25%	**51.** 4.78%	**52.** 5.5%
53. 78.9%	**54.** 54.8%	**55.** 23.56%	**56.** 17.33%
57. 31.3%	**58.** 74.9%	**59.** 50.2%	**60.** 63.9%
61. 0.403%	**62.** 0.158%	**63.** 0.374%	**64.** 0.273%
65. 0.145%	**66.** 0.963%	**67.** 2.672%	**68.** 38.439%

For more practice: **Skill Bank,** page 433.

Writing Percents as Fractions

MODEL

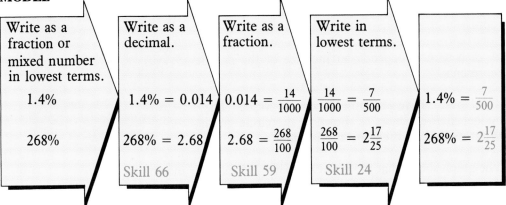

Write as a fraction or mixed number in lowest terms.

1.4%

268%

Write as a decimal.

$1.4\% = 0.014$

$268\% = 2.68$

Skill 66

Write as a fraction.

$0.014 = \frac{14}{1000}$

$2.68 = \frac{268}{100}$

Skill 59

Write in lowest terms.

$\frac{14}{1000} = \frac{7}{500}$

$\frac{268}{100} = 2\frac{17}{25}$

Skill 24

$1.4\% = \frac{7}{500}$

$268\% = 2\frac{17}{25}$

 GET READY. Write as a decimal.

1. 12%   **2.** 8.6%   **3.** 37.2%   **4.** 592%   **5.** 0.15%

6. 35%   **7.** 4.9%   **8.** 88%   **9.** 250%   **10.** 0.07%

Write as a fraction or mixed number in lowest terms.

11. 0.16   **12.** 0.35   **13.** 2.14   **14.** 0.082   **15.** 0.125

16. 3.45   **17.** 0.412   **18.** 0.36   **19.** 1.75   **20.** 0.225

 NOW USE SKILL 67. Write as a fraction or mixed number in lowest terms.

21. 6%   **22.** 8%   **23.** 3%   **24.** 5%   **25.** 9%

26. 10%   **27.** 15%   **28.** 20%   **29.** 24%   **30.** 75%

31. 25%   **32.** 34%   **33.** 56%   **34.** 60%   **35.** 55%

36. 40%   **37.** 50%   **38.** 70%   **39.** 80%   **40.** 90%

41. 64%   **42.** 7.8%   **43.** 84%   **44.** 9.5%   **45.** 36%

46. 100%   **47.** 12.5%   **48.** 450%   **49.** 180%   **50.** 745%

51. 1.5%   **52.** 245%   **53.** 290%   **54.** 2.65%   **55.** 815%

56. 600%   **57.** 554%   **58.** 725%   **59.** 800%   **60.** 10.6%

For more practice: **Skill Bank,** page 433.

Circle Graphs

A *circle graph* is one way of presenting information. A circle graph is divided into sections. The circle represents the whole quantity, and each of the sections represents a specific part of the whole. The size of each section depends on the size of the part it represents.

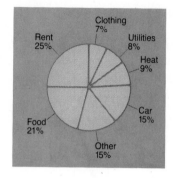

The Howells have a monthly income of $1200. They budget their expenses as shown in the circle graph. What percent of their income is spent on rent?

The graph shows that the Howells spend 25% of their income for rent.

 GET READY. Use the circle graph above.

1. Which is the smallest expense in the Howells' budget?

2. What percent of their income is spent on heat?

3. Which of the expenses shown are equal?

4. Which two expenses add up to almost 50% of their income?

5. What percent of the Howells' income is spent on rent, food, heat, and utilities?

6. Do the Howells spend more on heat or on utilities?

 NOW USE APPLICATION A31.

The manager of Jason's Stationery drew a circle graph to show how the store uses every dollar of income.

7. What is the store's biggest expense?

8. How much of every dollar is spent on salaries?

9. Rent and utilities and advertising expenses add up to how much of every dollar?

10. All categories in the graph, except for profit, are expenses. Out of every dollar, how much goes for expenses?

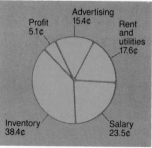

Roger keeps careful records of the family food budget. He uses a circle graph to show what percent is spent in each category.

11. What is their biggest food expense?

12. Do they spend more for vegetables or for fruit?

13. Which two items together take more than half of their budget?

14. What percent of their budget is spent on dairy products and eggs?

15. What percent of the budget is spent on all groceries except meat?

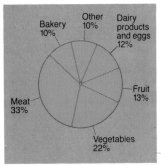

Sally Perreira wants to organize her time better. To analyze how she spends an average day, she uses a circle graph.

16. How much time does Sally spend at school and doing homework?

17. Does Sally spend more time on sports or watching television?

18. How many hours a day does Sally sleep?

19. Which two activities take the same amount of time?

20. Which two activities together take more than half the day?

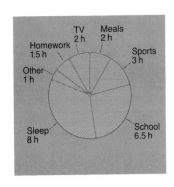

For more practice: **Application Bank,** page 460.

257

Small World

You plan to begin providing child care in your home. You are making plans for safety and recreation equipment and staffing needs. You expect to have eight children in your program.

PROBLEM: How much is your initial investment in equipment? How much will your weekly payroll be for helpers?

Solution Plan: Determine the cost of equipment you will make yourself and equipment that you will buy. Figure how much you will pay your assistants.

WHAT DO YOU WANT TO KNOW? State regulations require you to have a fire extinguisher, a first aid kit, and a fenced play area.

1. You buy a fire extinguisher for $18.95 and a first aid kit for $13.99. How much do you spend on these items?

2. You will have a rectangular play area that is 18 feet by 20 feet. You estimate that you can build the fence around it for 75¢ per foot. How much will the fence cost?

3. You will buy a climbing gym. It is available already assembled, or as a kit. The assembled gym costs $259.50. The kit costs $189.95. You decide to buy the kit. How much do you save by buying the kit?

You are going to build a large sandbox. You need boards for the sides, corner braces, and sand.

4. The sandbox requires four 2″ × 12″ boards, each 6′ long. The boards cost $1.02 per foot of length. How much do the boards cost?

5. You need four corner braces. Corner braces are two for $1.49. How much do the braces cost?

6. The sandbox will be 6′ wide, 6′ long, and 1′ deep. How many cubic feet of sand do you need?

7. Sand is sold by the cubic yard. One cubic yard is 27 cubic feet. You can only buy sand in whole cubic yards. How many cubic yards of sand must you buy?

8. Sand costs $31.80 per cubic yard, including delivery. How much do you spend on sand?

WHAT MORE DO YOU WANT? You want to find how much it will cost to hire helpers.

9. You hire Kara to work 7:30 A.M to 3:00 P.M. on Monday, Wednesday, and Thursday. How many hours does Kara work each week?

10. You hire Gil to work from 2:00 P.M. to 6:00 P.M. Monday through Friday. How many hours does Gil work each week?

NOW SOLVE THE PROBLEM.

11. How much is your total expense for equipment? Include all safety and play equipment and materials.

12. You pay your helpers $5.50 an hour. How much is your total weekly payroll for your helpers?

Quiz

Write as a decimal.

1. 93%   **2.** 6.78%

Write as a fraction or mixed number in lowest terms.

3. 48%   **4.** 260%

Application A31

5. Which age group had the highest attendance for the seminar?

6. Which age group had the lowest attendance for the seminar?

7. What percent of the participants were between 18 and 45?

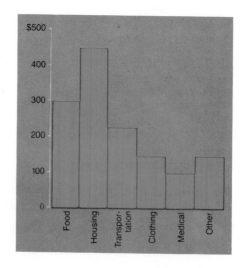

Seminar attendance

26–35 yrs 31%

18–25 yrs 33%

Under 18 yrs 10%

36–45 yrs 22%

Over 45 years 4%

Problem Solving

8. You hire two assistants to work from 1:00 to 4:00 Monday through Friday. You hire three aides to work from 7:00 to 12:30 on Monday, Thursday, and Friday. You pay all workers $5.80 an hour. How much is your weekly payroll?

CONSUMER NOTE

Family Budget

The City Housing Office uses a bar graph to show an average monthly budget for a family of four.

1. What is the biggest monthly expense?

2. Find the monthly food cost.

3. Which is higher, the cost of transportation or the cost of clothing?

4. Find the monthly medical expense.

5. Which expenses are over $200?

6. How much is the total monthly budget?

7. How much is the total annual budget?

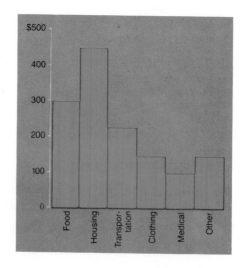

Skill 58 **Write as a decimal. Round to the nearest hundredth, when necessary.** (page 234)

1. $\frac{4}{9}$ 2. $\frac{2}{5}$ 3. $\frac{7}{9}$ 4. $\frac{3}{5}$ 5. $\frac{5}{7}$

Write as a decimal. Round to the nearest thousandth, when necessary.

6. $\frac{3}{7}$ 7. $\frac{5}{8}$ 8. $\frac{4}{7}$ 9. $\frac{2}{3}$ 10. $\frac{8}{9}$

Skill 59 **Write as a fraction.** (page 235)

11. 0.75   12. 0.875   13. 0.2   14. 0.6   15. 0.125

Skill 60 **Write as a decimal. Round to the nearest thousandth, when necessary.** (page 236)

16. $1\frac{1}{2}$ 17. $2\frac{1}{8}$ 18. $4\frac{3}{5}$ 19. $6\frac{8}{9}$

20. $5\frac{5}{7}$ 21. $9\frac{2}{5}$ 22. $3\frac{1}{3}$ 23. $12\frac{3}{4}$

Skill 61 **Write as a mixed number in lowest terms.** (page 237)

24. 9.4   25. 3.04   26. 7.35   27. 11.175

Skill 64 **Write as a percent.** (page 246)

28. 0.041   29. 0.26   30. 0.93   31. 4.37

32. 0.073   33. 0.83   34. 7.4   35. 0.61

Skill 65 **Write as a percent.** (page 247)

36. $\frac{3}{5}$ 37. $\frac{9}{20}$ 38. $\frac{7}{10}$ 39. $\frac{13}{40}$

40. $8\frac{1}{2}$ 41. $7\frac{3}{4}$ 42. $5\frac{9}{10}$ 43. $4\frac{3}{8}$

Skill 66 **Write as a decimal.** (page 254)

44. 14%   45. 5.9%   46. 0.28%   47. 165%

48. 38%   49. 7.5%   50. 4.26%   51. 408%

Skill 67 **Write as a fraction or mixed number in lowest terms.** (page 255)

52. 3%   53. 5%   54. 90%   55. 75%

56. 36%   57. 240%   58. 125%   59. 614%

Applications Review

Application A29 (pages 238–239)

Carr's Rent-a-Car keeps a record of the number of cars rented every month.

1. How many cars were rented in April?

2. Which month shows the highest number of rentals?

3. Which months show the same number of rentals?

4. Which month shows the lowest number of rentals?

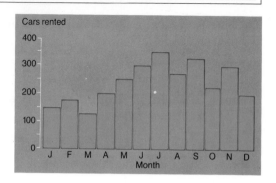

Application A30 (pages 248–249)

Garments manufactured by Ace Crafts are inspected every two hours. The percent of defects is recorded.

5. What percent of the garments inspected at 10:00 were defective?

6. How many inspections show the defects to be less than 4%?

7. Which inspection shows the highest percent of defects?

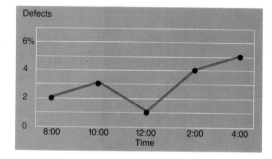

Application A31 (pages 256–257)

Grover's Appliance Center prepares a monthly graph of appliance sales.

8. Which appliance shows the highest sales record this month?

9. What percent of the total sales is radios?

10. Did the store sell more dishwashers or refrigerators?

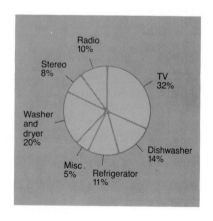

Unit Test

**Skill 58 Write as a decimal.
Round to the nearest hundredth.**

1. $\frac{13}{25}$

2. $\frac{3}{8}$

**Skill 60 Write as a decimal.
Round to the nearest hundredth.**

5. $5\frac{1}{2}$

6. $8\frac{5}{8}$

Skill 62

9. $\frac{11}{20} = \frac{?}{100}$

Skill 63

10. $\frac{66}{100} = \underline{\ ?\ }\%$

Skill 66 Write as a decimal.

13. 8%

14. 25.3%

**Skill 59 Write as a fraction in
lowest terms.**

3. 0.10

4. 0.32

**Skill 61 Write as a mixed number
in lowest terms.**

7. 7.4

8. 5.625

Skill 64

11. $0.524 = \underline{\ ?\ }\%$

Skill 65

12. $5\frac{4}{5} = \underline{\ ?\ }\%$

**Skill 67 Write as a fraction or
mixed number in lowest terms.**

15. 65%

16. 275%

Application A29

17. What is the total enrollment at
Valley High School?

18. How many more students are there
in 9th grade than in 10th grade?

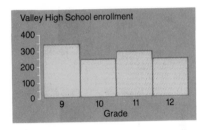

Application A30

19. In which year did Levi's earnings
increase the most?

20. How much more did Levi earn
in 1980 than in 1979?

21. When did Levi first earn more
than $20,000?

Application A31

22. Which two activities take the most of
Annette's day?

23. Do leisure and work together take as
much time as school?

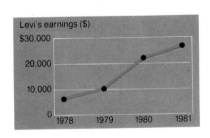

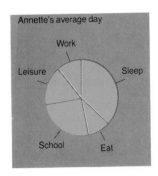

263

9

Percent Skills

Applying Percents

Problem Solving

You will use these skills and applications in this unit. Which do you already know? Work each problem.

Skill 68 **Write as a number sentence.**

1. 16% of 75 is 12

2. 35% of 92 is 32.2

Skill 69 **Identify the percent, base, and percentage.**

3. 6% of 82 is 4.92.   Percent: __?__   Base: __?__   Percentage: __?__

Skill 70

4. What is 35% of 60?

Skill 71

5. 7 is 10% of what number?

Skill 72

6. 6 is what percent of 24?

7. What percent of 420 is 357?

Skill 73 **Write as a proportion.**

8. 30 is 75% of 40.

9. 22 is what percent of 83?

Application A32

10. You owe $238.50 on your charge card account. The interest charge is 1.5% of the unpaid balance. How much do you pay in interest?

Application A33

11. World population is now about 4 billion. In 30 years it is expected to increase to around 10 billion. What would be the percent increase, if this were to happen?

Application A34

12. During August, City Motors discounts all cars by 7%. What is the sale price of a car whose original price was $6740?

Check your answers. If you had difficulty with any skill or application, be sure to study the corresponding lesson in this unit.

Writing Number Sentences

MODEL

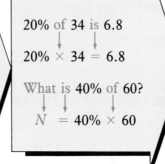

Write as a number sentence.

20% of 34 is 6.8

What is 40% of 60?

20% of 34 is 6.8

20% × 34 = 6.8

What is 40% of 60?

N = 40% × 60

20% × 34 = 6.8

N = 40% × 60

 GET READY. Complete the number sentence.

1. 55% of 24 is 13.2.
55% × __?__ = 13.2

2. What percent of 80 is 20?
__?__% × 80 = 20

3. 30 is 25% of 120.
__?__ = 25% × 120

4. What is 17% of 90?
__?__ = 17% × 90

NOW USE SKILL 68. Write as a number sentence.

5. 46% of 85 is 39.1.

6. 93% of 66 is 61.38.

7. 13% of 25 is 3.25.

8. 45% of 90 is 40.5.

9. 30 is 50% of 60.

10. 20 is 25% of 80.

11. 24 is 30% of 80.

12. 35 is 50% of 70.

13. 36 is what percent of 67?

14. 81 is what percent of 115?

15. What is 10% of 99?

16. What percent of 43 is 12?

17. What percent of 69 is 31?

18. What is 34% of 167?

19. What is 55% of 148?

20. What percent of 275 is 193?

21. 16.5% of 18 is 2.97.

22. What is 12% of 75?

23. 48% of 79 is what number?

24. What is 36% of 800?

25. What is 150% of 88?

26. 7.4% of 1708 is what number?

27. What is 15% of 90?

28. 4.2% of what number is 13.5?

For more practice: **Skill Bank,** page 434.

Identifying Percent, Base, and Percentage

MODEL

Identify the percent, base, and percentage.	Write as a number sentence.	Formula:	Percent Percentage Base
50% of 60 is 30	$50\% \times 60 = 30$	Percent × Base =	$50\% \times 60 = 30$
What is 6% of 192?	$N = 6\% \times 192$ Skill 68	Percentage	$N = 6\% \times 192$ Percent Base Percentage

 GET READY. Write as a number sentence.

1. 10% of 73 is 7.3.

2. 64 is 50% of 128.

3. What is 25% of 152?

4. 32 is what percent of 496?

 NOW USE SKILL 69. Identify the percent.

5. 48 is 60% of 80.

6. What is 13% of 768?

7. What percent of 812 is 139?

8. 17 is 20% of 85.

Identify the base.

9. 90% of 55 is 49.5.

10. What percent of 145 is 29?

11. 76 is 25% of 304.

12. 56 is what percent of 156?

Identify the percentage.

13. 40% of 82 is 32.8.

14. 91 is what percent of 213?

15. What is 39% of 77?

16. What is 14% of 319?

Identify the percent, base, and percentage.

17. 3% of 94 is 2.82.
   Percent: __?__   Base: __?__
   Percentage: __?__

18. 60 is what percent of 300?
   Percent: __?__   Base: __?__
   Percentage: __?__

For more practice: **Skill Bank,** page 434.

Finding the Percentage

MODEL

| Find the percentage.

What is 26% of 70? | Write as a number sentence.

$N = 26\% \times 70$

Skill 68 | Write 26% as a fraction in lowest terms.

$N = 26\% \times 70$

$N = \frac{26}{100} \times 70$
Skill 67

$N = \frac{13}{50} \times 70$

$N = 18\frac{1}{5}$ | 26% of 70 is $18\frac{1}{5}$.

Answer: $18\frac{1}{5}$ |

 GET READY. Write as a fraction with denominator 100.

1. 75%　　　**2.** 39%　　　**3.** 86%　　　**4.** 21%　　　**5.** 99%

Write the answer in lowest terms.

6. $\frac{32}{100} \times 50$　　**7.** $\frac{45}{100} \times 20$　　**8.** $\frac{72}{100} \times 5$　　**9.** $\frac{55}{100} \times 9$　　**10.** $\frac{60}{100} \times 30$

 NOW USE SKILL 70. Write the answer as a mixed number in lowest terms or a whole number.

11. What is 2% of 50?

12. 5% of 60 is what number?

13. 3% of 400 is what number?

14. 8% of 50 is what number?

15. What is 6% of 700?

16. 4% of 900 is what number?

17. 5% of 40 is what number?

18. What is 10% of 120?

19. What is 20% of 615?

20. 10% of 200 is what number?

21. What is 15% of 480?

22. 22% of 1400 is what number?

23. 20% of 75 is what number?

24. What is 75% of 104?

25. 15% of 260 is what number?

26. 20% of 490 is what number?

MODEL

| Find the percentage.

40% of 28 is what number? | Write as a number sentence.

40% × 28 = N

Skill 68 | Write 40% as a decimal.

40% × 28 = N
0.40 × 28 = N
Skill 66

11.2 = N | 40% of 28 is 11.2.

Answer: 11.2 |

 GET READY. Write as a decimal.

27. 23%   **28.** 67%   **29.** 7%   **30.** 1.5%   **31.** 350%

Multiply.

32. 0.32 × 60   **33.** 0.25 × 35   **34.** 0.03 × 167   **35.** 0.75 × 278

NOW USE SKILL 70. Write the answer as a decimal.

36. 4% of 35 is what number?

37. 3% of 740 is what number?

38. 7% of 40 is what number?

39. 5% of 150 is what number?

40. What is 9% of 120?

41. What is 6% of 735?

42. 5.5% of 326 is what number?

43. 2.7% of 290 is what number?

44. 3.6% of 4890 is what number?

45. 8.4% of 145 is what number?

46. 20% of 60 is what number?

47. What is 75% of 40?

48. What is 40% of 85?

49. 80% of 140 is what number?

50. 65% of 400 is what number?

51. 25% of 748 is what number?

52. What is 12% of 650?

53. What is 30% of 190?

54. What is 90% of 110?

55. 15% of 260 is what number?

56. 40% of 75 is what number?

57. What is 26% of 800?

58. 12.5% of 60 is what number?

59. 36.2% of 250 is what number?

60. What is 105% of 66?

61. 400% of 980 is what number?

62. 230% of 75 is what number?

63. 150% of 670 is what number?

For more practice: **Skill Bank,** page 435.

Interest

When you deposit money in a bank or make a personal loan you are letting the bank or individual use your money. The amount you earn for the use of your money is called *interest*. The amount of money deposited or loaned is called *principal*. The *interest rate* is expressed as a percent of the principal.

To find interest, use the formula:

Interest earned = Principal × Interest rate (%)

Bob Jones borrowed $500 from his brother. Bob agreed to repay the $500 loan plus 8% interest. How much interest will Bob pay?

MODEL

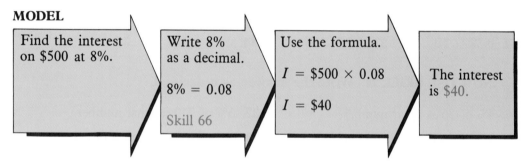

| Find the interest on $500 at 8%. | Write 8% as a decimal. $8\% = 0.08$ Skill 66 | Use the formula. $I = \$500 \times 0.08$ $I = \$40$ | The interest is $40. |

The interest on Bob's loan is $40.

 GET READY. Write as a decimal.

1. 12%	**2.** 4%	**3.** 7.5%	**4.** 30%
5. 75%	**6.** 3.8%	**7.** 0.6%	**8.** 0.2%

Multiply.

9. $176.00 × 0.06	**10.** $308.00 × 0.21	**11.** $296.50 × 0.18	**12.** $1462.45 × 0.2
13. $756.00 × 0.025	**14.** $318.25 × 0.16	**15.** $946.50 × 0.18	**16.** $532.00 × 0.095

NOW USE APPLICATION A32. **Round to the nearest cent. Each account receives an interest payment of 7%.**

	17.	18.	19.	20.	21.
Principal	$625.00	$508.50	$190.25	$7150.50	$8225.84
Interest earned	?	?	?	?	?
Total repaid	?	?	?	?	?

Round to the nearest cent.

22. Anna Chen borrowed $356 from her credit union. She agreed to make an interest payment of 12%.

  a. How much interest did Anna pay?

  b. How much money did Anna owe the credit union in all?

23. The Michaels' $10,000 certificate of deposit matures today. The certificate yields 14% interest.

  a. How much interest did the certificate earn?

  b. What is the total value of the certificate at maturity?

24. The Grace Department Store charges 1.5% interest on the outstanding balance of a charge account. Roger Bantor owes $65.25 this month.

  a. How much is Roger's finance charge (interest)?

  b. How much does Roger owe the department store in all?

25. The Maiers' property taxes were $1360.70, payable in March. Since they didn't pay until April, the county charged them 2% interest.

  a. How much was the penalty for late payment?

  b. How much did they pay the county in all?

26. Beverly has a NOW account. She is paid interest of 0.5% of her average monthly balance. Last month her average balance was $686.23. How much interest did she receive?

For more practice: **Application Bank,** page 461.

Fringe Benefits

You run a small business. Besides paying each employee a salary, you also provide them with fringe benefits, such as Social Security, health insurance, and a savings plan.

MONTHLY PAYROLL			
Data entry clerk	$1088	Bookkeeper	$1152
Sales manager	1585	Assembler	980
Service technician	1364		

PROBLEM: What is your total monthly cost for salaries and fringe benefits?

Solution Plan: Find the monthly payroll. Figure the cost of each fringe benefit. Add to find the total of salaries and fringe benefits.

WHAT DO YOU KNOW? You know each salary.

1. Find the total monthly payroll.

WHAT DO YOU WANT TO KNOW? You want to know the cost of Social Security and health insurance.

2. As an employer, you are required to pay an amount equal to 6.70% of all salaries to Social Security. How much do you pay each month to Social Security?

   The health insurance premium is $50.10 for individuals and $123.95 for the family plan. The employer pays 80% of the premium and the employee pays 20%.

3. Three employees have family health coverage and two have individual coverage. As employer, how much do you pay a month for health insurance?

4. How much do the employees pay per month, in total, for health insurance?

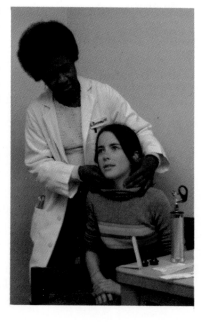

WHAT MORE DO YOU WANT? You want to know the cost of any other fringe benefits the company provides.

You provide group life insurance for your employees. The *monthly* premium is 30¢ per $1000 of your *annual* payroll rounded to the nearest $1000.

5. How much is your annual payroll?

6. Round the annual payroll to the nearest $1000.

7. What is your monthly premium for group life insurance?

You offer a savings plan to employees. The company contributes an amount, rounded to the nearest dollar, equal to 2% of monthly salary for participating employees. The sales manager, service technician, and assembler chose to participate in the savings plan.

8. Find the savings plan contribution for:
   a. sales manager   **b.** assembler
   c. service technician

9. How much is your total monthly contribution for the savings plan?

NOW SOLVE THE PROBLEM. Give the monthly amounts to complete the report.

	SALARY & BENEFITS—MONTHLY REPORT	
10.	Salary	_?_
11.	Social Security. . . . . _?_	
12.	Health (80%) . . . . _?_	
13.	Group Life . . . . _?_	
14.	Savings Plan . . . . _?_	
15.	Total Non-Salary . . . . _?_	
16.	Total Salary and Benefits. . . . . _?_	

Skill 68 Write as a number sentence.

1. 60% of 150 is 90.

2. What percent of 53 is 10.6?

Skill 69 Identify percent, base, and percentage.

3. 15% of 60 is 9.   Percent: __?__   Base: __?__   Percentage: __?__

Skill 70 Write the answer as a whole number or as a decimal.

4. 16% of 1500 is what number?

5. What is 38% of 240?

Application A32

6. City Savings raised its interest rate from 6% to 7%. How much more will a $3500 savings deposit earn in 1 year?

Problem Solving

7. You have a monthly payroll of $7500. You pay 6.70% of the payroll amount to Social Security. You also pay 75% of the health insurance premium of $585. How much do you pay a month for salary and benefits?

CONSUMER NOTE

East Bay Motors is selling the new PD-4. The $6425 sticker price includes a base price of $5305, options totaling $960, and a destination charge of $160. East Bay paid 80% of the base price, 75% of the options, and the full destination charge when buying the car from the manufacturer. How much did East Bay pay for each PD-4?

$$\text{Dealer's Cost} = \frac{80\% \text{ of}}{\text{Base Price}} + \frac{75\% \text{ of}}{\text{Options Price}} + \frac{\text{Destination}}{\text{Charge}}$$

$$\text{Dealer's Cost} = (0.80 \times \$5305) + (0.75 \times \$960) + \$160 = \$5124$$

East Bay Motors paid $5124 for each PD-4.

Jerry Cohen made East Bay an offer of $500 less than the sticker price.

1. What is Jerry's offer?

2. Jerry's offer is how much more than the dealer's cost?

Compound Interest

When you put money in a savings account, the bank pays you for the use of your money. The money you earn is called **interest.** The amount of interest earned depends on your deposit, or principal. The interest rate is expressed as a yearly rate. However, interest can be paid (or compounded) daily, monthly, quarterly, or annually. The interest earned is often added to your principal. Future interest based on the new total (principal + interest) is called **compound interest.**

Find the interest earned on a $600 deposit, at 18% annual interest, compounded quarterly for 9 months. Use the formula:

$$\text{Interest earned} = \text{Principal} \times \text{Rate} \times \text{Time}$$

KEY-IN	DISPLAY	
600	600	original principal = $600
$\times$	600	
.18	$.18$	rate = 18%
$\times$	108	
.25	$.25$	time for each payment = $\frac{1}{4}$ year = 0.25
$=$	27	1st quarter interest, $27
$+$	27	
600	600	
$=$	627	new principal for second quarter
$\times$	627	
.18	$.18$	
$\times$	112.86	
.25	$.25$	
$=$	28.215	2nd quarter interest, $28.215
$+$	28.215	
627	627	
$=$	655.215	new principal for third quarter
$\times$	655.215	
.18	$.18$	
$\times$	117.9387	
.25	$.25$	
$=$	29.484675	3rd quarter interest, $29.484675

The total amount of interest is the sum of the interest payments for the three quarters: $27.00 + $28.215 + $29.484675, or $84.70. Your new balance at the end of 9 months is $684.70, the sum of your original $600 deposit and three interest payments totalling $84.70.

Finding the Base

MODEL

Find the base.	Write as a number sentence.	Write 15% as a fraction in lowest terms.	Divide.	
15% of what number is 9?	$15\% \times N = 9$ Skill 68	$15\% \times N = 9$ $\dfrac{15}{100} \times N = 9$ $\dfrac{3}{20} \times N = 9$ Skill 67	$N = 9 \div \dfrac{3}{20}$ $N = 60$	15% of 60 is 9. Answer: 60

 GET READY. Write in lowest terms.

1. $\dfrac{48}{100}$ 2. $\dfrac{64}{100}$ 3. $\dfrac{50}{100}$ 4. $\dfrac{26}{100}$ 5. $\dfrac{95}{100}$

Complete the statement.

6. $\dfrac{4}{5} \times N = 12$

$N = \underline{\ ?\ } \div \dfrac{4}{5}$

7. $\dfrac{1}{2} \times N = 8$

$N = 8 \div \underline{\ ?\ }$

8. $\dfrac{7}{10} \times N = 6.3$

$N = 6.3 \underline{\ ?\ } \dfrac{7}{10}$

 NOW USE SKILL 71. Write the answer as a mixed number in lowest terms or a whole number.

9. 8 is 10% of what number?

10. 15% of what number is 30?

11. 9% of what number is 270?

12. 6% of what number is 24?

13. 108 is 45% of what number?

14. 240 is 16% of what number?

15. 12 is 15% of what number?

16. 360 is 80% of what number?

17. 126 is 9% of what number?

18. 30% of what number is 63?

19. 15% of what number is 810?

20. 198 is 11% of what number?

21. 5 is 15% of what number?

22. 3 is 40% of what number?

23. 50% of what number is $\dfrac{1}{8}$?

24. 3 is 24% of what number?

25. 54 is 120% of what number?

26. 250% of what number is 16?

MODEL

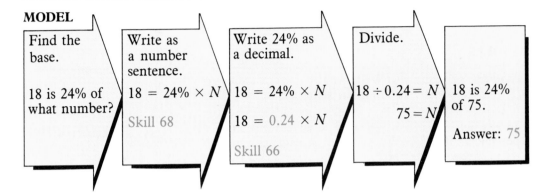

Find the base.

18 is 24% of what number?

Write as a number sentence.

$18 = 24\% \times N$

Skill 68

Write 24% as a decimal.

$18 = 24\% \times N$

$18 = 0.24 \times N$

Skill 66

Divide.

$18 \div 0.24 = N$

$75 = N$

18 is 24% of 75.

Answer: 75

 GET READY. Complete the statement.

27. $0.65 \times N = 780$
$N = 780 \div$ ___?___

28. $0.34 \times N = 195$
$N =$ ___?___ $\div 0.34$

29. $24 = 0.6 \times N$
$24 \div$ ___?___ $= N$

Divide to hundredths. Round to the nearest tenth.

30. $9 \div 0.25$

31. $21 \div 0.32$

32. $27 \div 0.81$

 NOW USE SKILL 71. Round to the nearest tenth when necessary.

33. 12 is 15% of what number?

34. 14% of what number is 56?

35. 6% of what number is 36?

36. 98 is 2% of what number?

37. 213 is 5% of what number?

38. 16% of what number is 54?

39. 25% of what number is 173?

40. 70% of what number is 420?

41. 818 is 50% of what number?

42. 284 is 32% of what number?

43. 60% of what number is 231?

44. 30% of what number is 153?

45. 30% of what number is 186?

46. 15% of what number is 45?

47. 4 is 12% of what number?

48. 440 is 16% of what number?

49. 5% of what number is 176?

50. 48% of what number is 80?

51. 105 is 7% of what number?

52. 256 is 20% of what number?

53. 19 is 125% of what number?

54. 84 is 240% of what number?

55. 300% of what number is 414?

56. 450 is 160% of what number?

57. 150% of what number is 105?

58. 300% of what number is 460?

For more practice: **Skill Bank,** page 435.

Finding the Percent

MODELS

Find the percent.	Write as a number sentence.	Divide.	
26 is what percent of 50?	$26 = N \times 50$	$26 \div 50 = N$ $0.52 = N$ $52\% = N$ ↑ Skill 64	26 is 52% of 50. Answer: 52%

Find the percent.	Write as a number sentence.	Divide.	
What percent of 72 is 9?	$N \times 72 = 9$	$N = 9 \div 72$ $N = 0.125$ $N = 12.5\%$ ↑ Skill 64	12.5% of 72 is 9. Answer: 12.5%

▶ **GET READY. Divide. Round the answer to the nearest hundredth.**

1. $17 \div 63$ **2.** $34 \div 963$ **3.** $74 \div 89$ **4.** $27 \div 44$

5. $120 \div 85$ **6.** $14 \div 56$ **7.** $200 \div 250$ **8.** $36 \div 54$

Write as a percent.

9. 0.4 **10.** 0.12 **11.** 0.06 **12.** 0.005 **13.** 1.07

14. 0.63 **15.** 0.55 **16.** 7.36 **17.** 0.096 **18.** 1.23

NOW USE SKILL 72.

19. 2 is what percent of 4?

20. What percent of 12 is 3?

21. What percent of 20 is 5?

22. 9 is what percent of 45?

23. 9 is what percent of 12?

24. What percent of 25 is 8?

25. 4 is what percent of 16?

26. What percent of 35 is 7?

27. What percent of 12 is 6?

28. 8 is what percent of 10?

29. 12 is what percent of 32?

30. What percent of 56 is 7?

31. What percent of 96 is 60?

32. 56 is what percent of 64?

33. 18 is what percent of 144?

34. 3 is what percent of 24?

35. 72 is what percent of 192?

36. What percent of 96 is 84?

37. What percent of 168 is 105?

38. What percent of 224 is 42?

39. What percent of 400 is 2?

40. What percent of 1500 is 3?

41. 4 is what percent of 1600?

42. 8 is what percent of 3200?

43. 21 is what percent of 2625?

44. 42 is what percent of 6000?

45. What percent of 3750 is 15?

46. What percent of 2000 is 6?

47. 100 is what percent of 20?

48. 228 is what percent of 38?

49. What percent of 46 is 138?

50. What percent of 54 is 378?

51. 75 is what percent of 15?

52. 60 is what percent of 24?

53. What percent of 176 is 440?

54. What percent of 576 is 1008?

55. 120 applicants.
45 accepted.
What percent of
applicants are accepted?

56. 96 cars in stock.
24 sold.
What percent of cars in
stock are sold?

57. $450 wages.
$81 saved.
What percent of
wages is saved?

58. 170-page novel.
85 pages read.
What percent of the
novel is read?

59. 36 games played.
27 games won.
What percent of the
games are won?

60. 60 employees.
9 absent.
What percent of the
employees are present?

For more practice: **Skill Bank,** page 436.

Percent Change

When a quantity changes from one value to another, the change can be expressed as a percent of the original amount. To find the percent change, use the formula:

$$\text{Percent change} = \frac{\text{Amount of change}}{\text{Original amount}}$$

The Emmerts' rent will be $483 in September. They now pay $420. What is the percent increase in their rent?

MODEL

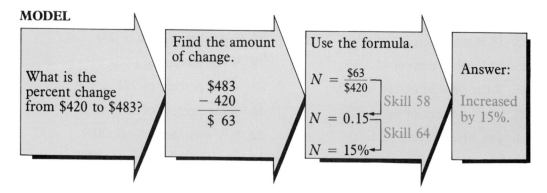

What is the percent change from $420 to $483?

Find the amount of change.

$\begin{array}{r} \$483 \\ -\ 420 \\ \hline \$\ 63 \end{array}$

Use the formula.

$N = \dfrac{\$63}{\$420}$ Skill 58

$N = 0.15$

  Skill 64

$N = 15\%$

Answer:

Increased by 15%.

Valley Senior High School has 256 graduates this year. Last year there were 320 graduates. What is the percent decrease?

MODEL

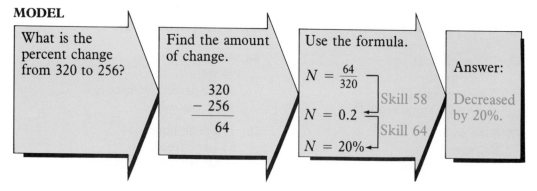

What is the percent change from 320 to 256?

Find the amount of change.

$\begin{array}{r} 320 \\ -\ 256 \\ \hline 64 \end{array}$

Use the formula.

$N = \dfrac{64}{320}$ Skill 58

$N = 0.2$

  Skill 64

$N = 20\%$

Answer:

Decreased by 20%.

 GET READY. Write as a decimal.

1. $\dfrac{16}{40}$ **2.** $\dfrac{8}{25}$ **3.** $\dfrac{96}{120}$ **4.** $\dfrac{124}{200}$

NOW USE APPLICATION A33. **When necessary, round answers to the nearest whole percent.**

5. The town of Sailor's Point had a population of 10,000 ten years ago. Today the population is 15,600. By what percent has the population increased?

6. United Sales Corporation showed a $1,675,000 profit last quarter. This quarter profits are down by $335,000. What is the percent decrease in profits?

7. In Prairie County 381,000 voters turned out in this year's election. Last election 300,000 people voted. What is the percent increase in voter turnout?

8. After installing a chimney damper, the Murphys saved $216 compared to last year's fuel bill of $1200. By what percent did they cut their fuel bill?

9. Before Dave bought a new car, his insurance premium was $180. He now pays $396. By what percent did the cost of his insurance go up?

10. The Turners used to average $50 a week on groceries. They now average $56. What is the percent increase in their grocery budget?

11. The price of a monthly pass for commuters on the Shore Line Railroad increased from $52 to $63. What was the percent increase?

12. The Census Bureau keeps a record of the number of tourists who travel to and from the United States each year.

   a. In 1980 8 million tourists came to the United States. What percent of change is this from 1965?

   b. What was the percent change in U.S. tourists travelling abroad between 1975 and 1980?

   c. On the average, each tourist spent $350 while visiting the United States for one week in 1965. In 1980 the average was $630 for one week. What is the percent change in the amount spent by each tourist?

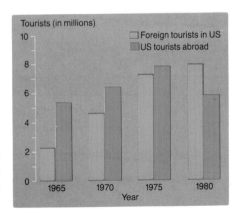

For more practice: **Application Bank,** page 461.

Tax Hike

You have learned that your community is adopting 100% valuation as the new basis for real estate taxes. You want to know how this will affect your monthly house payment.

Problem: What is your monthly house payment with the 100% valuation system?

Solution Plan: Use your old tax bill to figure the market value of your house. Find the tax under the new system, and the monthly amount of the new tax.

WHAT DO YOU KNOW? You have your old tax bill. The bill gives tax rates and amounts for the current year.

*TAX RATE PER $1000	**OFFICE OF THE COLLECTOR OF TAXES**	REAL ESTATE TAX FISCAL YEAR ENDING JUNE 30.

LOCATION DESCRIPTION	ASSESSED VALUE	REAL ESTATE TAX
19 MAIN STREET 1744100	TOTAL 27,000	1998.00

*SCHOOL	*GENERAL	*TOT. TAX	MAP	LOT	PLOT	BOOK	PAGE
29.70	44.30	74.00	2042	000C	0030	48560	4940

1. What is the assessed value of your house?

2. What is the present total tax rate?

3. What is your present real estate tax?

WHAT DO YOU WANT TO KNOW? You want to know the market value of your house. You call the assessor's office and find that the assessed value is 30% of the market value.

4. Find the market value of your house.

WHAT MORE DO YOU WANT? You want to know your new tax amount. In the new system, you pay taxes on the full market value. The new tax rate will be $26. This is the tax amount per $1000 of the market value, per year.

5. The tax rate is stated per thousand dollars. How many thousands of dollars is the value of your house?

6. What is your tax at the new rate?

NOW SOLVE THE PROBLEM.

7. Under the new system, what is your tax per month?

8. Under the old system, what was your tax per month?

9. You pay your real estate tax as part of your monthly house payment. How much will your monthly payment go up?

10. Your monthly statement from the bank shows your total house payment before the system changed. How much is your new monthly payment, with the increased tax amount?

ACCOUNT NUMBER	MAGNOLIA NATIONAL BANK		PAYMENT SHOULD BE MADE ON OR BEFORE DUE DATE	TOTAL PAYMENT DUE
4-126614				
DUE DATE	INTEREST DUE	PRINCIPAL DUE	ESCROW DUE	MISCELLANEOUS INSURANCE DUE	
06/01	650.02	121.46	166.50	0.00	937.98

Skill 71

1. 30 is 60% of what number?

2. 78% of what number is 117?

Skill 72

3. 55 is what percent of 88?

4. What percent of 350 is 70?

Application A33

5. The price of gold fell from $550 per ounce to $484. By what percent did the price decrease?

Problem Solving

6. The assessed value of your house is $48,000. The tax rate is $50 per $1000 of the assessed value. You pay your real estate tax in 12 equal monthly payments. How much do you pay for real estate taxes each month?

EXTRA!

Nutrition

Before eating her yogurt, Ellen read the nutrition information on the back of the carton.

1. Ellen has an average of 4 servings of yogurt a week. How many Calories are there in 4 servings?

2. In each serving there are 6 grams of protein, which is 15% of the U.S. RDA. What is the U.S. RDA for protein?

3. According to Ellen's doctor, she should eat a total of 1700 Calories a day. One carton of yogurt is what percentage of this total?

4. How many servings of yogurt would be needed to get the full RDA of calcium?

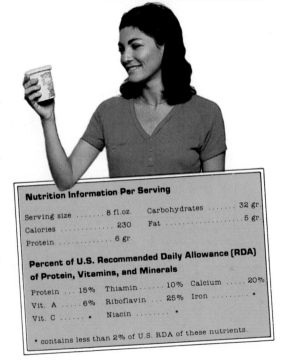

Nutrition Information Per Serving

Serving size .... 8 fl.oz.	Carbohydrates .... 32 gr.
Calories 230	Fat .... 5 gr.
Protein .... 6 gr	

Percent of U.S. Recommended Daily Allowance (RDA) of Protein, Vitamins, and Minerals

Protein ... 15%	Thiamin 10%	Calcium 20%
Vit. A 6%	Riboflavin ... 25%	Iron .... *
Vit. C .... *	Niacin .... *	

* contains less than 2% of U.S. RDA of these nutrients.

Annual Percentage Rate

You want to borrow $1000. You plan to repay the loan in 24 payments of $48 each. You use a table to find the annual percentage rate (APR). You can use the APR to compare types of loans available from different banks.

To find the APR for your loan, figure out the finance charge and the finance charge per $100 of the loan amount.

Total repaid = Number of payments × Amount of each payment
$$= 24 \times \$48$$
$$= \$1152$$

Finance charge = Total repaid − Loan amount
$$= \$1152 - \$1000$$
$$= \$152$$

$$\text{Finance charge per \$100} = \frac{\text{Finance charge}}{\text{Loan amount}} \times \$100$$
$$= \frac{\$152}{\$1000} \times \$100$$
$$= 0.152 \times \$100$$
$$= \$15.20$$

Use the table to find the amount closest to $15.20 in the 24-payment row. The heading for that column is your APR.

APR	12.5%	13%	13.5%	14%	14.5%	15%
Term	Finance charge per $100					
12	6.90	7.18	7.46	7.74	8.03	8.31
24	13.54	14.10	14.66	15.23	15.80	16.37

The APR for your loan is 14%.

Find the APR.

1. $750 loan.
   12 payments of $67.50.

2. $1500 loan.
   24 payments of $72.50.

3. $430 loan.
   12 payments of $38.50.

4. One bank offers a $2000 loan with 24 payments of $96.50 each. Another bank offers a $2000 loan with 12 payments of $180.51 each. Which bank has the lower APR?

285

Using Proportions

MODEL

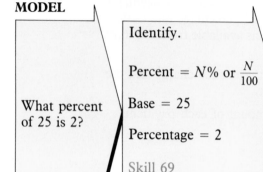

What percent of 25 is 2?

Identify.

Percent = $N\%$ or $\dfrac{N}{100}$

Base = 25

Percentage = 2

Skill 69

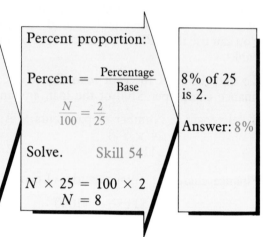

Percent proportion:

Percent = $\dfrac{\text{Percentage}}{\text{Base}}$

$\dfrac{N}{100} = \dfrac{2}{25}$

Solve.   Skill 54

$N \times 25 = 100 \times 2$
$N = 8$

8% of 25 is 2.

Answer: 8%

GET READY. Identify.

1. What percent of 9 is 4?
Percent: __?__   Percentage: __?__
Base: __?__

2. 6 is what percent of 15?
Percent: __?__   Percentage: __?__
Base: __?__

NOW USE SKILL 73. Write as a proportion.

3. What percent of 72 is 9?

4. What percent of 46 is 18?

5. 51 is what percent of 93?

6. 3 is what percent of 22?

7. 13 is what percent of 65?

8. What percent of 91 is 14?

9. 17 is what percent of 221?

10. 54 is what percent of 176?

11. What percent of 276 is 180?

12. What percent of 242 is 66?

Answer the question.

13. What percent of 8 is 2?

14. What percent of 40 is 12?

15. 46 is what percent of 115?

16. What percent of 54 is 27?

17. 48 is what percent of 64?

18. What percent of 20 is 15?

19. 12 is what percent of 120?

20. 70 is what percent of 175?

21. What percent of 3600 is 1368?

22. 306 is what percent of 408?

MODEL

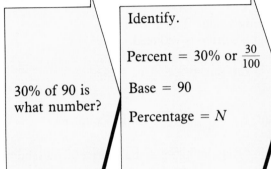

Identify.

Percent = 30% or $\frac{30}{100}$

Base = 90

Percentage = N

Percent proportion:

Percent = $\frac{\text{Percentage}}{\text{Base}}$

$\frac{30}{100} = \frac{N}{90}$

Solve.   Skill 54

$30 \times 90 = 100 \times N$
$27 = N$

30% of 90 is what number?

30% of 90 is 27.

Answer: 27

 GET READY. Write as a fraction with denominator 100.

23. 42%   **24.** 55%   **25.** 163%   **26.** 8%   **27.** 17%

Identify.

28. 40% of 68 is what number?
    Percent: __?__   Percentage: __?__
                     Base: __?__

29. What number is 32% of 90?
    Percent: __?__   Percentage: __?__
                     Base: __?__

30. 15% of what number is 53?
    Percent: __?__   Percentage: __?__
                     Base: __?__

31. 77 is 43% of what number?
    Percent: __?__   Percentage: __?__
                     Base: __?__

NOW USE SKILL 73. Write as a proportion.

32. 30% of 120 is what number?

33. 80 is 24% of what number?

34. What is 16% of 109?

35. What is 42% of 730?

36. 130 is 10% of what number?

37. What is 60% of 200?

Answer the question.

38. 27% of what number is 54?

39. What is 45% of 120?

40. What is 65% of 400?

41. 80% of what number is 280?

42. 35 is 35% of what number?

43. What is 100% of 70?

44. 48% of what number is 384?

45. What is 20% of 660?

For more practice: **Skill Bank,** page 436.

Discounts

Stores often have items specially priced during a sale. The *discount*, or *markdown*, is the amount by which the regular price is reduced. The discount rate is the discount expressed as a percent of the regular price. The price of the item after the discount is the sale price. Use these formulas:

Discount rate × Regular price = Discount $\dfrac{\text{Discount}}{\text{Regular price}}$ = Discount rate

Regular price − Discount = Sale price   Regular price − Sale price = Discount

The Linen Kloset is taking 20% off its entire inventory this week. The regular price for a king size comforter is $67.50. How much is the discount? What is the sale price?

MODEL

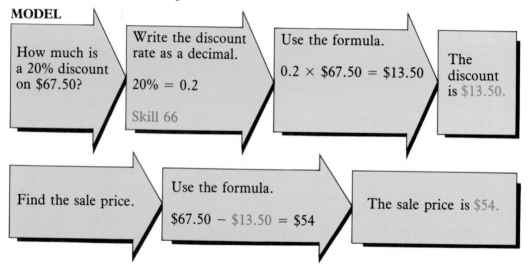

How much is a 20% discount on $67.50?

Write the discount rate as a decimal.

20% = 0.2

Skill 66

Use the formula.

0.2 × $67.50 = $13.50

The discount is $13.50.

Find the sale price.

Use the formula.

$67.50 − $13.50 = $54

The sale price is $54.

 NOW USE APPLICATION A34.

1. Fischer's is taking 25% off all summer dresses. Gina chose one with a regular price of $53.
   a. How much is a 25% discount on this dress?
   b. What is the sale price?

2. Continental Plastics allows its customers a 2% discount if payment is received within 10 days. If Maxie's pays their $7500 bill two days after receiving their shipment, how much is the discount?

3. At the Lane Furnishings clearance sale all items are marked 50% to 75% off. Terry bought a chair with a regular price of $123.
   a. What is the lowest the discount could be?
   b. What is the highest the discount could be?
   c. If the chair gets the full 75% discount, how much will Terry pay?

4. Bay Motors is taking 12% off last year's Royal Coupe. The list price was $7432. What is the discount? the sale price?

A dinnerware set is on sale. The regular price for each set is $68.50. The sale price is $51.30. What is the discount? What is the discount rate?

MODEL

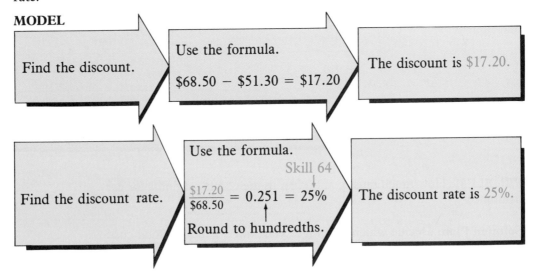

Find the discount. → Use the formula. $68.50 − $51.30 = $17.20 → The discount is $17.20.

Find the discount rate. → Use the formula. Skill 64 $\frac{\$17.20}{\$68.50} = 0.251 = 25\%$ Round to hundredths. → The discount rate is 25%.

Round percents to the nearest whole percent.

5. The Boutique has winter coats on sale. The $88 coat is selling for $59.75. What is the discount?

6. Jim used a 40¢ off coupon to buy a bottle of fabric softener. The regular price is $3.20. What is the discount rate?

7. Leechmont has a special this week. They are selling the VM electric typewriter at $152.99. The regular price is $169.99.
   a. How much is the discount?   b. What is the discount rate?

8. Sarah bought a pair of irregular shoes for $16.99 at Farley's shoe sale. The pre-sale price was $67.99.
   a. How much is the discount?   b. What is the discount rate?

For more practice: **Application Bank,** page 461

Auto Insurance Premiums

You are buying insurance for your car. Your insurance agent explains the coverages available. You must decide which coverages to buy. The agent shows you the Coverage Selection Form and tells you about the payment plan.

COVERAGE SELECTION FORM

COMPULSORY INSURANCE	LIMITS	DEDUCTIBLE	PREMIUM
1 Bodily Injury To Others	$10,000 per person $20,000 per accident	None	$48.00
2 Personal Injury Protection	$ 2,000 per person	None	$17.00
		$ 2,000	$ 8.00
3 Bodily Injury Caused By An Uninsured Auto	$10,000 per person $20,000 per accident	None	$ 7.00
4 Damage To Someone Else's Property	$10,000 per accident	None	$107.00
OPTIONAL INSURANCE			
5 Optional Bodily Injury To Others	$100,000 per person $300,000 per accident	None	$69.00
6 Collision	Actual Cash Value	$ 100	$213.00
		$ 200	$198.00
7 Comprehensive	Actual Cash Value	$ 100	$160.00
		$ 200	$154.00
8 Towing and Labor	$ 25	None	$ 4.00

PROBLEM: How much must you budget in the coming months for car insurance?

Solution Plan: Decide which coverages you want. Find the total cost, including any discounts that apply to you. Figure the amounts you need for car insurance in your budget.

WHAT DO YOU KNOW? You know which coverages are required by law (compulsory).

1. You have a good medical insurance program that you get at work, so you take the $2000 deductible on Coverage 2. How much is your premium for Coverage 2?

2. What is the total of your premiums for compulsory insurance, Coverages 1–4?

WHAT DO YOU WANT TO KNOW? You must decide which optional coverages you want.

The agent explains that with $200 deductible you must pay the first $200 of damage in an accident, and the insurance company pays the rest. With $100 deductible you pay only the first $100.

3. You decide to take $200 deductible on collision. How much is the premium?

4. Since there is very little difference in the premium for $100 deductible and $200 deductible on comprehensive, you decide to take $100 deductible. How much is the premium?

5. You also decide to take Coverages 5 and 8. How much is the premium for these coverages?

6. You get a 10% discount on your comprehensive coverage for having an anti-theft device. How much is your discount?

7. Find your total insurance bill, including all premiums and the discount.

NOW SOLVE THE PROBLEM.

The company has a 3 month payment plan. You pay 30% of the total the first month, 35% the second month, and the remaining balance the third month. You make your first payment in July.

8. How much do you budget for car insurance in the given month?
  a. July   **b.** August   **c.** September

Quiz

Skill 73 **Write as a proportion.**

1. 35% of 40 is 14.

2. 15 is what percent of 385?

3. 25 is 75% of what number?

4. 52 is 80% of 65.

5. What is 16% of 55?

6. What is 85% of 220?

Application A34

7. You own the Stereo Store. In August you offer a 10% discount on all stereos. In September you offer a 30% discount on the unsold stock. A stereo regularly sells for $675. What is the sale price in August? in September?

Problem Solving

8. You receive an $11 discount from the listed premium for personal injury protection, because you select $1000 deductible. How much is your total premium for compulsory insurance, including the discount?

COMPULSORY INSURANCE	Premium
Bodily injury to others	$ 52.00
Personal injury protection	$ 20.00
Bodily injury by uninsured auto	$ 6.00
Damage to someone else's property	$130.00

CALCULATOR DISPLAYS

Use your calculator to find the answer to the math exercise. Invert the calculator and read the message to complete the sentence.

1. Farms in Iowa have very rich __?__.
100 + 1597 + 2577 + 2415 + 416

2. The __?__ Desert is hot.
115 + 512 + 78 + 1056 + 45

3. The puzzle will __?__ your mind!
66,666 + 99,999 + 66,666 + 143,277

4. Don't get too near a __?__.
220,376 ÷ 652

5. The frog rests on the log in the __?__.
686 − 78

6. She sells sea __?__.
8,795,563 − 8,218,218

7. Hamburger __?__ while it's cooking.
2,151,484 + 1,145,658 + 218,279 + 321,386 + 1,535,408

292

Skills Review

Skill 69 (page 267)

1. 7% of 80 is 5.6.
Percent: ?   Base: ?
Percentage: ?

2. 15% of 320 is 48.
Percent: ?   Base: ?
Percentage: ?

3. 74 is 50% of 148.
Percent: ?   Base: ?
Percentage: ?

4. 798 is 28% of 2850.
Percent: ?   Base: ?
Percentage: ?

Skill 70 **Write the answer as a whole or mixed number.** (page 268)

5. 2% of 600 is what number?

6. What is 14% of 350?

7. What is 35% of 74?

8. 40% of 46 is what number?

Write the answer as a decimal. Round to the nearest tenth. (page 269)

9. What is 25% of 190?

10. 5.5% of 340 is what number?

11. 12.7% of 700 is what number?

12. What is 30% of 84?

Skill 71 (pages 276–277)

13. 18% of what number is 9?

14. 24 is 60% of what number?

15. 72 is 45% of what number?

16. 100% of what number is 346?

Skill 72 (pages 278–279)

17. What percent of 25 is 5?

18. 36 is what percent of 288?

19. 90 is what percent of 24?

20. What percent of 480 is 24?

Skill 73 (pages 286–287)

Write as a proportion.

21. What percent of 55 is 11?

22. What is 11% of 395?

23. 62% of what number is 60?

24. What percent of 317 is 20?

Answer the question.

25. What is 20% of 140?

26. 15% of what number is 60?

27. 36 is 50% of what number?

28. 102 is what percent of 120?

Applications Review

Application A32 (pages 270–271)

1. You owe $387.90 on your Passport charge card account. The finance charge (interest) is 1.5% of the unpaid balance. How much is your finance charge this month? Round to the nearest cent.

2. You owe $55,600 on your mortgage. The monthly interest rate is 1.25% of the unpaid balance of the mortgage. How much is your interest payment this month?

3. A bank advertises that its "current effective yield" on savings certificates is 12.9%. This means that their interest payments are equivalent to one annual interest payment of 12.9%. Find the interest on a certificate for $3500 for one year.

Application A33 (pages 280–281)

4. This winter the DeLuca family cut their heating oil consumption from 1200 gallons to 1056 gallons. However, the price of oil was raised from $1.20 a gallon to $1.38 a gallon. Which percent change was larger, the decrease in oil consumption or the increase in the price of oil?

5. The table gives census information about the area and population of the United States. Answer the following questions to the nearest whole percent.

Year	Area (square miles)	Population
1800	900,000	5,300,000
1850	3,000,000	23,200,000
1900	3,000,000	76,000,000

a. By what percent did the area of the United States increase between 1800 and 1850?

b. By what percent did the population increase between 1800 and 1850?

c. What was the percent change in the population between 1850 and 1900?

Application A34 (pages 288–289)

6. Granite Builder's Supply gives contractors a 2% discount for payment within two weeks. Capitol Contractors Inc. has a bill for $3975. How much of a discount will Capitol receive for paying promptly?

7. A $120 suit is on sale for $84. What is the percent discount rate?

8. A new car has a list price of $7500. The dealer advertises a rebate program. You receive a rebate of 6% of the list price. What is the cost of the car with the rebate?

Unit Test

Skill 68

Write as a number sentence.

1. 30% of 30 is 9.

2. What is 16% of 85?

Skill 69

Identify.

3. 9% of 76 is 6.84.
Percent: __?__   Base: __?__
Percentage: __?__

Skill 70

Write the answer as a mixed number in lowest terms or a whole number.

4. What is 25% of 18?

5. 60% of 8 is what number?

Write the answer as a decimal.

6. What is 23% of 120?

7. 52% of 2520 is what number?

Skill 71

Write the answer as a mixed number in lowest terms or a whole number.

8. 25% of what number is 103?

9. 80% of what number is 10?

Write the answer as a whole number or a decimal rounded to the nearest tenth.

10. 15 is 12% of what number?

11. 38% of what number is 20?

Skill 72

12. 84 is what percent of 240?

Skill 73 **Write as a proportion.**

13. 62% of 850 is 527.

Application A32

14. You have $700 in a savings account. Your sister has $1500 in a savings account. If both accounts pay 2% interest this quarter, how much more does her account earn than yours?

Application A33

15. *New Life* magazine has a circulation of 80,000. The circulation manager predicts a circulation of 150,000 in 3 years. What percent increase in circulation would this be?

Application A34

16. You bought a pair of shoes for $36. The original price was $45. What is the discount? What is the discount rate?

Skills

Skill 52 **Write the ratio as a fraction.** (page 202)

1. 5 out of 8 dresses   **2.** 6 out of 7 calculators   **3.** 50 days out of a year

Skill 53 **Are the ratios equal?** (page 203)

4. $\frac{5}{6}$; $\frac{15}{18}$ **5.** $\frac{4}{7}$; $\frac{12}{14}$ **6.** $\frac{4}{9}$; $\frac{8}{16}$ **7.** $\frac{14}{24}$; $\frac{21}{36}$

Skill 54 **Solve the proportion.** (pages 204–205)

8. $\frac{1}{6} = \frac{4}{t}$ **9.** $\frac{2}{5} = \frac{6}{a}$ **10.** $\frac{3}{7} = \frac{m}{21}$ **11.** $\frac{f}{9} = \frac{12}{18}$

Skill 56 **Write the equal rates as a proportion. Solve.** (pages 214–215)

12. 35 ounces in 7 jars = x ounces in 14 jars

13. 10 records in 2 files = n records in 4 files

Skill 57 **Find the unit rate.** (pages 222–223)

14. $8 for 4 plants   **15.** 156 miles in 3 hours

Skill 58 **Write as a decimal. Round to the nearest hundredth.** (page 234)

16. $\frac{3}{10}$ **17.** $\frac{2}{25}$ **18.** $\frac{29}{50}$ **19.** $\frac{5}{12}$

Write as a decimal. Round to the nearest thousandth.

20. $\frac{9}{20}$ **21.** $\frac{3}{40}$ **22.** $\frac{7}{16}$ **23.** $\frac{8}{9}$

Skill 59 **Write as a fraction in lowest terms.** (page 235)

24. 0.6   **25.** 0.36   **26.** 0.05   **27.** 0.075

Skill 60 **Write as a decimal.** (page 236)

Round to the nearest hundredth.   **Round to the nearest thousandth.**

28. $3\frac{3}{4}$ **29.** $2\frac{9}{16}$ **30.** $8\frac{1}{8}$ **31.** $2\frac{5}{6}$

Skill 61 **Write as a mixed number in lowest terms.** (page 237)

32. 7.5   **33.** 9.02   **34.** 11.6   **35.** 10.005

Skill 62 **Write as a fraction with a denominator of 100.** (page 244)

36. $\frac{9}{20} = \frac{?}{100}$ **37.** $\frac{6}{25} = \frac{?}{100}$ **38.** $\frac{4}{5} = \frac{?}{100}$

Skill 63 (page 245)

39. $\frac{17}{100} = \underline{\ ?\ }\%$ **40.** $\frac{55}{100} = \underline{\ ?\ }\%$ **41.** $\frac{210}{100} = \underline{\ ?\ }\%$

Skill 64 **Write as a percent.** (page 246)

42. 0.25   **43.** 0.51   **44.** 0.017   **45.** 0.002

Skill 65 **Write as a percent.** (page 247)   Skill 66 **Write as a decimal.** (page 254)

46. $\frac{7}{10}$ **47.** $\frac{17}{20}$ **48.** 9%   **49.** 13%

Skill 67 **Write as a fraction or mixed number in lowest terms.** (page 255)

50. 80%   **51.** 45%   **52.** 36%   **53.** 850%

Skill 68 **Write as a number sentence.** (page 266)

54. 33% of 50 is 16.5   **55.** What is 17% of 73?

Skill 69 **Identify the percent, base, and percentage.** (page 267)

56. 5% of 56 is 2.8
Percent: __?__   Base: __?__
Percentage: __?__

57. 60% of 105 is 63
Percent: __?__   Base: __?__
Percentage: __?__

Skill 70 (pages 268–269)

58. 20% of 42 is what number?   **59.** What is 23% of 200?

Skill 71 (pages 276–277)

60. 13 is 20% of what number?   **61.** 240 is 32% of what number?

Skill 72 (pages 278–279)

62. What percent of 20 is 5?   **63.** 36 is what percent of 75?

Skill 73 **Write as a proportion.** (pages 286–287)

64. 8 is 25% of 32.   **65.** 17 is what percent of 55?

Applications

Application A25 (pages 206–207)

1. You punch a small hole in the bottom of a bag containing a mixture of 50 red marbles and 100 green marbles. A marble falls out of the hole.

   a. What is the probability that the marble is green?

   b. What is the probability that the marble is red?

Application A26 (pages 216–217)

2. Three out of every 10 students at your high school are enrolled in a computer course. If there are 2500 students in your school, how many are enrolled in a computer course?

Application A27 (page 224)

3. Ernie Jones bought 3 suits for $150.33, $200, and $175.69. What was the mean price of the suits?

4. The scores on a quiz were 88, 75, 67, 91, 75, 84, 76, and 92. What was the mean score?

Application A28 (page 225)

5. Find the median and range: 56, 43, 51, 47, 65.

6. Find the mode and median: 18, 23, 17, 19, 21, 18, 20.

Application A29 (page 239)

7. Which car gets the highest mileage?

8. Which car gets the lowest mileage?

9. How many more miles per gallon does the Puma get compared to the Deluxe?

10. Which cars get less than 25 MPG?

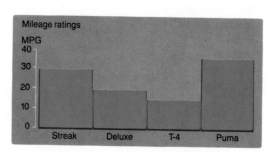

Application A30 (pages 248–249)

11. In which years did the budget increase?

12. In which year did the budget increase the most?

13. In which years was the budget less than $30,000,000?

14. In which years was the budget more than $50,000,000?

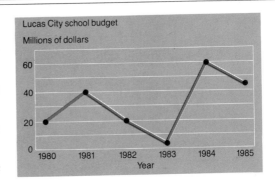

Lucas City school budget

Millions of dollars

Application A31 (page 256)

15. Which is the biggest expense?

16. Which is the smallest expense?

17. Which costs more, maintenance or concessions?

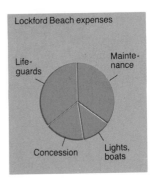

Lockford Beach expenses

Application A32 (pages 270–271)

18. You have $400 in a savings account. You receive an interest payment of 6%.
    a. How much interest do you get?
    b. How much do you have in your account in total?

19. You owe $438 on your charge card account. Your interest payment (finance charge) is 1.5% of your balance. How much interest do you pay?

Application A33 (pages 280–281)

20. Nanette accelerates her car from 30 miles per hour to 45 miles per hour. What is the percent increase in her speed?

21. The population of Deadrock City fell from 24,000 in 1880 to 6,000 in 1920. By what percent did the population decline?

Application A34 (pages 288–289)

22. The regular cost of a sheet of plywood is $20. You buy it on sale for $17. What is the discount rate?

23. A mattress with a regular price of $159.95 is on sale for 20% off.
    a. How much is the discount?   **b.** What is the sale price?

10

You will use these skills and applications in this unit. Which do you already know? Work each problem.

Skill 74 Write a positive or negative number.

1. 235 ft below sea level

2. 14° above zero

Skill 75

Write > or < in place of __?__.

3. 3 __?__ ⁻6

4. ⁻5 __?__ ⁻2

Skill 76

5. ⁻9 + ⁻5

6. ⁻11 + ⁻14

Skill 77

7. 18 + ⁻7

8. ⁻16 + 12

Skill 78

9. 6 − 10

10. ⁻4 − 9

Skill 79

11. 3 − ⁻8

12. ⁻5 − ⁻2

Skill 80

13. ⁻3 × ⁻7

14. ⁻84 ÷ ⁻7

Skill 81

15. ⁻12 × 9

16. 120 ÷ ⁻15

Skill 82 Simplify.

17. 6 + 4 × 7

18. 9 − 6 ÷ 3

Skill 83 Simplify.

19. ⁻10 × (⁻5 + 3)

Application A35 Write a sum of positive or negative numbers. Then solve the problem.

20. A diver swimming at a depth of 30 ft rose 14 ft. What is the diver's depth?

Application A36 Write a difference of positive or negative numbers. Then solve the problem.

21. Before making a payment you owed $42. After the payment you owed $25. How much did you reduce your debt?

Application A37

22. The wind chill factor yesterday was ⁻21°F. Today the wind chill factor is ⁻5°F. How many degrees warmer does it seem today?

Check your answers. If you had difficulty with any skill or application, be sure to study the corresponding lesson in this unit.

Positive and Negative Numbers

Think of a boat at sea level.
Call its position 0.

Think of a helicopter 30 meters above sea level.
Call its position "positive 30." Write it $^+30$.

Think of a diver 20 meters below sea level.
Call this position "negative 20."
Write it $^-20$.

MODEL

Write a positive or negative number.

10 points won	$^+10$
20 dollars in debt	$^-20$
250 meters climbed	$^+250$

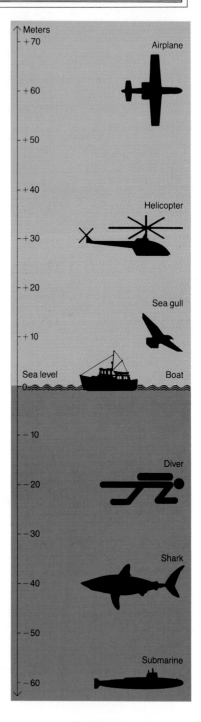

 GET READY. Name the position by a positive or negative number. Use the scale.

1. Sea gull   **2.** Airplane

3. Submarine   **4.** Shark

5. Helicopter   **6.** Diver

 NOW USE SKILL 74. Write a positive or negative number.

7. 20° above zero   **8.** 10° below zero

9. A $50 profit   **10.** A 75¢ loss

11. 15 kg lost   **12.** 5 lb gained

13. 150 m below sea level   **14.** $25 owed

15. A well 500 ft deep   **16.** A kite 11 m high

For more practice: **Skill Bank,** page 437.

Using a Number Line

You can write positive numbers without the + sign: $^+5 = 5$.
You must always write the − sign with a negative number: $^-5$.

Number Line

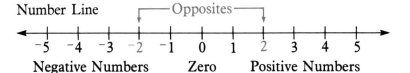

Negative Numbers   Zero   Positive Numbers

MODELS

Write > or < in place of ? .

$4 \stackrel{?}{=} {}^-4$
$^-4 \stackrel{?}{=} {}^-2$

Look at the number line.

4 is to the right of $^-4$.
$^-4$ is to the left of $^-2$.

$4 > {}^-4$
$^-4 < {}^-2$

Write the opposite of $^-4$.

Find $^-4$ on the number line.
What number is the same
distance from 0 but on the
opposite side?

The opposite
of $^-4$ is 4.

 GET READY. Write > or < in place of ?. Use the number line.

1. $3 \underline{\underline{?}} 2$ **2.** $^-2 \underline{\underline{?}} 1$ **3.** $^-1 \underline{\underline{?}} 0$ **4.** $2 \underline{\underline{?}} {}^-5$

Write the opposite of the number.

5. 4   **6.** $^-5$ **7.** 12   **8.** $^-6$

 NOW USE SKILL 75. Write > or < in place of ?.

9. $4 \underline{\underline{?}} 8$ **10.** $^-4 \underline{\underline{?}} {}^-8$ **11.** $^-4 \underline{\underline{?}} 8$ **12.** $4 \underline{\underline{?}} {}^-8$

13. $0 \underline{\underline{?}} 9$ **14.** $^-9 \underline{\underline{?}} 0$ **15.** $6 \underline{\underline{?}} {}^-6$ **16.** $^-8 \underline{\underline{?}} {}^-6$

Write the opposite.

17. $^-1$ **18.** 16   **19.** $^-2$ **20.** $^-3$ **21.** 11

For more practice: **Skill Bank,** page 437.

Adding Numbers with the Same Sign

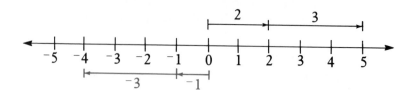

MODEL

Add.	Look at the number line.	
2 + 3 ⁻1 + ⁻3	The black arrows show 2 + 3. The red arrows show ⁻1 + ⁻3.	2 + 3 = 5 ⁻1 + ⁻3 = ⁻4

To add a positive number, move to the right.

To add a negative number, move to the left.

 GET READY. Use the number line when necessary.

1. 1 + 4	**2.** 2 + 2	**3.** 4 + 1
4. 3 + 2	**5.** 1 + 1	**6.** 0 + 4
7. ⁻3 + ⁻1	**8.** ⁻2 + ⁻2	**9.** ⁻1 + ⁻4
10. 0 + ⁻4	**11.** ⁻1 + ⁻2	**12.** ⁻3 + ⁻2

 NOW USE SKILL 76. Use a number line when necessary.

13. 3 + 7	**14.** 4 + 8	**15.** ⁻9 + ⁻8
16. ⁻3 + ⁻5	**17.** 7 + 9	**18.** ⁻8 + ⁻7
19. 7 + 6	**20.** ⁻7 + ⁻6	**21.** 8 + 8
22. ⁻8 + ⁻8	**23.** ⁻9 + ⁻9	**24.** 10 + 7

25. ⁻16 + ⁻2	**26.** ⁻5 +⁻9	**27.** ⁻8 +⁻6

For more practice: **Skill Bank,** page 438.

Adding Numbers with Different Signs

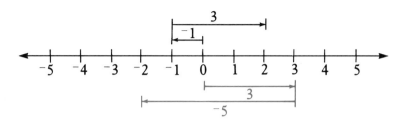

MODEL

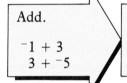

Add.	Look at the number line.	
$^-1 + 3$ $3 + {}^-5$	The black arrows show $^-1 + 3$. The red arrows show $3 + {}^-5$.	$^-1 + 3 = 2$ $3 + {}^-5 = {}^-2$

To add a positive number, move to the right.

To add a negative number, move to the left.

> **GET READY. Add. Use the number line.**

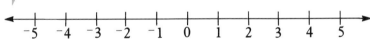

1. $4 + {}^-3$ **2.** $5 + {}^-1$ **3.** $3 + {}^-2$

4. $3 + {}^-4$ **5.** $1 + {}^-5$ **6.** $2 + {}^-3$

> **NOW USE SKILL 77. Use a number line if you wish.**

7. $7 + {}^-3$ **8.** $8 + {}^-6$ **9.** $9 + {}^-5$

10. $6 + {}^-6$ **11.** $4 + {}^-4$ **12.** $^-2 + 2$

13. $^-7 + 3$ **14.** $^-5 + 8$ **15.** $^-9 + 4$

16. $10 + {}^-5$ **17.** $12 + {}^-7$ **18.** $^-15 + 8$

19. $12 + {}^-4$ **20.** $7 + {}^-14$ **21.** $^-17 + 7$

22. $18 + {}^-9$ **23.** $13 + {}^-6$ **24.** $7 + {}^-16$

25. $^-6 + 4$ **26.** $^-9 + 5$ **27.** $8 + {}^-2$

For more practice: **Skill Bank,** page 438.

Using Addition

Positive and negative numbers can be used to represent many situations. For example, positive numbers can be used for gains, temperatures above zero, or elevations above sea level. Negative numbers can be used for losses, temperatures below zero, or elevations below sea level.

An airplane was flying at 30,000 feet. It descended 8000 feet, and then climbed 14,000 feet. What was the final altitude?

Write a sum of positive and negative numbers to represent the information. Then add to solve the problem.

MODEL

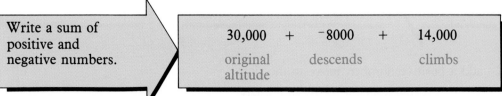

Write a sum of positive and negative numbers.	30,000 + ⁻8000 + 14,000
	original altitude descends climbs

Add.	Solve the problem.
30,000 + ⁻8000 + 14,000 = 36,000	The final altitude was 36,000 feet.

 GET READY.

1. 6 + ⁻10   2. 3 + ⁻3   3. ⁻5 + ⁻3

4. 9 + ⁻3   5. ⁻15 + ⁻10   6. 11 + ⁻16

7. 9 + ⁻7 + ⁻16   8. 3 + ⁻8 + 2   9. 100 + ⁻55 + 20

Write a positive or negative number.

10. 200 ft below sea level   11. a profit of $8000

12. 15° below zero   13. a loss of 6 yards

NOW USE APPLICATION A35. Write a sum of positive or negative numbers to represent each problem. Then add to solve the problem.

14. At 4:00 A.M. the temperature was ⁻16°C. By 2:00 P.M. the temperature had risen 12°. What was the temperature then?

15. The Dead Sea is about 1300 feet below sea level. A balloon starts from the shore and ascends 1500 feet. What is the altitude of the balloon?

16. A submarine cruising at a depth of 150 m rises 60 m. What is the depth of the submarine?

17. Kimo's checking account has a balance of $238. He writes checks for $25 and $42, and makes a deposit of $36. What is his new balance?

18. The ball starts 4 yards behind the first down marker. On the next two plays there is a gain of 8 yards followed by a loss of 6 yards. Where is the ball after these plays?

19. You start at an elevation of 7400 feet. After hiking to the top of the pass you have gained 2200 feet in elevation. Then you descend to the valley, dropping down 3600 feet in elevation. What is the elevation of the valley?

20. Maria and Bernie push a car 75 feet up a hill. They let go and the car rolls 34 feet back down the hill before it can be stopped. Then they push it 30 feet more up the hill. How far is the car from the starting point when they finish?

21. Garcia and Granoff, Inc. opened for business in 1979. They made a profit of $50,000 in 1979, a loss of $35,000 in 1980, a loss of $125,000 in 1981, and a profit of $85,000 in 1982. What is their net profit or loss for the four years?

For more practice: **Application Bank,** page 462.

Tide's In

You are a shipping agent. You are planning to load a ship in West Harbor on August 7. You want to know what the depth of the water will be during the time from the first high tide to the next low tide so you can plan your work carefully. Average water depth is 12 ft at your pier at low tide. You have learned that high and low tides occur about 6 hours apart. You consult a tide manual, where you find a cross-reference listing.

	Time of High Tide		Rise above low tide (ft)
Hingham	Same as	Boston	9.5
Lynn	10 min after	''	9.2
Richmond Is.	15 min before	''	8.9
Tucker	40 min after	''	2.6
West Harbor	45 min after	''	6.0
York	10 min before	''	7.4

PROBLEM: How deep is the water at your pier during the time involved?

Solution Plan: Find the time of the first high tide. Figure the water depth at one-hour intervals between high tide and the following low tide.

WHAT DO YOU KNOW? You know certain facts about your pier and the harbor.

1. What is the average depth at low tide?

2. How much does the water rise between low and high tide?

3. Suppose high tide was at 1:30 in Boston. When would high tide be in West Harbor?

WHAT DO YOU WANT TO KNOW? You want to know when high and low tides occur. You look up the Boston tide table.

August											
Day		1	2	3	4	5	6	7	8	9	10
High Tide	a.m.	2:15	3:09	4:06	5:07	6:11	7:15	8:17	9:14	10:04	10:52
	p.m.	2:46	3:41	4:38	5:36	6:38	7:38	8:36	9:30	10:18	11:04

4. When is the first high tide in Boston on August 7?

5. What time is the first high tide in West Harbor?

6. About when is the next low tide after this in West Harbor?

7. What is the depth of the water at high tide at your pier in West Harbor?

NOW SOLVE THE PROBLEM. Find the water depth at your pier.

There is a "rule of thumb" widely used in navigation. The rule tells how much of the total rise or fall of the tide will occur during each hour of the 6 hour tide cycle.

According to this rule, the water will fall $\frac{1}{12}$ of the total high tide rise in the first hour after high tide.

$$\frac{1}{12} \times 6\,\text{ft} = \frac{1}{2}\,\text{ft}$$

total rise   fall in 1st hour

You figure that during the first hour after high tide the water level will drop $\frac{1}{2}$ ft.

Complete the table.   **8.** **9.** **10.** **11.** **12.**

Hour after high tide		1st	2nd	3rd	4th	5th	6th
Part of fall		$\frac{1}{12}$	$\frac{2}{12}$	$\frac{3}{12}$	$\frac{3}{12}$	$\frac{2}{12}$	$\frac{1}{12}$
Decrease in depth (ft)		$\frac{1}{2}$	?	?	?	?	?
Water depth (ft)	18	$17\frac{1}{2}$	?	?	?	?	?

13. For safety, you want to load only while the water is at least 15 ft deep. About when will the water reach this depth?

Quiz

Skill 74

Write a positive or negative number.

1. 1 mile high

2. 200 feet deep

Skill 75

Write > or < in place of __?__.

3. ⁻3 __?__ ⁻1

4. ⁻5 __?__ 2

Skill 76

5. 8 + 2

6. ⁻9 + ⁻15

Skill 77

7. ⁻6 + 15

8. ⁻17 + 11

Application A35 Write a sum of positive or negative numbers to represent the problem. Then solve the problem.

9. You have $370 in a savings account. You withdraw $90, and then deposit $50. How much do you have in your account?

Problem Solving

10. Water depth at low tide is 8 ft. There is a total rise of 12 feet between low tide and high tide. Of the total rise, $\frac{1}{4}$ takes place in the first two hours after low tide. Low tide occurs at 6:30 P.M. What is the water depth at 8:30 P.M.?

CALCULATOR DISPLAYS

Some calculators have a key that can change the sign of a number. The "+/−" key will change a number from positive to negative or from negative to positive.

Here is how you can use a calculator to add 34 + ⁻62.

KEY-IN	DISPLAY	
34	3 4	
⊞	3 4	
62	6 2	
⊻	- 6 2	The +/− key changes 62 to ⁻62.
⊟	- 2 8	The negative sign may appear on either the left or the right.

Use a calculator with a +/− key, if possible.

1. 38 + ⁻75

2. ⁻300 + 126

3. ⁻250 + ⁻780

4. 96 + ⁻18

5. ⁻1281 + 847

6. ⁻3889 + 5280

Graduated Income Tax

You are required by law to pay federal income tax and, in many states, state income tax. The federal income tax and some state income taxes are *graduated* taxes. That is, the tax rate depends on your income. The higher your income, the higher your tax rate is. You pay taxes on your *taxable income*. This is the amount of your income that is left after you have subtracted all deductions.

Claire Gerhardt earned $18,750 last year at the *Daily News*. Her deductions total $2460. What is Claire's taxable income? How much state income tax will Claire pay?

First, find Claire's taxable income.

$18,750	Gross income
− 2,460	Deductions
$16,290	Taxable income

Use the table to figure Claire's tax.

Tax Rate Table Your tax is	
1% of the first $2000	1% of $2000: $20.00
2% of the next $2000	2% of $2000: $40.00
3% of the next $2000	3% of $2000: $60.00
4% of the next $2000	4% of $2000: $80.00
5% of the next $2000	5% of $2000: $100.00
6% of amount over $10,000	6% of $6290: $377.40
	Total tax $677.40

Claire must pay $677.40 in state income tax.

Use the tax rate table to find the income tax.

	1.	**2.**	**3.**	**4.**
Gross income	$16,850	$11,500	$25,281	$9,830
Deductions	$ 3,370	$ 3,741	$ 6,566	$1,420
Taxable income	?	?	?	?
Income tax	?	?	?	?

5. Roberto earned $23,310 last year. He had $5815 in deductions.
   a. Find his taxable income.
   b. Find the income tax he must pay.
   c. How much would Roberto's tax be if he could pay only 3% on his entire taxable income?

Subtracting a Positive Number

To subtract a number, add its opposite.

MODEL

Subtract.	To subtract, add the opposite.	
$4 - 3$	$4 - 3 \rightarrow 4 + {}^-3 = 1$	$4 - 3 = 1$
$1 - 2$	$1 - 2 \rightarrow 1 + {}^-2 = {}^-1$	$1 - 2 = {}^-1$
${}^-2 - 5$	${}^-2 - 5 \rightarrow {}^-2 + {}^-5 = {}^-7$	${}^-2 - 5 = {}^-7$

 GET READY. Write the opposite.

1. 4   **2.** 1   **3.** 6   **4.** 9

5. 13   **6.** 8   **7.** 17   **8.** 36

Complete the statement.

9. $4 - 6 = 4 + \underline{?}$

10. ${}^-9 - 4 = {}^-9 + \underline{?}$

11. $8 - 3 = 8 + {}^-3 = \underline{?}$

12. ${}^-8 - 3 = {}^-8 + {}^-3 = \underline{?}$

13. ${}^-3 - 8 = {}^-3 + {}^-8 = \underline{?}$

14. ${}^-6 - 1 = {}^-6 + {}^-1 = \underline{?}$

15. $27 - 9 = 27 + {}^-9 = \underline{?}$

16. $5 - 12 = 5 + {}^-12 = \underline{?}$

17. $3 - 24 = 3 + {}^-24 = \underline{?}$

18. $34 - 9 = 34 + {}^-9 = \underline{?}$

 NOW USE SKILL 78.

19. $3 - 9$ **20.** ${}^-5 - 8$ **21.** $9 - 11$

22. $7 - 2$ **23.** ${}^-12 - 4$ **24.** $5 - 8$

25. $6 - 9$ **26.** ${}^-4 - 7$ **27.** ${}^-13 - 7$

28. $10 - 7$ **29.** ${}^-5 - 8$ **30.** $11 - 13$

31. $4 - 12$ **32.** $15 - 20$ **33.** ${}^-3 - 6$

34. ${}^-9 - 3$ **35.** ${}^-2 - 8$ **36.** ${}^-1 - 7$

37. $3 - 7$ **38.** $6 - 2$ **39.** ${}^-6 - 2$

40. $13 - 29$ **41.** ${}^-21 - 9$ **42.** $100 - 250$

For more practice: **Skill Bank**, page 439.

Subtracting a Negative Number

To subtract a number, add its opposite.

MODEL

Subtract.	To subtract, add the opposite.	
$4 - {}^-3$	$4 - {}^-3 \rightarrow 4 + 3 = 7$	$4 - {}^-3 = 7$
${}^-2 - {}^-5$	${}^-2 - {}^-5 \rightarrow {}^-2 + 5 = 3$	${}^-2 - {}^-5 = 3$
${}^-4 - {}^-3$	${}^-4 - {}^-3 \rightarrow {}^-4 + 3 = {}^-1$	${}^-4 - {}^-3 = {}^-1$

 GET READY. Write the opposite.

1. ${}^-2$ **2.** ${}^-6$ **3.** ${}^-5$ **4.** ${}^-10$

5. ${}^-15$ **6.** ${}^-20$ **7.** ${}^-36$ **8.** ${}^-12$

Complete the statement.

9. $5 - {}^-3 = 5 + \underline{?}$

10. ${}^-7 - {}^-2 = {}^-7 + \underline{?}$

11. $8 - {}^-3 = 8 + 3 = \underline{?}$

12. ${}^-8 - {}^-3 = {}^-8 + 3 = \underline{?}$

13. $3 - {}^-8 = 3 + 8 = \underline{?}$

14. ${}^-3 - {}^-8 = {}^-3 + 8 = \underline{?}$

15. ${}^-6 - {}^-3 = {}^-6 + 3 = \underline{?}$

16. ${}^-10 - {}^-8 = {}^-10 + 8 = \underline{?}$

 NOW USE SKILL 79.

17. $8 - {}^-6$ **18.** ${}^-9 - {}^-4$ **19.** $7 - {}^-8$

20. ${}^-7 - 10$ **21.** $8 - {}^-11$ **22.** $13 - 4$

23. $7 - {}^-3$ **24.** ${}^-4 - {}^-12$ **25.** $3 - {}^-5$

26. ${}^-6 - {}^-9$ **27.** ${}^-9 - {}^-3$ **28.** ${}^-4 - {}^-10$

29. $10 - {}^-7$ **30.** $4 - {}^-8$ **31.** ${}^-12 - {}^-4$

32. $16 - 5$ **33.** ${}^-11 - {}^-4$ **34.** ${}^-11 - 4$

35. ${}^-4 - 11$ **36.** ${}^-10 - {}^-5$ **37.** ${}^-4 - {}^-11$

38. $15 - 8$ **39.** ${}^-15 - {}^-8$ **40.** ${}^-3 - {}^-5$

41. $23 - 30$ **42.** ${}^-36 - 19$ **43.** ${}^-28 - {}^-32$

For more practice: **Skill Bank,** page 439.

Multiplying and Dividing—Same Sign

Multiplication: The product of two numbers with the same sign is positive.

MODEL

Multiply.

3×4
$^-2 \times ^-5$

Both numbers have the same sign.
The product is positive.

$3 \times 4 = 12$
$^-2 \times ^-5 = 10$

Division: The quotient of two numbers with the same sign is positive.

MODEL

Divide.

$12 \div 3$
$^-8 \div ^-4$

Both numbers have the same sign.
The quotient is positive.

$12 \div 3 = 4$
$^-8 \div ^-4 = 2$

 GET READY.

1. 7×9
2. 6×8
3. 5×12
4. $16 \div 4$
5. $42 \div 6$
6. $48 \div 12$

NOW USE SKILL 80.

7. 8×5
8. $^-8 \times ^-5$
9. $^-5 \times ^-8$
10. $^-6 \times ^-7$
11. 6×7
12. 5×6
13. $^-7 \times ^-8$
14. $^-5 \times ^-9$
15. $^-10 \times ^-2$
16. $^-12 \times ^-16$
17. $^-13 \times ^-7$
18. $^-9 \times ^-15$

19. $24 \div 3$
20. $^-24 \div ^-3$
21. $^-32 \div ^-8$
22. $50 \div 5$
23. $^-56 \div ^-8$
24. $^-63 \div ^-9$
25. $^-16 \div ^-8$
26. $^-40 \div ^-5$
27. $^-64 \div ^-8$
28. $^-96 \div ^-6$
29. $^-144 \div ^-9$
30. $^-108 \div ^-18$

For more practice: **Skill Bank,** page 440.

Multiplying and Dividing—Different Signs

Multiplication: The product of two numbers with different signs is negative.

MODEL

| Multiply. $3 \times {}^-4$ $^-2 \times 5$ | The numbers have different signs. The product is negative. | $3 \times {}^-4 = {}^-12$ $^-2 \times 5 = {}^-10$ |

Division: The quotient of two numbers with different signs is negative.

MODEL

| Divide. $14 \div {}^-2$ $^-18 \div 3$ | The numbers have different signs. The quotient is negative. | $14 \div {}^-2 = {}^-7$ $^-18 \div 3 = {}^-6$ |

 NOW USE SKILL 81.

1. $6 \times {}^-2$ **2.** $^-6 \times 2$ **3.** $4 \times {}^-3$

4. $3 \times {}^-4$ **5.** $12 \times {}^-2$ **6.** $^-12 \times 2$

7. $7 \times {}^-3$ **8.** $3 \times {}^-7$ **9.** $^-9 \times 4$

10. $^-4 \times 9$ **11.** $7 \times {}^-8$ **12.** $^-4 \times 7$

13. $6 \times {}^-8$ **14.** $8 \times {}^-6$ **15.** $9 \times {}^-7$

16. $^-20 \times {}^-6$ **17.** $15 \times {}^-7$ **18.** $^-12 \times 11$

19. $7 \times {}^-23$ **20.** $^-13 \times {}^-6$ **21.** $8 \times {}^-35$

22. $16 \div {}^-2$ **23.** $^-16 \div 2$ **24.** $56 \div {}^-8$

25. $^-63 \div 9$ **26.** $^-48 \div 6$ **27.** $^-100 \div 10$

28. $^-15 \div 15$ **29.** $36 \div {}^-4$ **30.** $49 \div {}^-7$

31. $^-120 \div 8$ **32.** $96 \div {}^-8$ **33.** $^-364 \div 14$

34. $^-250 \div {}^-25$ **35.** $^-75 \div 3$ **36.** $^-680 \div {}^-20$

For more practice: **Skill Bank,** page 440.

Using Subtraction

Positive and negative numbers can be used to represent the change in a quantity. If the change is positive it indicates an increase. If the change is negative it indicates a decrease.

To find the change in a quantity, use the formula:

New value − Original value = Change

At noon the temperature was ⁻2°C. At midnight the temperature was ⁻15°C. How much did the temperature change? Did it increase or decrease?

Write a difference to represent the situation. Use the sign of the change to determine whether there is an increase or a decrease.

MODEL

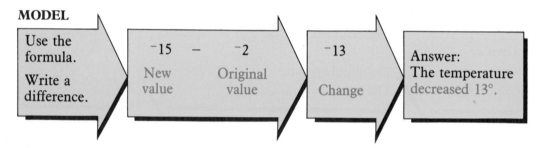

Use the formula. Write a difference.

⁻15 − ⁻2
New value   Original value

⁻13
Change

Answer: The temperature decreased 13°.

 GET READY.

1. 4 − ⁻3 **2.** ⁻10 − 5 **3.** ⁻8 − ⁻6 **4.** 7 − 12

5. ⁻16 − ⁻45 **6.** 36 − 28 **7.** 212 − ⁻116 **8.** ⁻37 − ⁻25

Write a positive or negative number.

9. 5 inches above normal

10. 3 degrees below average

11. a loss of $5,000,000

12. 2600 feet above sea level

13. a gain of 5 yards

14. a descent of 300 feet

 NOW USE APPLICATION A36. Write a difference of positive or negative numbers to represent the situation. Then subtract to solve the problem. State whether the change is an increase or a decrease.

15. In 1980 Comox Corp. had a loss of $3 million. In 1982 Comox had a profit of $12 million. What was the change in profitability?

16. A play began 3 yards behind the first down marker. The play ended 11 yards behind the first down marker. How many yards were gained or lost?

17. The temperature in a laboratory vessel went from the freezing point of ethanol, at 175°F below zero, to the freezing point of water, at 32°F. What was the change in the temperature?

18. A hiker walked from Pete's Park, at an elevation of 8762 feet, to Deep Valley, which is 534 feet below sea level. What was the change in elevation?

19. A submarine at a depth of 30 m dived to a depth of 73 m. What was the change in the sub's depth?

20. Two years ago the total rainfall was 11 inches below normal. Last year the rainfall was 4 inches below normal. How much more rain was there last year than the year before?

21. When you woke up, your temperature was four degrees above normal. When you checked at lunch, your temperature was one degree below normal. What was the change in your temperature?

22. Last year the average reading score at Central Elementary School was 14 points above the national average. This year the average score was 6 points below the national average. How much better or worse were the scores this year?

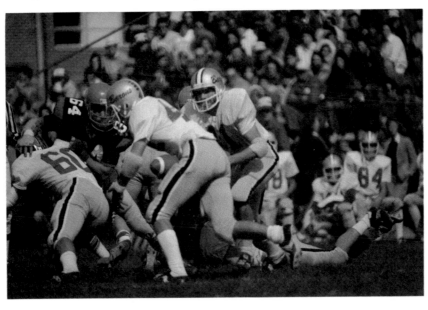

For more practice: **Application Bank,** page 462.

Time of Your Life

You are preparing a report for the Life Insurance Institute. You want to know how life expectancy (the average number of years a person is expected to live) has changed over the years. You also want to find the probability that a person who is born now will reach a certain age. You look up statistics to answer these questions.

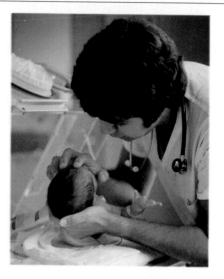

Life Expectancy at Birth (in years)

Year		Year	
1900	47.3	1960	69.7
1920	54.1	1980	73.8
1940	62.9		

PROBLEM: How has the average life expectancy changed? What is the probability of a person reaching a given age?

Solution Plan: Use statistics to answer the questions. Calculate the probabilities from the statistics you find.

WHAT DO YOU KNOW? You know the life expectancy over the years since 1900.

1. How much did life expectancy increase between 1900 and 1940?

2. How much did life expectancy increase between 1940 and 1980?

3. How much did life expectancy increase between 1900 and 1980?

Complete the table showing the increase in life expectancy over each 20-year period.

	4.	5.	6.	7.
Years	1900–1920	1920–1940	1940–1960	1960–1980
Increase	?	?	?	?

WHAT DO YOU WANT TO KNOW? You want to find the probability of reaching certain ages. You find more statistics and analyze them.

The table shows the number, out of 100,000 to begin with, who are living at the given age.

Age	Number	Age	Number
0	100,000	50	91,356
10	98,289	60	83,397
20	97,610	70	67,879
30	96,282	80	41,923
40	94,740		

The probability that a person born now will live to age 50 is 0.91, to the nearest hundredth.

$$\frac{\text{Number living at 50}}{\text{Total number}} = \frac{91,356}{100,000}$$

Application A25   0.91 ←

Complete the table.
Find the probability of reaching the given age.

	8.	**9.**	**10.**	**11.**	**12.**	**13.**	**14.**
Age	10	20	30	40	50	60	70	80
Probability	?	?	?	?	0.91	?	?	?

NOW SOLVE THE PROBLEM. Complete the report.

VITAL STATISTICS REPORT
PREPARED FOR LIFE INSURANCE INSTITUTE
SUMMARY OF FINDINGS

I. Life Expectancy

15.   20-year period with greatest increase: ?
16.   Total increase, 1900 to 1980: ?
17.   Percent increase from 1900 to 1980: ?

II. Age Probabilities

Probability is more than 0.9 for all
18.   ages up to and including age ?

Probability is more than 0.5 for all
19.   ages up to and including age ?

Probability that you will not reach
20.   age 40 is ?

Quiz

1. $3 - 7$ **2.** $^-6 - 4$ **3.** $^-9 - 4$ **4.** $^-8 - 9$

5. $3 - ^-7$ **6.** $^-4 - ^-3$ **7.** $^-6 - ^-12$ **8.** $5 - ^-2$

9. $^-9 \times ^-3$ **10.** $^-72 \div ^-8$ **11.** $7 \times ^-8$ **12.** $^-28 \div 4$

Write a difference of positive or negative numbers. Then solve the problem.

13. Last month your weight was 7 lb below average. This month your weight is 2 lb above average. How much did your weight change?

14. In 1900 life expectancy was 47.3 years. In 1960 life expectancy was 69.7 years. What was the percent change in life expectancy from 1900 to 1960, to the nearest whole percent?

CALCULATOR DISPLAYS

Timothy Frankland is buying a $450 scuba diving tank and regulator. The sales tax on the purchase is 6%. How much is the sales tax?

KEY-IN	DISPLAY
450	450.
✕	450.
6	6.
%	27.

On some calculators you may need to press the "=" key after the "%" key.

Find the sales tax to the nearest cent.

1. $39.95 toaster oven. Tax rate is 5%.

2. $225.50 set of tires. Tax rate is 4.5%.

3. $359.99 television. Tax rate is 4%.

Credit Counselor

Julia Watanabe is a counselor for a consumer credit counseling service. Julia gives advice to people who have debts they can't handle. She advises them on how to keep their use of credit within a reasonable limit.

The Merricks' monthly take-home pay is $1960. They currently have debts requiring monthly payments of $521. Julia recommends that they keep a 20% credit limit. This means that their monthly loan and credit payments should be no more than 20% of their take-home pay.

$$\text{Credit limit} = 20\% \text{ of take-home pay}$$

$$\text{Credit limit} = 0.2 \times \$1960 = \$392.$$

The Merricks are exceeding a 20% credit limit by a considerable amount. Julia advises them to reduce their spending by cutting their use of credit cards.

1. The Tremonts have monthly credit payments of $173. Their take-home pay is $1125. If they buy a car with monthly payments of $142, will they exceed their 20% limit?

2. Gerry Ashton has a take-home income of $1430. His car payments and other credit expenses total $429 a month.
   a. Is he within a 25% credit limit?
   b. If not, by how much must he reduce his monthly payments to reach a 25% limit?

Order of Operations

When there are more than two numbers to combine, follow this rule:

Step 1. Do all multiplications and divisions. Work in order from left to right.

Step 2. Then do all additions and subtractions. Work in order from left to right.

MODELS

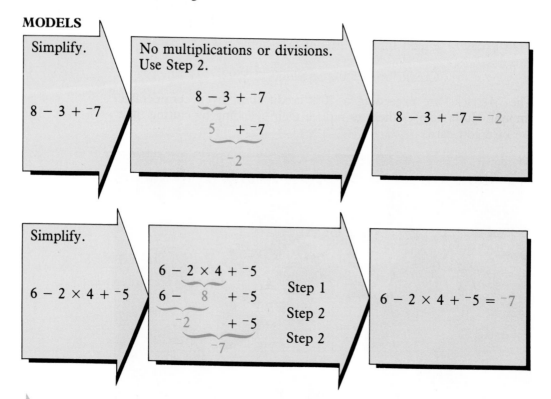

Simplify.

$8 - 3 + {}^-7$

No multiplications or divisions. Use Step 2.

$8 - 3 + {}^-7$

$5 \quad + {}^-7$

$^-2$

$8 - 3 + {}^-7 = {}^-2$

Simplify.

$6 - 2 \times 4 + {}^-5$

$6 - 2 \times 4 + {}^-5$

$6 - \quad 8 \quad + {}^-5$ Step 1

$^-2 \quad + {}^-5$ Step 2

$^-7$ Step 2

$6 - 2 \times 4 + {}^-5 = {}^-7$

NOW USE SKILL 82. Simplify.

1. $3 + 5 \times 2$

2. $6 \times 2 - 4$

3. $2 - 2 \times 2 + 4$

4. $^-3 \times {}^-3 + 5$

5. $5 + {}^-2 + 6$

6. $5 + {}^-2 \times 6$

7. $^-2 + 2 \times 7$

8. $^-3 \times 0 + 4$

9. $6 \div 2 \times 3$

10. $5 \times 2 + 3 \div 3$

11. $8 - 8 \div 2 + {}^-2$

12. $3 + 9 - 6 \div 2$

13. $4 + 4 \times 4 \div 4$

14. $4 + 4 \div 4 - 4$

15. $4 \times 4 + 4 \div 4$

For more practice: **Skill Bank,** page 441.

Grouping Symbols

Parentheses
↓　　↘
$2 \times (6 + 4)$

Brackets
↙　　　↘
$[2 \times (6 + 4)] - 3$

First work inside parentheses, then inside brackets.

MODEL

Simplify.

$[2 \times (6 + 4)] - 3$

$[2 \times (6 + 4)] - 3$

$[2 \times \underbrace{\quad 10 \quad}] - 3$

$\underbrace{20} \qquad -3$

$\underbrace{17}$

$[2 \times (6 + 4)] - 3 = 17$

 GET READY. Complete the statement.

1. $2 + (6 - 2) = 2 + \underline{}$

2. $(7 \times 2) - 4 = \underline{} - 4$

3. $5 \times (8 - 3) = 5 \times \underline{}$

4. $5 - (3 \times 2) = 5 - \underline{}$

5. $[2 + (6 - 2)] - 3 = [2 + \underline{}] - 3$

6. $[(9 - 4) - 3] \times 2 = [\underline{} - 3] \times 2$

 NOW USE SKILL 83. Simplify.

7. $9 + (4 - 3)$

8. $(10 - 2) \times 6$

9. $8 - (4 \times 2)$

10. $6 - (3 - 3)$

11. $(3 + 3) \div 2$

12. $12 \div (8 - 5)$

13. $(3 - 5) \times (2 + 6)$

14. $20 \div (4 - 9)$

15. $^-4 - (16 + 3)$

16. $7 \times [(16 \div 2) - 5]$

17. $[12 \times (9 - 4)] + 20$

18. $3 + [(4 + 1) \times 3]$

19. $[(6 - 5) \times 4] + 3$

20. $6 - [5 \times (4 + 3)]$

21. $6 - [(5 \times 4) + 3]$

22. $3 + (9 - {}^-4)$

23. $8 \times [(6 + {}^-5) - 1]$

24. $[({}^-2 \times {}^-2) + 3] \div 7$

25. $15 \div [(8 - 2) \div 2]$

26. $[({}^-10 + 1) \times 9] \div {}^-3$

27. $24 - [({}^-8 \times {}^-3) - 12]$

For more practice: **Skill Bank,** page 441.

Charts with Positive and Negative Numbers

Charts sometimes use positive and negative numbers to give information.

The world is divided into time zones. When communicating within the United States or to another part of the world, you often pass from one time zone into another.

You live in the Eastern Standard Time (EST) zone. At 10:00 P.M. (EST) you telephone your cousin in American Samoa. Use the chart to find what time it is in American Samoa when you call.

Location	Time Difference U.S. Time Zone			
	EST	CST	MST	PST
American Samoa	−6	−5	−4	−3
Singapore	$12\frac{1}{2}$	$13\frac{1}{2}$	$14\frac{1}{2}$	$15\frac{1}{2}$
Nigeria	6	7	8	9
Denver	−2	−1	0	1
Los Angeles	−3	−2	−1	0
New York	0	1	2	3
San Francisco	−3	−2	−1	0

MODEL

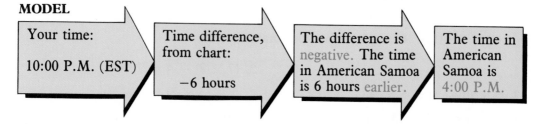

Your time: 10:00 P.M. (EST) → Time difference, from chart: −6 hours → The difference is negative. The time in American Samoa is 6 hours earlier. → The time in American Samoa is 4:00 P.M.

 NOW USE APPLICATION A37.

1. Aisha wants to talk to her father who is in Nigeria. Aisha is calling from within the Mountain Standard Time (MST) zone. She wants to call when it is 6:00 P.M. in Nigeria. What time (MST) should Aisha place her call?

2. You are flying from New York to Singapore. You leave New York at 9:30 A.M. (EST) and arrive in Singapore after traveling 24 hours. What time is it in Singapore when you arrive?

3. Barbara leaves New York at 10:00 A.M. (EST). She arrives at 1:00 P.M. (PST) in Los Angeles. How long did her flight take?

The wind chill factor measures the cooling effect of the wind. In cold weather, your body loses more heat when the wind is blowing than when it is calm. It feels colder than what the temperature shows. Suppose, for example, that the wind chill factor is ⁻15°F. This means that present conditions feel as cold as a temperature of ⁻15°F on a day with no wind.

Wind Speed (mph)	Thermometer Reading in Farenheit Degrees							
	50	40	30	20	10	0	−10	−20
	Equivalent Calm-air Temperature							
Calm	50	40	30	20	10	0	−10	−20
5	48	37	27	16	6	−5	−15	−26
10	40	28	16	4	−9	−24	−33	−46
15	36	22	9	−5	−18	−32	−45	−58
20	32	18	4	−10	−25	−39	−53	−67
25	30	16	0	−15	−29	−44	−59	−74
30	28	13	−2	−18	−33	−48	−63	−79
35	27	11	−4	−20	−35	−51	−67	−82
40	26	10	−6	−21	−37	−53	−69	−86

The temperature is 20°F and there is a 15 mile-per-hour wind. What is the wind chill factor?

MODEL

Read down the temperature column. Read across the wind speed row. The wind chill factor is −5°F.

 NOW USE APPLICATION A37.

1. Today's weather report indicates a high temperature of 40°F. There will be no wind. What will the wind chill factor be?

2. On New Year's Day the temperature was −10°F, and there was a 20-mile-per-hour wind. What was the wind chill factor?

3. On a day when the wind was 25 miles per hour the temperature dropped from 30°F to 10°F. How much colder did it feel after the change in temperature?

For more practice: **Application Bank,** page 462.

Moving Experience

You are moving from Milwaukee to Omaha. It is 520 miles from your old residence to the new one. You plan to rent a 20-foot truck and to keep it four days. You want to know how much your moving expenses will be.

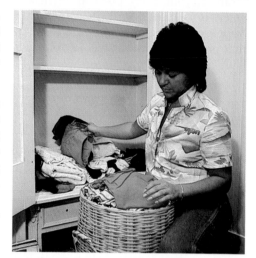

There are two rental rates. You can rent on a local basis, driving from Milwaukee to Omaha and then back, returning the truck in Milwaukee. Or you can rent one-way, driving from Milwaukee to Omaha and leaving the truck in Omaha. If you rent one-way, however, you will fly back to Milwaukee, since you have further business to take care of in Milwaukee before you leave for good. The air fare is $55.

You get the rental rates from the truck rental company.

     Local: $45 a day plus 18¢ a mile.
One-way: $317 for 4 days, including 500 miles.
                 25¢ extra for each mile over 500.

PROBLEM: Which rate will cost less, local or one-way? What will the total cost be?

Solution Plan: Figure all costs, including mileage charges, gas, and plane fare. Add in any extras you'll need for packing and moving.

WHAT DO YOU KNOW?

1. How far will you drive if you rent one-way?

2. How far will you drive if you rent local?

3. What is the basic charge for 4 days on the one-way rental?

4. What is the basic charge for 4 days on the local rental?

WHAT DO YOU WANT TO KNOW?
You want to know the mileage charges.

5. How much extra will you pay for mileage on the one-way rate?

6. How much will you pay for mileage on the local rate?

WHAT MORE DO YOU WANT TO KNOW? You need to include the cost of gas, air fare back to Milwaukee, and extras.

7. The truck averages 8 miles per gallon. Gas sells for about $1.60 a gallon. What is the cost per mile for gas?

8. If you choose the one-way rental, how much will gas cost?

9. If you choose the local rental, how much will gas cost?

Pads rent for $8 a dozen, and hand trucks for $7. You decide to get 24 pads and a hand truck.

10. How much will extras (pads and hand truck) cost?

NOW SOLVE THE PROBLEM.

11. Find the total cost using the one-way basis. Include the basic charge, mileage, gas, air fare, and extras.

12. Find the total cost using the local basis. Include the basic charge, mileage, gas, and extras.

13. Which rate gives you the lower total cost?

14. How much money can you save by using the less expensive method?

Skill 82 **Simplify.**

1. $1 + 5 \times 2$ **2.** $3 \times 6 - 4$ **3.** $8 - 2 \times 5$

Skill 83

4. $12 \div (6 - 2)$ **5.** $(9 - 4) \times 5$ **6.** $[5 - (3 \times 2)] \times 2$

Application A37

7. It's 8:00 A.M. EST in Atlanta. What time is it in Tahiti?

8. It's 5:00 P.M. CST in Dallas. What time is it in Taiwan?

Time Difference U.S. Time Zone				
Country	EST	CST	MST	PST
Tahiti	−5	−4	−3	−2
Taiwan	13	14	15	16

Problem Solving

9. You are moving 575 miles. You will need a truck for 5 days. The local rental rate is $50 a day plus 20¢ a mile. The one-way rate is $350 for 5 days including 500 miles, plus 25¢ extra for each mile over 500. Which rate will cost less?

COMPUTER BITS

A department store has a clearance sale on several items. The chart below lists each item, the original price, and the percent discount.

The sale prices will be figured by a computer.

ITEM	PRICE	% DISCOUNT
Shirt	18.50	20%
Pants	26.00	30%
Shoes	36.50	25%
Hat	9.25	15%
Dress	32.75	20%

Flowcharting shows the steps in the solution of a problem. The flowchart shows the process of reading an item, its price, and the percent discount, and then computing the sale price. The word STOP tells the computer that there is no more data for processing.

ACTIVITY Find the sale price for each item.

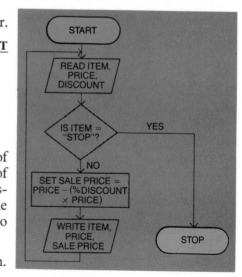

Skills Review

Skill 74 **Write as a positive or negative number.** (page 302)

1. $12 loss

2. 6 under par

3. 10 extra chairs

4. 8 m higher

5. 109 points won

6. 5 days behind

Skill 75 **Write > or < in place of ?.** (page 303)

7. 5 $\underline{?}$ 9

8. $^-3$ $\underline{?}$ $^-8$

9. $^-7$ $\underline{?}$ 7

10. $^-12$ $\underline{?}$ $^-11$

Write the opposite.

11. 4

12. $^-15$

13. $^-26$

14. 74

Skill 76 (page 304)

15. 6 + 8

16. $^-5 + {}^-9$

17. $^-13 + {}^-4$

18. 4 + 3

19. $^-17 + {}^-2$

20. $^-42 + {}^-15$

Skill 77 (page 305)

21. 5 + $^-2$

22. $^-12 + 7$

23. 8 + $^-6$

24. $^-18 + 7$

25. $^-6 + 16$

26. 3 + $^-28$

Skill 78 (page 312)

27. 9 − 2

28. 5 − 13

29. 11 − 14

30. $^-14 − 3$

31. $^-10 − 6$

32. $^-18 − 9$

Skill 79 (page 313)

33. $^-9 − {}^-12$

34. 4 − $^-6$

35. $^-7 − {}^-3$

36. 27 − $^-15$

37. $^-84 − {}^-86$

38. $^-62 − {}^-10$

Skill 80 (page 314)   Skill 81 (page 315)

39. 8 × 6

40. $^-5 × {}^-2$

43. $^-3 × 20$

44. 7 × $^-8$

41. 16 ÷ 8

42. $^-45 ÷ {}^-9$

45. 28 ÷ $^-4$

46. $^-64 ÷ 8$

Skill 82 **Simplify.** (page 322)

47. 4 + 7 × 8

48. 6 × 3 − 9

49. 10 + 10 ÷ 5

Skill 83 **Simplify.** (page 323)

50. (8 − 11) × $^-8$

51. [(28 ÷ 7) + 6] ÷ 5

52. 15 ÷ (3 − 8)

Application A35 **Write a sum of positive or negative numbers to represent each problem. Then add to solve the problem.** (pages 306–307)

1. The normal level of water in a reservoir is 30 feet. During a heavy rain the level rises 15 feet. Then it drops 10 feet during a drought. What is the final level of water in the reservoir?

2. A building is 34 stories high. You get on an elevator that is not working properly. It goes up to the 20th floor, then down 7 floors, up 2 floors, up 7 floors, down 6 floors, and up 1 floor. On which floor did the elevator end up?

Application A36 **Write a difference of positive or negative numbers to represent the situation. Then subtract to solve the problem. State whether the change is an increase or a decrease.** (pages 316–317)

3. An airplane tracks a submarine across the Atlantic. The airplane is flying at an altitude of 300 feet. The submarine is cruising at a depth of 100 feet. A monitor is dropped from the plane to the submarine. How far does the monitor travel?

4. Last year Hopestill Inc. had a profit of $9 million. This year Hopestill predicts it will lose $2 million. What is the predicted difference in Hopestill's earnings?

5. The original price of a truck is $700 above dealer's cost. Later it is sold at $1000 below dealer's cost. How much was the truck marked down from the original price?

Application A37
(pages 324–325)

6. What is the wind chill factor when the wind is 20 mph and the thermometer reads 10°F?

7. The wind chill factor is ⁻45°F. The wind speed is 15 mph. What is the thermometer reading?

Wind Speed (mph)	Actual Thermometer Reading					
	20	10	0	−10	−20	−30
	Wind Chill Factor					
Calm	20	10	0	−10	−20	−30
5	16	6	−5	−15	−26	−36
10	4	−9	−21	−33	−46	−58
15	−5	−18	−36	−45	−58	−72
20	−10	−25	−39	−53	−67	−82
25	−15	−29	−44	−59	−74	−88
30	−18	−33	−48	−63	−79	−94
35	−20	−35	−49	−67	−82	−98
40	−21	−37	−53	−69	−85	−100

8. The thermometer reads 10°F. The wind increases from 15 mph to 25 mph. What is the change in the wind chill factor?

Unit Test

Skill 74

Write a positive or negative number.

1. a profit of $570

2. 25 ft below the surface

Skill 75   Skill 76   Skill 77

Write > or < in place of __?__.

5. $^-6 + ^-11$ **7.** $^-19 + 13$

3. $^-4$ _?_ $^-9$ **4.** $^-3$ _?_ 1 **6.** $^-15 + ^-25$ **8.** $^-12 + 20$

Skill 78   Skill 79

9. $^-5 - 11$ **10.** $7 - 14$ **11.** $^-7 - ^-10$ **12.** $2 - ^-8$

Skill 80   Skill 81

13. $^-6 \times ^-9$ **14.** $^-48 \div ^-4$ **15.** $8 \times ^-13$ **16.** $^-98 \div 7$

Skill 82   Skill 83

Simplify.   **Simplify.**

17. $9 + 3 \times 7 - 2$ **18.** $^-8 \div (4 + ^-2)$ **19.** $(3 - 8) \times ^-6$

Application A35

Write a sum of positive or negative numbers. Then solve the problem.

20. An airplane at an altitude of 22,000 ft descended 3000 ft. What was the final altitude?

Application A36

Write a difference of positive or negative numbers. Then solve the problem.

21. At noon the temperature was 8°F. At 10:00 P.M. the temperature was $^-5$°F. What was the change in temperature?

Application A37

22. In Yaktaw the time is two hours earlier than here. In Corwick the time is three hours later than here. When it's 6:00 P.M. in Corwick, what time is it in Yaktaw?

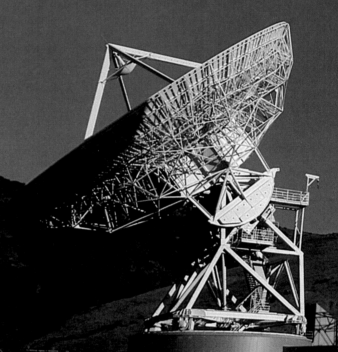

11

Variable Expressions

Translating Words into Symbols

Problem Solving

You will use these skills and applications in this unit.
Which do you already know? Work each problem.

Skill 84 State the meaning of the variable expression.

1. $c + 3$ **2.** $x - 7$

Skill 85 State the meaning of the expression.

3. $5b$ **4.** $4 \cdot 11$

Skill 86 Find the value of the expression. Use $x = 2$, $y = 4$.

5. $x - 2$ **6.** $4x + 7y$

Skill 87 Simplify.

7. $9n + 0$ **8.** $\dfrac{3f}{f}$

Skill 88 Simplify by combining like terms.

9. $4c - 6c$ **10.** $3n - 4 + 2n$

Skill 89 Write with exponents.

11. $3 \times 3 \times 3$ **12.** $(r \cdot s)(r \cdot r)$

Skill 90 Find the value of the expression. Use $a = 3$, $b = 2$.

13. $a^2 - b^2$ **14.** $5b^2$

Skill 91 Find the solution.

15. $3n = 6$ **16.** $a + 8 = 3$

Skill 92 Check these values for x: $^-1$, 0, 1, 2, 3. Which are solutions?

17. $x > 0$ **18.** $x < 1$ **19.** $x - 2 > 0$

Application A38 Complete the statement with a variable expression.

20. The cost of 3 shirts at x dollars each is __?__ dollars.

21. Monique is 16 years old; n years from now she will be __?__ years old.

Application A39 Write a formula for the perimeter.

22.

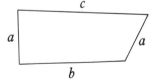

Application A40 Select the equation that fits the problem.

23. A tennis racket costs $50. The cost of the racket and a pair of shoes is $80.

a. $50 = s - 80$ **b.** $50 + s = 80$

Check your answers. If you had difficulty with any skill or application be sure to study the corresponding lesson in this unit.

Variable Expressions

A letter can sometimes stand for a number. That letter is called a **variable.** If the letter x stands for 7, then the value of x is 7. An expression containing a variable is called a **variable expression.**

MODEL

State the meaning of each expression.	Meaning:
1. $n + 1$	**1.** Add 1 to the value of n.
2. $5 \times V$	**2.** Multiply the value of V by 5.
3. $\dfrac{9}{y}$	**3.** Divide 9 by the value of y.
4. $2 \times (w - 6)$	**4.** Subtract 6 from the value of w. Then multiply the result by 2.
5. $a + b$	**5.** Add the value of b to the value of a.

 GET READY. Complete the statement.

1. $\dfrac{y + 1}{3}$

Add __?__ to the value of __?__.
Then divide by __?__.

2. $2 \times (m + 4)$

Add __?__ to the value of __?__.
Then multiply the result by __?__.

 NOW USE SKILL 84. State the meaning of the variable expression.

3. $16 - a$

4. $16 \times y$

5. $2 + w$

6. $\dfrac{b}{8}$

7. $z - 4$

8. $b \times c$

9. $\dfrac{12}{s}$

10. $n + r$

11. $5 \times (d + 4)$

12. $x + (3 \times y)$

13. $\dfrac{m + f}{4}$

14. $\dfrac{g - 18}{r}$

15. $18 - (w + 6)$

16. $\dfrac{i + j}{j}$

17. $\left(\dfrac{k + 9}{7}\right) \times 2$

For more practice: **Skill Bank,** page 442.

Multiplication Symbols

Different ways to write the product of a and b:

$$a \times b \quad a \cdot b \quad ab \quad a(b) \quad (a)b \quad (a)(b)$$
$$\uparrow \qquad \uparrow \qquad \uparrow \qquad \uparrow \qquad \uparrow \qquad \uparrow$$
times   raised   no   parentheses
sign   dot   sign

You can eliminate the raised dot and the parentheses only with a variable. You can write $3 \cdot a = 3a$. You cannot write $3 \cdot 4 = 34$.

MODEL

State the meaning of each expression.	Meaning:
1. $4 \cdot 5$	**1.** Multiply 5 by 4.
2. $7a$	**2.** Multiply the value of a by 7.
3. $9(x)$	**3.** Multiply the value of x by 9.
4. $(b)5$	**4.** Multiply 5 by the value of b.
5. $(m)(c)$	**5.** Multiply the value of c by the value of m.
6. $3(a + b)$	**6.** Add the value of b to the value of a. Then multiply the result by 3.

 GET READY. Complete the statement.

1. $4y$
Multiply the value of __?__ by 4.

2. bz
Multiply the value of __?__ by the value of __?__.

3. $m + 2d$
Multiply the value of __?__ by 2. Then add the result to the value of __?__.

4. $6(m - n)$
Subtract the value of __?__ from the value of __?__. Then multiply the result by __?__.

 NOW USE SKILL 85. State the meaning of the expression.

5. $9n$

6. $3 \cdot 8$

7. mn

8. $(10x)y$

9. $4(x + 1)$

10. $11(x - y)$

11. $(a + b)3$

12. $(11m)c$

13. $16(2m - 3)$

For more practice: **Skill Bank,** page 442.

Evaluating Variable Expressions

You can find the *value* of a variable expression when you know the
values of the variables in the expression.

MODEL

Find the value of the expression.

Expression	Values of variables
$5a$	$a = 2$
$x + 2$	$x = 3$
$a + b$	$a = 1, b = 8$
$\frac{x}{z}$	$x = 12, z = 2$

Substitute.

$5a \longrightarrow 5 \cdot 2$
$x + 2 \longrightarrow 3 + 2$
$a + b \longrightarrow 1 + 8$
$\frac{x}{z} \longrightarrow \frac{12}{2}$

$5a = 10$
$x + 2 = 5$
$a + b = 9$
$\frac{x}{z} = 6$

 GET READY.

1. $8 - 10$ **2.** $4 \cdot 16$ **3.** $\frac{27}{9}$

4. $^-2 + {}^-9$ **5.** $7 \cdot {}^-7$ **6.** $3 + {}^-3$

 **NOW USE SKILL 86. Find the value of the expression.
Use $x = 3$, $y = 2$.**

7. $x + 1$ **8.** $x - 1$ **9.** $10 + x$ **10.** $10 - x$

11. $x + 10$ **12.** $x - 10$ **13.** $y + 8$ **14.** $8 + y$

15. $8 - y$ **16.** $y - 8$ **17.** $\frac{x}{3}$ **18.** $\frac{2}{y}$

19. $3x$ **20.** $2y$ **21.** $x + y$ **22.** $x - y$

23. xy **24.** yx **25.** $\frac{3}{x}$ **26.** $\frac{y}{4}$

27. $9x$ **28.** $\frac{x}{9}$ **29.** $\frac{9}{x}$ **30.** $x + y + 1$

31. $x - y + 1$ **32.** $x + y - 1$ **33.** $x - y - 1$ **34.** $9 - x - y$

MODEL

Find the value of the expression. Use $a = 5, b = 8, c = {}^-1$.	Substitute.	
$2(a + b)$	$2(a + b) \rightarrow 2(5 + 8)$	$2(a + b) = 26$
$2a + b$	$2a + b \rightarrow 2 \cdot 5 + 8$	$2a + b = 18$
$\dfrac{a + c}{2}$	$\dfrac{a + c}{2} \rightarrow \dfrac{5 + {}^-1}{2}$	$\dfrac{a + c}{2} = 2$

 GET READY. Find the value of the expression.

35. $3(4 + {}^-2)$

36. $5(5 - 2)$

37. $3 \cdot 2 + {}^-2$

38. $\dfrac{{}^-3 + 6}{3}$

39. $2 \cdot 3 + 8 \cdot 3$

40. $8(3 - 4)$

 NOW USE SKILL 86. Find the value of the expression. Use $x = 4, y = 5, z = {}^-1$.

41. $x + y + z$

42. $y - z$

43. $x - y + z$

44. $x + y - z$

45. $3(x + y)$

46. $3(x - y)$

47. $3x + y$

48. $3x - y$

49. $3x + z$

50. $\dfrac{x + z}{3}$

51. $\dfrac{x + y}{3}$

52. $\dfrac{4y}{z}$

53. $3x - z$

54. $5y - z$

55. $8 - 3x$

56. $8 + 3x$

57. $7 + 2z$

58. $7 - 2z$

59. xz

60. xyz

61. yz

62. $xy({}^-2)$

63. $x(y + 1)$

64. $y(x + 1)$

65. $z(4 + x)$

66. $4x - y$

67. $2z + x + y$

68. $\dfrac{y + 2z}{x - 1}$

69. $\dfrac{4(x + y)}{6}$

70. $\dfrac{3(x - y)}{z}$

71. $x(6 - z)$

72. $(x + z)y$

73. $(2z + 4y)z$

74. $(xyz) - 2$

75. $37 - xy$

76. $12 + 5y + 7z$

For more practice: **Skill Bank,** page 442.

From Words to Variable Expressions

You can translate word sentences into variable expressions. The table below lists some words commonly used to describe an operation.

Operation	Words that describe the operation
Addition	add, sum, plus, increase, more
Subtraction	subtract, difference, remainder, less
Multiplication	multiply, times, product, double, twice
Division	divide, quotient, ratio

MODEL

Write a variable expression.

1. the sum of a and b
2. the difference of m and n
3. twice a number x
4. the quotient of x divided by y
5. 3 more than c

1. $a + b$
2. $m - n$
3. $2x$
4. $\dfrac{x}{y}$
5. $c + 3$

 GET READY. Write the meaning of the expression.

1. $x + y$ 2. $x - y$ 3. $2y$ 4. $\dfrac{a}{b}$

5. $2x + y$ 6. $x - 2y$ 7. $2a + 1$ 8. $2(a + 1)$

 NOW USE APPLICATION A38. Complete the third column.

	Variable	Operation	Variable Expression
9.	x	add 3	$x + 3$
10.	x	add z	?
11.	a	subtract 4	?
12.	a	subtract b	?
13.	c	multiply by 2	?
14.	b	double it	?
15.	c	add twice a	?

Write as a variable expression.

16. 12 increased by m

17. t less 9

18. the difference of n and s

19. 7 more than w

20. 4 added to x

21. 10 plus x

22. x added to y

23. 4 increased by y

24. the sum of x, y, and z

25. 5 more than n

26. x subtracted from 10

27. 10 subtracted from x

28. the difference between x and y

29. 9 less than n

30. x decreased by 4

31. x multiplied by 10

32. the product of x and y

33. twice n

34. m doubled

35. the sum of x and twice y

36. n times m

37. the quotient of m divided by b

38. the ratio of x to y

39. 2 more than twice x

40. twice the sum of n and 1

41. the product of y and twice x

Complete the statement with a variable expression.

42. a. The cost of a radio at \$15 and a toaster at \$20 is $(15 + 20)$ dollars.

    b. The cost of a radio at \$15 and a toaster at x dollars is __?__ dollars.

43. a. The cost of 3 bikes at \$90 each is (3×90) dollars.

    b. The cost of 3 bikes at y dollars each is __?__ dollars.

44. a. Tim is 12 years old; in 3 years he will be $(12 + 3)$ years old.

    b. Tim is 12 years old; in n years he will be __?__ years old.

45. a. Anna is 15 years old; 4 years ago she was $(15 - 4)$ years old.

    b. Anna is 15 years old; b years ago she was __?__ years old.

46. a. A cone costs 60¢; a sundae costs twice as much, or (2×60) cents.

    b. A cone costs x cents; a sundae costs twice as much or __?__ cents.

47. a. The price of a record was \$3; it increased \$1. The new price is $(3 + 1)$ dollars.

    b. The price of a record was \$3; it increased y dollars. The new price is __?__ dollars.

For more practice: **Application Bank**, page 463.

Hot Stuff

You have noticed that snow can take a long time to melt in the spring. You know that glaciers do not melt away, even during a warm summer. You have observed that water can be boiled fairly quickly but it takes much longer for the water to actually boil away. You decide to investigate the heat-related behavior of water. To focus your work, you will trace some changes in a 5000 g piece of ice.

PROBLEM: How much heat is needed to melt a 5000 g piece of ice, heat the water to boiling, and then change the water to steam?

Solution Plan: Look up some important facts about the amount of heat needed to warm water and to change ice to water and water to steam. Use these facts to figure the amount of heat needed for the problem.

WHAT DO YOU WANT TO KNOW? You want to know the amount of heat needed to cause changes in ice and water. You learn that it takes 80 calories (cal) of heat to melt 1 g of ice.

1. How much heat is needed to melt 5 g of ice?

2. How much heat is needed to melt 200 g of ice?

You find that it takes 1 cal of heat to raise the temperature of 1 g of water 1°C.

3. How much heat is needed to raise the temperature of 1 g of water from 5°C to 25°C?

4. How much heat is needed to raise the temperature of 100 g of water from 36°C to 37°C?

5. How much heat is needed to raise the temperature of 100 g of water from 36°C to 40°C?

According to a manual you read, it takes 540 more cal of heat to cause 1 g of water to evaporate, in addition to the heat it takes to bring the water to the boiling point.

6. How much heat does it take to evaporate 5 g of water at the boiling point?

7. How much heat does it take to evaporate 1000 g of water at the boiling point?

NOW SOLVE THE PROBLEM. Use the facts you have found about ice and water to answer the original question about a 5000 g piece of ice.

8. How much heat is needed to melt the ice?

Ice melts at 0°C and water boils at 100°C.

9. After the ice melts you have 5000 g of water. How much heat is needed to raise the temperature of the water from the melting point to the boiling point?

10. How much more heat is needed to evaporate the water, converting it to steam?

11. What is the total amount of heat needed to convert the 5000 g of ice to steam?

Quiz

Skill 84 State the meaning of the variable expression.

1. $\dfrac{8}{n}$ **2.** $6 - y$

Skill 85 State the meaning of the expression.

3. $(9m)n$ **4.** $2(7 \cdot x)$

Skill 86 Find the value of the expression. Use $x = 2$, $y = 4$.

5. $\dfrac{4}{y}$ **6.** $1 - 3x$ **7.** $y(x + 1)$

Application A38 Complete the statement with a variable expression.

8. Paul is 16 years old now; y years ago he was __?__ years old.

9. The cost of a shirt at \$14 and a tie at t dollars is __?__ dollars.

Problem Solving

10. It takes 80 calories of heat to melt 1 gram (g) of ice. How much heat is needed to melt a 3500 g chunk of ice?

EXTRA!

A block of wood is painted on all sides. Then it is cut into 8 cubes as shown. When you take the 8 cubes apart, how many sides of each cube are painted?

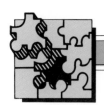 →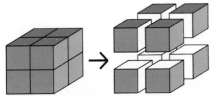

Answer:
3 sides of each
cube are painted

Each of the blocks below has been painted on all sides. When the blocks are cut into cubes and taken apart, how many sides of each cube are painted?

1.

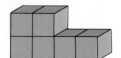

2.

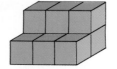

3.

__?__ cube 3 sides painted   __?__ cube 2 sides painted   __?__ cubes 3 sides painted
__?__ cubes 4 sides painted   __?__ cubes 3 sides painted   __?__ cubes 4 sides painted
__?__ cube 5 sides painted   __?__ cubes 4 sides painted

Changes in Value

Many items *depreciate*, or decrease in value, as they become older.

Craig Donahue paid $4800 for his car. He estimates that its present value is $3600. Find the depreciation. What percent of the original value is the depreciation?

KEY-IN	DISPLAY	
4800	*4800*	
⊟	*4800*	
3600	*3600*	
⊟	*1200*	⟶ Depreciation
⊞	*1200*	
4800	*4800*	
⊟	*0.25*	⟶ 0.25 = 25%

The depreciation is 25% of the original value. However, not all items depreciate. For example, antiques *appreciate*, or increase in value.

Norma Platte paid $18,000 for an antique porcelain vase. She recently sold the vase for $23,760. Find the appreciation. What percent of the original value is the appreciation?

KEY-IN	DISPLAY	
23760	*23760*	
⊟	*23760*	
18000	*18000*	
⊟	*5760*	⟶ Appreciation
⊞	*5760*	
18000	*18000*	
⊟	*0.32*	⟶ 0.32 = 32%

The appreciation is 32% of the original value.

Complete the table. Tell if the change is appreciation or depreciation.

	Original Cost	Present Value	Change	Percent of Original Value
1.	$3750	$3000	?	?
2.	$14,500	$21,170	?	?
3.	$46,800	$60,840	?	?
4.	$8450	$6929	?	?

343

Properties of Zero and One

Using 0	Using 1

Using 0

1. $3 + 0 = 3$ $a + 0 = a$
any number $+ 0 =$ the number

2. $3 \times 0 = 0$ $a \times 0 = 0$
any number $\times 0 = 0$

3. $\frac{0}{3} = 0$ $\frac{0}{a} = 0$
$0 \div$ any number $= 0$
(Never divide by 0.)

4. $5 - 5 = 0$ $a - a = 0$
any number $-$ itself $= 0$

Using 1

1. $3 \times 1 = 3$ $a \times 1 = a$
any number $\times 1 =$ the number

2. $\frac{3}{1} = 3 \div 1 = 3$ $\frac{a}{1} = a$
any number $\div 1 =$ the number

3. $\frac{8}{8} = 1$ $\frac{a}{a} = 1$
any number $\div$ itself $= 1$
(except 0)

MODEL

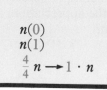

Simplify.

$n(6 - 6)$
$n(7 - 6)$
$\frac{4n}{4}$

$n(0)$
$n(1)$
$\frac{4}{4}n \longrightarrow 1 \cdot n$

0
n
n

 GET READY.

1. $^-6 + 0$

2. $^-6 \cdot 1$

3. $^-6 \cdot 0$

4. $\frac{0}{6}$

5. $7 - 7$

6. $\frac{4}{4}$

7. $\frac{4}{1}$

8. $100 \cdot 0$

NOW USE SKILL 87. Simplify.

9. $x + 0$

10. $3x + 0$

11. $\frac{12y}{1}$

12. $x - x$

13. $a \cdot 0$

14. $3a \cdot 0$

15. $3x \cdot 1$

16. $b \cdot 1$

17. $\frac{a}{a}$

18. $\frac{2n}{2}$

19. $\frac{0}{n}$

20. $\frac{2b}{b}$

21. $30n + 0$

22. $5 - 5 + y$

23. $0 \cdot 6n$

24. $4 + y - 4$

25. $\frac{0}{2a}$

26. $\frac{0}{24}$

27. $\frac{4n}{2 \cdot 2}$

28. $\frac{5}{5} \cdot a$

For more practice: **Skill Bank,** page 443.

Combining Like Terms

$2x + 3x + 3y$
↑   ↑   ↑
Terms of the
expression

$2x + 3x + 3y$
   ↑   ↑
Like terms
(same variable)

$2x + 3x + 3y$
     ↑   ↑
Unlike terms
(different variables)

You can combine like terms: $2x + 3x = 5x$
You cannot combine unlike terms: $2x + 3x + 3y = 5x + 3y$

It may be helpful to rewrite b as $1b$, x as $1x$, and so on.

MODEL

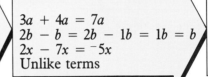

Simplify by combining like terms.

$3a + 4a$
$2b - b$
$2x - 7x$
$7a + 3b$

$3a + 4a = 7a$
$2b - b = 2b - 1b = 1b = b$
$2x - 7x = {}^-5x$
Unlike terms

$7a$
b
${}^-5x$
$7a + 3b$

GET READY. **Can you combine the terms? Write Yes or No.**

1. $2x + 5x$ **2.** $3a + a$ **3.** $3a + 1$ **4.** $5x - x$

5. $2c - 2d$ **6.** $1 + 10m$ **7.** $5n + 10m$ **8.** $2r - 5r$

NOW USE SKILL 88. **Simplify by combining like terms.**

9. $3x + 3x$ **10.** $4y + 2y$ **11.** $11a - 10a$ **12.** $7b - b$

13. $2x + 2y$ **14.** $5x - 3x$ **15.** $5a + 7a$ **16.** $9a - 11a$

17. $6x - 8x$ **18.** $6y - y$ **19.** $4m - 3m$ **20.** $3b - 7b$

21. $5b + 2$ **22.** $3r + 9r$ **23.** $8s + 7s$ **24.** $18x - 9x$

25. $15x - 7x$ **26.** $7n - 7n$ **27.** $14n + 7n$ **28.** $16x + 6x$

29. $2x + x + 3x$ **30.** $9y + y + 3y$ **31.** $4x + x + y$

32. $3n - 2n + n$ **33.** $4n - 3n + 1$ **34.** $9 + 8z - 7z$

35. $4r - 5r + r$ **36.** $6x - 7x + x$ **37.** $3x + 8x - 11x$

38. $8y + 2y - 9z$ **39.** $5y - y - 4y$ **40.** $x + 3x + 4$

For more practice: **Skill Bank,** page 443.

Exponents

$x \cdot x \cdot x = x^3$ ◀— exponent 3

3 factors the same

$(ab)(ab) = (ab)^2$ ◀— exponent 2

2 factors the same

MODELS

Write with an exponent.

5×5

$c \cdot c \cdot c \cdot c$

$(pq)(pq)(pq)$

2 factors the same

4 factors the same

3 factors the same

5^2

c^4

$(pq)^3$

Show the meaning.

3^4

$a \cdot b^2$

$(x + y)^2$

$3^4 = 3 \cdot 3 \cdot 3 \cdot 3$

$a \cdot b^2 = a \cdot b \cdot b$

$(x + y)^2 = (x + y)(x + y)$

GET READY. Simplify.

1. $3 \cdot 3 = 3^?$ **2.** $2 \cdot 2 \cdot 2 = 2^?$ **3.** $x \cdot x = x^?$ **4.** $5a \cdot a = 5a^?$

5. $4^2 = 4 \cdot ?$ **6.** $6^3 = ? \cdot ? \cdot ?$ **7.** $(xy)^2 = (?)(?)$ **8.** $3b^2 = 3 \cdot ? \cdot ?$

NOW USE SKILL 89. Write with exponents.

9. $5 \times 5 \times 5$ **10.** 9×9 **11.** $2 \cdot 2 \cdot 2 \cdot 2$ **12.** $5 \cdot 5 \cdot 6 \cdot 6$

13. $a \cdot a$ **14.** $c \cdot c \cdot c$ **15.** $(ab)(ab)$ **16.** $a \cdot b \cdot b$

17. $(y + 2)(y + 2)$ **18.** $x \cdot x \cdot y$ **19.** $3x \cdot x$ **20.** $(ab)(ab)(ab)$

21. $(x - 5)(x - 5)$ **22.** $2m \cdot n \cdot n$ **23.** $(rs)(rs)$ **24.** $6f \cdot g \cdot g$

25. $x \cdot x \cdot x \cdot y \cdot y$ **26.** $8x \cdot x \cdot y$ **27.** $4n \cdot n$ **28.** $(3y - 1)(3y - 1)$

29. $(3xy)(3xy)$ **30.** $(3x - 1)(3x - 1)$ **31.** $5x \cdot xy$ **32.** $(a \cdot a)(a \cdot b)$

33. $7x \cdot x \cdot y \cdot y$ **34.** $(3y - 5)(3y - 5)$ **35.** $3 \cdot x \cdot x \cdot y \cdot y \cdot y$ **36.** $(x - y)(x - y)$

37. $4(a - b)(a - b)$ **38.** $7a \cdot a \cdot b \cdot b \cdot b$ **39.** $6x \cdot x \cdot x \cdot y$ **40.** $4x \cdot x \cdot x \cdot x$

For more practice: **Skill Bank,** page 443.

Exponents and Evaluation

MODEL

Find the value
of the expression.
Use $x = 2, y = 3$.

xy^3

$x^2 - y^3$

$5x^3 - 4$

$2 \cdot 3^3 = 2 \cdot 3 \cdot 3 \cdot 3 = 2 \cdot 27$

$2^2 - 3^3 = 2 \cdot 2 - 3 \cdot 3 \cdot 3 = 4 - 27$

$5 \cdot 2^3 - 4 = 5 \cdot 2 \cdot 2 \cdot 2 - 4 = 40 - 4$

$xy^3 = 54$

$x^2 - y^3 = {}^-23$

$5x^3 - 4 = 36$

 GET READY. Find the value of the expression.

1. 2^3

2. $3 \cdot 2^3$

3. $3^2 \cdot 2^3$

4. $2^2 + 5^2$

5. $3 \cdot 2^2 - 3^2$

6. $5^2 \cdot 3^2$

7. $3 \cdot 4^2 + 2^2$

8. $5^2 \cdot 2 - 2$

 **NOW USE SKILL 90. Find the value of the expression.
Use $a = 1, b = 3, c = {}^-2$.**

9. a^2

10. b^2

11. c^2

12. $3a^2$

13. $4b^2$

14. $5c^2$

15. $3a^2 - b^2$

16. $4b^2 - c^2$

17. $(a - b)^2$

18. $(b - c)^2$

19. $c^3 + 1$

20. $b^2 - 1$

21. $c^2 + 3$

22. $b^2 + 4$

23. $\frac{1}{2}a^2$

24. $\frac{1}{3}b^2$

25. $\frac{c^2}{4}$

26. $b^2 - b$

27. $b^2 - a$

28. $c^2 + c$

Find the value. Use $x = 2, y = 5$.

29. x^2

30. y^2

31. x^2y^2

32. $3x^3$

33. $2y^2$

34. $6x^3y^2$

35. $(y + 1)^2$

36. $(x + 8)^2$

37. $(y - 1)^2$

38. $(x - 1)^2$

39. $3x^2y$

40. $5y^3$

41. $(xy)^2$

42. $(xy)^3$

43. $2xy^2$

44. $(5x)^2$

45. $(x + 7)^2$

46. $(x + 7)^3$

47. $(x + y)^2$

48. $(x - y)^3$

For more practice: **Skill Bank,** page 444.

Formulas

A **formula** is a short form of a rule. For example:

Rule: The perimeter of a square is four times the length of one side.

Formula: Let P stand for the perimeter. Let s stand for the length of one side. Then $P = 4s$.

MODEL

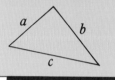

Write the formula for the perimeter of the triangle.

Rule:

The perimeter of a triangle is the sum of the lengths of the sides.

Formula:

$P = a + b + c$

> **GET READY. Select the correct answer.**

1. The perimeter of a figure is
   a. the distance around it.
   b. the distance across it.

3. The area of a rectangle is
   a. length plus width.
   b. length times width.

2. The volume of a box is
   a. length times width times height.
   b. length plus width plus height.

4. π is a Greek letter standing for
   a. 4.13.
   b. 3.14.

> **NOW USE APPLICATION A39. Write the formula for the perimeter P. If possible, simplify by combining like terms.**

5.

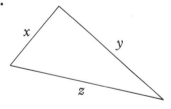

$P = \underline{\ ?\ }$

6.

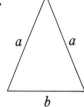

$P = \underline{\ ?\ }$

7.

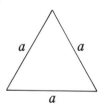

$P = \underline{\ ?\ }$

348

8. Rectangle

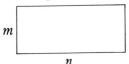

$P =$ __?__

9. Square

$P =$ __?__

10. Parallelogram

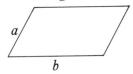

$P =$ __?__

11. Trapezoid

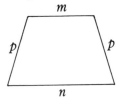

$P =$ __?__

12.

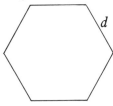

$P =$ __?__

13.

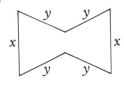

$P =$ __?__

14. Kite

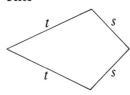

$P =$ __?__

15. Regular hexagon

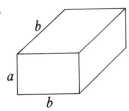

$P =$ __?__

16.

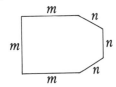

$P =$ __?__

Write the formula for the volume of the box. When possible, use exponents.

17.

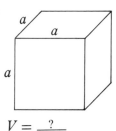

$V =$ __?__

18.

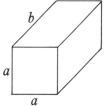

$V =$ __?__

19.

$V =$ __?__

Use the formula and the given values to find the area.

20. Square $A = s^2$

$s = 8$ in.
$A =$ __?__

21. Rectangle $A = lw$

$l = 13$ cm, $w = 6$ cm
$A =$ __?__

22. Triangle $A = \frac{1}{2}bh$

$b = 8.5$ cm, $h = 20$ cm
$A =$ __?__

For more practice: **Application Bank,** p. 463.

Homemaker

You are a homemaker. You provide many services for yourself and your family. You decide to try to find a value, in dollars, of the services you provide.

You keep a diary of your activities for two days. Besides the activities in the diary, you estimate that you are involved in child care an average of 5 hours a day. For each homemaking service you assign an hourly wage you would receive if you provided that service for hire.

Diary
(* indicates free time for personal use)

Monday

7:00–8:00	Breakfast, fix lunches
8:00–10:00	Laundry
10:00–11:30	Cleaning
12:30–3:00	Paint bedroom
3:00–5:00	Reading*
5:00–7:00	Dinner
7:00–8:30	Cleaning

Tuesday

6:30–7:30	Pay bills
7:30–8:00	Breakfast
8:00–9:30	Garden
9:30–11:00	Shopping
11:30–1:00	Lunch out*
1:00–3:30	Supervise repair work
3:30–5:30	Wash windows
5:30–6:30	Dinner
6:30–7:30	Cleaning

PROBLEM: Find the annual value of the services you provide as a homemaker.

Solution Plan: Find the average daily value of your services. Multiply to convert to an annual basis.

WHAT DO YOU KNOW? You know the time spent on each activity for the two-day period. Complete the table.

	Activity	Hours
1.	Meals	?
2.	Laundry	?
3.	Cleaning	?
4.	Maintenance/repair	?
5.	Financial (including shopping)	?
6.	Lawn/garden	?

WHAT DO YOU WANT TO KNOW? You want to figure the value of your services. You assign an hourly wage to each activity.

Meals $8
Laundry $5
Cleaning $6
Maintenance/repair $10
Financial $7
Lawn/garden $4
Child care $6

Find the value of each service for the two-day period.

7. Meals

8. Laundry

9. Cleaning

10. Maintenance/repair

11. Financial

12. Lawn/garden

13. Child care

NOW SOLVE THE PROBLEM.

14. Find the total value of all services for the two days.

15. Find the average value, per day, of the services you provide.

16. Use the average value per day to find the annual value of your services. Use 365 for the number of days in a year.

Skill 87 Simplify.

1. $d(4 - 3)$ 2. $14t(0)$

Skill 88 Simplify by combining like terms.

3. $6x + 8x$ 4. $7r - r$

Skill 89 Write with exponents.

5. $2u \cdot u \cdot v \cdot v$

6. $(4x - y)(4x - y)$

Skill 90 Find the value. Use $a = 5$ and $c = 2$.

7. $3a^2 + c^2$ 8. $\frac{c^3}{4}$

Application A39 Use the formula and the given values to find the area.

9. $A = lw$
$l = 9$ cm, $w = 6$ cm
$A = \underline{\ ?\ }$

10. $A = s^2$
$s = 14$ in.
$A = \underline{\ ?\ }$

Problem Solving

11. You want to find the value of your services. You assign the following hourly wages: child care $6, cleaning $7, meals $8. In one week you spend 35 hours on child care, 10 hours on cleaning, and 25 hours on meals. What is the value of your services that week?

CONSUMER NOTE

Universal Product Code

To help speed the consumer through the check-out counter, many supermarkets now use the Universal Product Codes (UPC) which are printed on items. An electronic "reader" in the counter scans the UPC, then signals a computer to search its memory for the item. In an instant the item is "rung up," and its name and price are automatically printed on the sales slip. After each sale, the computer keeps a running total of each product sold, so that the store manager can be advised at all times of the status of the inventory.

Each UPC uses lines and spaces of varying width as codes for the numbers beneath.

Code number of the manufacturer.   Code number of the product.

Federal Income Tax

To complete a federal tax return you need a Wage and Tax Statement (W-2 form). This statement shows your total wages and the amount withheld for federal and state taxes and social security (FICA).

Bryan Goh is preparing his federal income tax return. A portion of his W-2 form is shown below.

8 Employee's social security number 123-45-6789	9 Federal income tax withheld $2360.15	10 Wages, tips, other compensation $14,300	11 FICA tax withheld 950.95
12 Employer's name, address and ZIP code	13 FICA wages $14,300		14 FICA tips 0
	16 Employer's use		

Soundsystems
411 Gravel Street
Rapid City, So. Dakota

17 State income tax $1716	18 State wages, tips, etc. $14,300	19 Name of State So. Dakota
20 Local income tax 0	21 Local wages, tips, etc. $14,300	22 Name of locality Rapid City

Form **W-2 Wage and Tax Statement 1980**   Copy B To be filed with employee's FEDERAL tax return   This information is being furnished to the Internal Revenue Service.   Department of the Treasury Internal Revenue Service

Bryan uses the information in his W-2 form to complete Form 1040A, or the short form, for this year's tax return. Below is a section of Form 1040A Bryan must fill out. He copies the amounts for lines 7 and 12b from the W-2 form. Complete the form.

7 Wages, salaries, tips, etc. (Attach Forms W-2. See page 10 of Instructions)	**7**	?
8 Interest income (See pages 3 and 10 of Instructions)	**8**	238.84
9a Dividends __234 ¦ 65__ (See pages 3 and 10 of Instructions) 9b Exclusion __100 ¦ 00__ Subtract line 9b from 9a	**9c**	?
10a Unemployment compensation (insurance). Total received from Form(s) 1099-UC __0 ¦__		
b Taxable amount, if any, from worksheet on page 10 of Instructions	**10b**	0
11 Adjusted gross income (add lines 7, 8, 9c, and 10b). If under $10,000, see page 12 of Instructions on "Earned Income Credit"	**11**	?
12a Credit for contributions to candidates for public office. (See page 11 of Instructions) . . . . **12a** 0 IF YOU WANT IRS TO FIGURE YOUR TAX, PLEASE STOP HERE AND SIGN BELOW.		
b Total Federal income tax withheld (If line 7 is more than $25,900, see page 11 of Instructions) . . . . **12b** ?		
c Earned income credit (from page 12 of Instructions) . . . **12c** 0		
13 Total (add lines 12a, b, and c)	**13**	?
14a Tax on the amount on line 11. (See page 13 of Instructions; then find your tax in the Tax Tables on pages 15–26) . . . . **14a** 2,261 ¦ 00		
b Advance earned income credit (EIC) (from Form W–2) . . . **14b** 0		
15 Total (add lines 14a and 14b)	**15**	?
16 If line 13 is larger than line 15, enter amount to be **REFUNDED TO YOU** . . . . ▶	**16**	?
17 If line 15 is larger than line 13, enter **BALANCE DUE.** Attach check or money order for full amount payable to "Internal Revenue Service." Write your social security number on check or money order . ▶	**17**	

Please Attach Copy B of Forms — *Attach Payment Here*

Solving an Equation

To **solve** an equation, find the value of the variable that makes the equation a true statement.

Equation: $x + 3 = 9$ Check: $x + 3 = 9$
Solution: 6 $6 + 3 = 9$ True

MODEL

Find the solution.

$2n = 10$

$$2n = 10$$
Try $n = 3 \longrightarrow 2 \cdot 3 = 10$ false
Try $n = 4 \longrightarrow 2 \cdot 4 = 10$ false
Try $n = 5 \longrightarrow 2 \cdot 5 = 10$ TRUE

Solution is 5.

 GET READY. Find the value of the expression. Use $n = 4$.

1. $n + 1$ **2.** $5 + n$ **3.** $n - 3$ **4.** $10 + n$

5. $\frac{n}{2}$ **6.** $3n$ **7.** $10n$ **8.** $\frac{n}{4}$

NOW USE SKILL 91. Is the given value of the variable a solution of the equation? Write Yes or No.

9. $5n = 10$ $n = 2$ **10.** $3x = 27$ $x = 9$

11. $x + 1 = 4$ $x = 4$ **12.** $10 + y = 12$ $y = 2$

13. $x - 3 = 0$ $x = 3$ **14.** $\frac{x}{6} = 3$ $x = 2$

Find the solution.

15. $3x = 6$ **16.** $3n = 12$ **17.** $4n = 12$ **18.** $3n = 0$

19. $2n = 14$ **20.** $7x = 63$ **21.** $6y = 24$ **22.** $7y = 56$

23. $n + 1 = 3$ **24.** $n - 1 = 3$ **25.** $1 + n = 3$ **26.** $n + 2 = 5$

27. $x + 3 = 7$ **28.** $y + 2 = 11$ **29.** $n - 7 = 8$ **30.** $9 + x = 14$

31. $16 + y = 18$ **32.** $16 + y = 20$ **33.** $n - 6 = 15$ **34.** $x - 5 = 9$

35. $m + 8 = 6$ **36.** $x + 4 = 6$ **37.** $x + 6 = 4$ **38.** $x - 4 = 6$

39. $\frac{n}{3} = 2$ **40.** $\frac{x}{4} = 2$ **41.** $\frac{n}{5} = 1$ **42.** $\frac{n}{3} = 1$

43. $5x = 30$ **44.** $7x = {}^{-}28$ **45.** $9x = 81$ **46.** $x + 4 = 0$

For more practice: **Skill Bank**, page 444.

Solving an Inequality

The statements below are inequalities.

$$x + 5 < 9 \qquad x - 11 > 2 \qquad x + 3 < 4$$

To solve an inequality, find values of the variable that make the inequality a true statement.

MODEL

Inequality: $x + 3 < 4$ Check these values for x: $1, 0, ^-1, ^-2$	Try $x =$ 1	$\begin{array}{c} x + 3 < 4 \\ 1 + 3 < 4 \\ 4 < 4 \quad \text{false} \end{array}$	$0, ^-1,$ and $^-2$ are solutions of $x + 3 < 4$.
	Try $x =$ 0	$\begin{array}{c} 0 + 3 < 4 \\ 3 < 4 \quad \text{TRUE} \end{array}$	
	Try $x = ^-1$	$\begin{array}{c} ^-1 + 3 < 4 \\ 2 < 4 \quad \text{TRUE} \end{array}$	
	Try $x = ^-2$	$\begin{array}{c} ^-2 + 3 < 4 \\ 1 < 4 \quad \text{TRUE} \end{array}$	

 GET READY. Use > or < to make the statement true.

1. 5 _?_ 4   **2.** 0 _?_ 4   **3.** $^-3$ _?_ 4   **4.** $^-3$ _?_ $^-4$

5. 3 _?_ 10   **6.** $^-3$ _?_ 10   **7.** 3 _?_ $^-10$ **8.** $^-8$ _?_ $^-5$

 NOW USE SKILL 92. Check these values for x: $^-1, 0, 1, 2, 4$.
Which of these values are solutions of the inequality?

9. $x > 0$ **10.** $x < 4$ **11.** $x < 0$ **12.** $x > ^-1$

13. $x < 1$ **14.** $x > 1$ **15.** $x < 2$ **16.** $x > 2$

17. $x + 1 > 0$ **18.** $x + 1 < 0$ **19.** $x + 4 > 2$ **20.** $x + 3 < 4$

21. $x + 5 > 8$ **22.** $x - 5 < 8$ **23.** $x - 1 > 0$ **24.** $x + 2 > ^-1$

25. $2x < 4$ **26.** $3x < 4$ **27.** $3x > 4$ **28.** $4x > 1$

29. $x + 4 > 1$ **30.** $x + {}^-1 > 3$ **31.** $x - {}^-2 < 3$ **32.** $x + {}^-2 < 1$

33. $x - 3 > 2$ **34.** $x - 2 > 2$ **35.** $\frac{x}{2} > 0$ **36.** $\frac{x}{2} < 0$

37. $x + 2 > 0$ **38.** $x + 3 < 6$ **39.** $2x > 0$ **40.** $2x > 3$

For more practice: **Skill Bank,** page 444.

From Words to Equations and Inequalities

You can translate a statement from words into an equation or inequality by using mathematical terms and symbols. For example,

Word expression Twice a number is 30.

Equation $2 \cdot n = 30$
$$2n = 30$$

MODEL

Write as an equation.

A box costs 75¢. A box and ribbons cost $1.85. ➤ Let y be the cost of ribbons. ➤ $75 + y = 185$

 GET READY. Write as a variable expression.

1. a number plus 3

2. a number less 4

3. three times a number

4. five times a number, then add 8

 NOW USE APPLICATION A40. Select the equation or inequality that best fits the problem.

5. The cost of a radio is $15. The cost of the radio and a toaster together is $40. Let the cost of the toaster be n.

   a. $15 + n = 40$

   b. $n - 15 = 40$

   c. $2n = 40$

   d. $n + 40 = 15$

6. A sandwich costs 75¢. The sandwich and a drink cost $1.25.

   a. $n + 75n = 1.25$

   b. $75 + n = 1.25$

   c. $75 + n = 125$

   d. $2n = 125$

7. Two bookshelves cost $190. Both bookshelves cost the same amount.

   a. $2 \cdot 190$

   b. $2n = 190$

   c. $n = 190$

   d. $2n + 2n = 190$

8. Each side of a square measures s cm. The area is 49 cm².

   a. $2s = 49$

   b. $s + 2 = 49$

   c. $s^2 = 49$

   d. $s + s + s + s = 49$

9. Each side of an equilateral triangle measures b in. The perimeter is 27 in.

   a. $2b = 27$

   c. $b^2 = 27$

   b. $3b = 27$

   d. $4b = 27$

10. The temperature is below zero.

    a. $t > 0$

    c. $t = 0$

    b. $t < 0$

11. A number is less than $^-2$.

    a. $n > 2$

    c. $n < {}^-2$

    b. $n > {}^-2$

    d. $n = {}^-2$

12. A number plus 6 is greater than 12.

    a. $n + 6 > 12$

    c. $n + 12 < 6$

    b. $12 - 6 > n$

    d. $n - 12 > 6$

13. Tom's age is less than 20.

    a. $T > 20$

    c. $T - 20$

    b. $T = 20$

    d. $T < 20$

14. If Elli were twice her present age, she would be more than 30 years old.

    a. $E + 2 > 30$

    c. $\frac{E}{2} > 30$

    b. $2E > 30$

    d. $2E = 30$

15. Half of y is less than 24.

    a. $\frac{1}{2}(y) < 24$

    c. $2y > 24$

    b. $2y < 24$

    d. $y + y > 24$

16. The card cost more than 80¢.

    a. $n < 80$

    c. $n > 80$

    b. $n = 80$

    d. $n(80)$

17. Three years from now Ann will be 20 years old.

    a. $3A = 20$

    c. $A - 3 = 20$

    b. $A + 3 > 20$

    d. $A + 3 = 20$

For more practice: **Application Bank**, page 463.

Leverage

You bought some property with the intention of reselling it at a profit later. You want to take advantage of *leverage*. Leverage means using a small amount of your own money to control property of much greater value. You hope to make a good profit on a fairly small investment.

To use leverage effectively, you want to tie up as little of your own money as possible. This means you want to make a small down payment. You also want to resell as soon as possible so that you don't make very many monthly payments on the loan you take to finance your purchase of the property.

You bought the property two years ago at a price of $18,000. You were required to make a down payment of 10%, and you got a loan for the remainder of the price. Your monthly payments on the loan are $276.70.

PROBLEM: What is the return on your investment when you sell the property?

Solution Plan: Figure the total amount of money you have invested in the property. Find the profit on the sale.

WHAT DO YOU KNOW? You know the terms of the sale when you bought the property.

1. What was the price when you bought the property?

2. How much was your down payment?

3. How much is each monthly payment?

4. How long have you been making monthly payments?

WHAT DO YOU WANT TO KNOW? You want to know the total amount of money that you have spent on the property.

5. How much have you paid altogether in monthly payments?

You have had some other expenses involved with owning the property. You want to include these as part of your total investment amount.

You pay real estate taxes of $124 every six months. You hired a neighborhood resident for $35 a month to act as an informal caretaker, mowing the grass, shoveling the snow, and keeping an eye on the place.

6. How much have you paid in taxes for the property?

7. How much have you paid your caretaker?

8. How much money does your total investment amount to? Include the down payment and all other expenses.

NOW SOLVE THE PROBLEM. Property in your area increases in value. You sell your property for $28,000. The agent who handles the sale charges you a commission of 6% of the sale price.

9. How much is the agent's commission?

10. After repaying your loan from the bank, you clear $14,752 on the deal. From this you deduct the agent's commission and your total investment in the property to find your profit. How much is your profit?

11. Express your profit as a percent of your total investment. Round to the nearest whole percent.

Quiz

Skill 91 **Find the solution.**

1. $m + 7 = 15$ **2.** $12 - t = 4$ **3.** $7x = 63$

Skill 92 **Check these values for x: $^-1$, 0, 1, 2, 3. Which of these values are solutions of the inequality?**

4. $x < 2$ **5.** $x + 2 < 4$ **6.** $3x > 4$

Application A40 **Select the equation or inequality that best fits the problem.**

7. Two identical boxes weigh more than 7 pounds together.

    a. $2 + n = 7$ **b.** $2n > 7$ **c.** $n + 2n > 7$ **d.** $n + n = 7$

Problem Solving

8. You make a down payment of $1200. You make monthly payments of $273.50 for 3 years. You pay taxes of $420 a year for 3 years. How much is your total investment?

CONSUMER NOTE

Energy Guide Labels

Many major appliances now carry Energy Guide labels. The label estimates the yearly cost of using the appliance. It also compares the energy efficiency of the given model with similar models.

This label appears on your new refrigerator-freezer. The $60 estimated yearly operating cost is based on current average electric rates.

Your refrigerator cost $645 to purchase. Find the total cost of owning and operating it for 5 years. Assume that the overall electric rate during the five years is 8¢ per kilowatt hour.

Refrigerator-Freezer   Model 772

ENERGYGUIDE

Model with lowest energy cost $45	$60	Model with highest energy cost $88

How much will this model cost you to run yearly?

		Yearly Cost
Cost per kilowatt hour	2¢	$24
	4¢	$48
	6¢	$72
	8¢	$96
	10¢	$120
	12¢	$144

Skills Review

Skill 85 State the meaning of the variable expression. (page 334)

1. $20 - b$ **2.** $8x$ **3.** $6 + m$

4. $\dfrac{v}{a}$ **5.** $\dfrac{f - 2}{3}$ **6.** $6(v - m)$

Skill 86 Find the value of the expression. Use $x = 5$, $y = 3$. (pages 336–337)

7. $3x$ **8.** $x - y$ **9.** $15 + x - y$ **10.** $4xy$

11. $\dfrac{^-15}{x}$ **12.** $\dfrac{y + 1}{6}$ **13.** $\dfrac{2(x + y)}{8}$ **14.** $6(y - x)$

Skill 87 Simplify. (page 344)

15. $9 + 0$ **16.** $(8x + 1)0$ **17.** $\dfrac{x}{x}$ **18.** $y - y$

19. $(15 - 2)\dfrac{x}{x}$ **20.** $\dfrac{3}{3} \cdot a$ **21.** $\dfrac{8n}{2 \cdot 4}$ **22.** $(6 - 5)n$

Skill 88 Simplify. (page 345)

23. $9x + 2x$ **24.** $3x + 2y + x$ **25.** $7y - 11y$

26. $5y - y + 3y$ **27.** $23x + 6 - 15x$ **28.** $10y - 9y + 7y$

Skill 89 Write with exponents. (page 346)

29. $6 \times 6 \times 6$ **30.** $(xy)(xy)$ **31.** $a(a \cdot b)$

32. $4x \cdot x \cdot y \cdot y$ **33.** $7a \cdot a \cdot b$ **34.** $(4 - y)(4 - y)$

Skill 90 Find the value. Use $x = 2$, $y = 6$. (page 347)

35. x^3 **36.** $7x^2y$ **37.** $(2x)^2$ **38.** $(y - x)^2$

39. $x^2 - y^2$ **40.** $(xy)^2$ **41.** $3y^2$ **42.** $(x + 1)^3$

Skill 91 Find the solution. (page 354)

43. $6x = 18$ **44.** $a - 2 = 10$ **45.** $\dfrac{b}{5} = 1$ **46.** $m + 4 = {}^-3$

47. $b + 2 = 3$ **48.** $6n = {}^-24$ **49.** $x - 7 = 19$ **50.** $6y = 54$

Skill 92 Check these values for y: $^-4$, $^-3$, 3, 4. Which of these values are solutions of the inequality? (page 355)

51. $y > 1$ **52.** $y < {}^-1$ **53.** $4y > 0$

Applications Review

Application A38 **Write as a variable expression.** (pages 338–339)

1. 9 increased by t

2. x decreased by 10

3. the sum of 3 and twice n

4. twice x times y

Complete the statement with a variable expression.

5. The cost of 4 tickets at \$16 each is (4×16) dollars. The cost of m tickets at \$16 each is __?__ dollars.

6. Jacob is 3 years old; 8 years from now he will be $(3 + 8)$ years old. Jacob is 3 years old; v years from now he will be __?__ years old.

Application A39 **Write the formula.** (pages 348–349)

7.
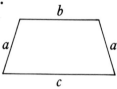

$P =$ __?__

8.
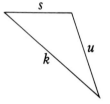

$P =$ __?__

9. Parallelogram
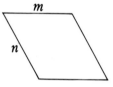

$P =$ __?__

10. Parallelogram
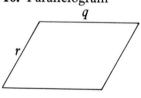

$P =$ __?__

11.
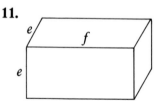

$V =$ __?__

12.
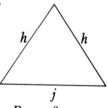

$P =$ __?__

Application A40 **Select the best answer.** (pages 356–357)

13. A rectangle is t cm long and a cm wide. The perimeter is 64 cm.
   a. $t + a = 64$ b. $t + t = 64$
   c. $a^2 = 64$ d. $2t + 2a = 64$

14. A ring costs \$7. A ring and a necklace cost \$15.
   a. $7 + n = 15$ b. $7 + 7n = 15$
   c. $7 + 2n = 15$ d. $7n = 15$

15. If twice as many guests come, there will be more than 40 people.
   a. $2 + g > 40$ b. $2g > 40$
   c. $2g < 40$ d. $g + g > 40$

16. Five years from now Sheila will be more than 20 years old.
   a. $n + 5 > 20$ b. $n + 5 < 20$
   c. $5n > 20$ d. $5n < 20$

Unit Test

Skill 84 **State the meaning of the variable expression.**

1. $4 \times h$

2. $2 - w$

Skill 85 **State the meaning of the expression.**

3. $(x + y)2$

4. $4(r - 3)$

Skill 86 **Find the value of the expression. Use $x = 3$, $y = 1$.**

5. $10 + y$

6. $3(zx - y)$

Skill 87 **Simplify.**

7. $\dfrac{0}{4x}$

8. $(9 - 8)\dfrac{p}{p}$

Skill 88 **Simplify by combining like terms.**

9. $7k - 3k - 2k$

10. $3r - 5r$

Skill 89 **Write with exponents**

11. $(ab)(ab)$

12. $5 \cdot 5 \cdot 5 \cdot 5$

Skill 90 **Find the value of the expression. Use $a = {}^{-}3$, $b = 2$, $c = 1$.**

13. $(a + b)^2$

14. $(c + 4)^2$

Skill 91 **Find the solution.**

15. $7g = 42$

16. $y - 6 = 2$

Skill 92 **Check these values for x: $^{-}2$, $^{-}1$, 0, 1, 2. Which of these values are solutions of the inequality?**

17. $x + 2 < 4$

18. $\dfrac{x}{2} < 0$

19. $2x < 1$

Application A38 **Complete the statement with a variable expression.**

20. The quotient of x divided by y is __?__.

21. The cost of a pen at 89¢ and a ruler at n cents is __?__ cents.

Application A39 **Write the formula for the volume of the box.**

22.

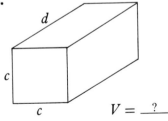

$V = \underline{\ ?\ }$

23.

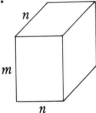

$V = \underline{\ ?\ }$

Application A40 **Select the best answer.**

24. Five years from now Sean will be more than 20 years old.

    a. $5n = 20$ **b.** $5n > 20$ **c.** $n + 5 > 20$ **d.** $n + 20 = 5$

12

Equations

Coordinates and Graphs

Problem Solving

Unit Preview

You will use these skills and applications in this unit.
Which do you already know? Work each problem.

Skill 93 Solve the equation.

1. $x - 7 = 12$ 2. $20 = n - 18$

Skill 94 Solve the equation.

3. $m + 8 = 17$ 4. $25 = y + 15$

Skill 95 Solve the equation.

5. $7y = 49$ 6. $9x = 72$

Skill 96 Solve the equation.

7. $\frac{n}{4} = 10$ 8. $\frac{x}{6} = {}^-8$

Skill 97 Solve the equation.

9. $n - 15 = 20$

Skill 98 Solve the equation.

10. $2x + 7 = 27$

Skill 99 Is the given ordered pair a solution of the equation?

11. $2x + y = 13;\ (5, 3)$ 12. $4x = y;\ (0, 4)$

Skill 100 Find the value of _y_.

13. $3x + y = 15;\ x = 4$ 14. $x - 2y = 2;\ x = 10$

Applications A41 and A42
Graph the solution on a number line.

15. $x + 7 = 12$ 16. $n - 8 = 0$ 17. $y < {}^-3$

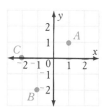

Application A43 Give the coordinates of the point.

18. A 19. B 20. C

Application A44 Complete the table. Mark the point for each ordered pair. Then draw a line through the points to graph the equation.

21. $x + y = {}^-2$

x	y	(x, y)
⁻2	0	(?, ?)
0	?	(?, ?)
2	?	(?, ?)

Check your answers. If you had difficulty with any skill or application, be sure to study the corresponding lesson in this unit.

Solving Equations by Addition

An equation is like a scale with two equal weights on each side. To keep the balance, you must treat both sides alike.

To solve an equation, find the value of the variable by isolating the variable. To do this, select an operation that leaves the variable alone on one side of the equation.

Rule: To solve a subtraction equation, add the same number to both sides of the equation.

MODEL

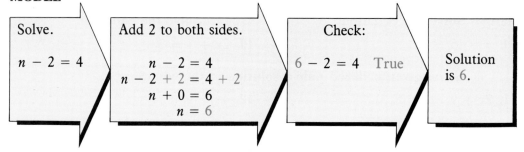

Solve.	Add 2 to both sides.	Check:	Solution is 6.
$n - 2 = 4$	$n - 2 = 4$ $n - 2 + 2 = 4 + 2$ $n + 0 = 6$ $n = 6$	$6 - 2 = 4$ True	

 GET READY. Complete the second equation.

1. $x - 1 = 6$
$x - 1 + \underline{\ ?\ } = 6 + \underline{\ ?\ }$
$x = 7$

2. $x - 3 = 5$
$x - 3 + \underline{\ ?\ } = 5 + \underline{\ ?\ }$
$x = 8$

 NOW USE SKILL 93. Solve the equation.

3. $n - 4 = 11$ **4.** $n - 9 = 3$ **5.** $n - 10 = 6$

6. $n - 15 = 0$ **7.** $x - 6 = {}^-20$ **8.** $x - 9 = 36$

9. $y - 3 = 9$ **10.** $z - 11 = 0$ **11.** $x - 7 = 11$

12. $y - 7 = 42$ **13.** $y - 15 = 30$ **14.** $y - 18 = 12$

15. $n - 8 = {}^-7$ **16.** $n - 18 = 72$ **17.** $n - 5 = 36$

18. $m - 8 = 9$ **19.** $x - 16 = 21$ **20.** $x - 12 = 18$

21. $n - 35 = 5$ **22.** $n - 19 = 18$ **23.** $x - 17 = {}^-7$

24. $y - 19 = 21$ **25.** $y - 14 = 4$ **26.** $x - 12 = 34$

In an equation, $y - 3 = 4$ and $4 = y - 3$ mean the same thing.

MODELS

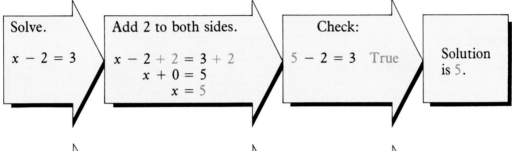

Solve.

$x - 2 = 3$

Add 2 to both sides.

$x - 2 + 2 = 3 + 2$
$x + 0 = 5$
$x = 5$

Check:

$5 - 2 = 3$ True

Solution is 5.

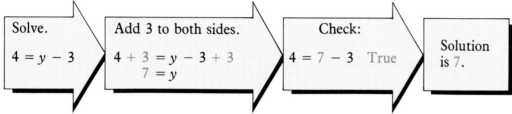

Solve.

$4 = y - 3$

Add 3 to both sides.

$4 + 3 = y - 3 + 3$
$7 = y$

Check:

$4 = 7 - 3$ True

Solution is 7.

 GET READY. Complete the second equation.

27. $5 = x - 1$
$5 + \underline{\quad?\quad} = x - 1 + \underline{\quad?\quad}$
$6 = x$

28. $x - 4 = 3$
$x - 4 + \underline{\quad?\quad} = 3 + \underline{\quad?\quad}$
$x = 7$

 NOW USE SKILL 93. Solve the equation.

29. $3 = m - 1$ **30.** $4 = x - 5$ **31.** $7 = x - 2$ **32.** $5 = x - 12$

33. $^-7 = x - 14$ **34.** $0 = y - 17$ **35.** $8 = x - 15$ **36.** $9 = y - 18$

37. $15 = x - 12$ **38.** $5 = x - 25$ **39.** $6 = y - 26$ **40.** $17 = n - 17$

41. $13 = y - 17$ **42.** $14 = n - 21$ **43.** $^-5 = n - 17$ **44.** $0 = n - 100$

45. $n - 4 = 3$ **46.** $n - 6 = 2$ **47.** $x - 4 = 4$ **48.** $y - 8 = 0$

49. $4 = n - 2$ **50.** $3 = x - 6$ **51.** $5 = n - 5$ **52.** $0 = n - 5$

53. $6 = n - 1$ **54.** $8 = n - 3$ **55.** $4 = x - 9$ **56.** $5 = x - 8$

57. $8 = n - 8$ **58.** $0 = x - 6$ **59.** $9 = y - 9$ **60.** $10 = y - 1$

61. $6 = y - 2$ **62.** $7 = y - 3$ **63.** $7 = y - 8$ **64.** $6 = n - 10$

For more practice: **Skill Bank,** page 445.

Solving Equations by Subtraction

Rule: To solve an addition equation, subtract the same number from both sides of the equation.

MODEL

| Solve. | Subtract 8 from both sides. | Check: | Solution is 4. |

$$n + 8 = 12$$

$$n + 8 = 12$$
$$n + 8 - 8 = 12 - 8$$
$$n + 0 = 4$$
$$n = 4$$

$$4 + 8 = 12 \quad \text{True}$$

 GET READY. Complete the second equation.

1. $n + 3 = 9$
$n + 3 - \underline{\ ?\ } = 9 - \underline{\ ?\ }$
$n = 6$

2. $x + 2 = 7$
$x + 2 - \underline{\ ?\ } = 7 - \underline{\ ?\ }$
$x = 5$

3. $x + 2 = 1$
$x + 2 - \underline{\ ?\ } = 1 - \underline{\ ?\ }$
$x = {}^-1$

4. $y + 3 = 12$
$y + 3 - \underline{\ ?\ } = 12 - \underline{\ ?\ }$
$y = 9$

5. $y + 5 = 5$
$y + 5 - \underline{\ ?\ } = 5 - \underline{\ ?\ }$
$y = 0$

6. $m + 2 = 4$
$m + 2 - \underline{\ ?\ } = 4 - \underline{\ ?\ }$
$n = 2$

 NOW USE SKILL 94. Solve the equation.

7. $x + 4 = 10$ **8.** $x + 9 = 12$ **9.** $x + 6 = 10$ **10.** $x + 5 = 6$

11. $n + 6 = 11$ **12.** $n + 9 = 16$ **13.** $y + 3 = 8$ **14.** $y + 11 = 12$

15. $x + 7 = 15$ **16.** $z + 7 = 16$ **17.** $n + 14 = 16$ **18.** $m + 8 = 8$

19. $n + 17 = 17$ **20.** $x + 9 = 18$ **21.** $n + 6 = 14$ **22.** $n + 7 = 14$

23. $x + 3 = 2$ **24.** $n + 4 = 1$ **25.** $n + 5 = 4$ **26.** $y + 1 = 2$

27. $n + 9 = 2$ **28.** $x + 13 = 12$ **29.** $x + 14 = 57$ **30.** $y + 14 = 12$

31. $x + 17 = 27$ **32.** $x + 27 = 17$ **33.** $x + 11 = 11$ **34.** $x + 11 = 0$

35. $y + 10 = 10$ **36.** $y + 15 = 0$ **37.** $n + 21 = 0$ **38.** $n + 70 = 70$

MODELS

Solve.	Subtract 4 from both sides.	Check:	Solution is $^-1$.
$3 = x + 4$	$3 - 4 = x + 4 - 4$ $^-1 = x + 0$ $^-1 = x$	$3 = {}^-1 + 4$ True	

Solve.	Subtract $^-2$ from both sides.	Check:	Solution is 6.
$n + {}^-2 = 4$	$n + {}^-2 - {}^-2 = 4 - {}^-2$ $n + 0 = 6$ $n = 6$	$6 + {}^-2 = 4$ True	

 GET READY. Simplify.

39. $3 - 5$	**40.** $6 - 7$	**41.** $10 - 15$	**42.** $4 - 4$
43. $^-3 - {}^-3$	**44.** $5 + {}^-2$	**45.** $^-1 + 5$	**46.** $5 - {}^-2$
47. $16 - 11$	**48.** $4 - {}^-9$	**49.** $^-6 - 12$	**50.** $^-5 - {}^-9$
51. $4 - {}^-4$	**52.** $11 - 16$	**53.** $^-8 - {}^-5$	**54.** $^-6 - 6$
55. $^-23 - {}^-23$	**56.** $^-8 - {}^-7$	**57.** $^-7 - {}^-8$	**58.** $6 - {}^-6$

 NOW USE SKILL 94. Solve the equation.

59. $5 = n + 1$	**60.** $10 = n + 6$	**61.** $15 = n + 8$	**62.** $16 = 3 + n$
63. $17 = 9 + n$	**64.** $2 = x + 10$	**65.** $3 = x + 9$	**66.** $5 = x + 10$
67. $14 = 8 + x$	**68.** $17 = 9 + m$	**69.** $11 = n + 3$	**70.** $0 = x + 1$
71. $0 = y + 9$	**72.** $9 = y + 9$	**73.** $13 = x + 13$	**74.** $3 = x + 11$
75. $x + {}^-1 = 8$	**76.** $x + {}^-2 = 2$	**77.** $x + {}^-2 = {}^-2$	**78.** $^-1 = x + 1$
79. $^-4 = n + 2$	**80.** $^-8 = n + 7$	**81.** $n + {}^-7 = 8$	**82.** $n + {}^-7 = {}^-8$
83. $n + {}^-8 = 4$	**84.** $n + {}^-16 = {}^-4$	**85.** $n + {}^-3 = 10$	**86.** $n + 11 = {}^-17$
87. $0 = m + {}^-4$	**88.** $x + {}^-5 = {}^-1$	**89.** $^-2 = n + 10$	**90.** $x + {}^-8 = {}^-5$

For more practice: **Skill Bank,** page 445.

Graphing the Solution of an Equation

MODEL

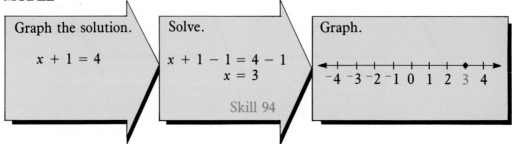

Graph the solution.

$$x + 1 = 4$$

Solve.

$$x + 1 - 1 = 4 - 1$$
$$x = 3$$

Skill 94

Graph.

 GET READY. Show these points on a number line.

1. 4   **2.** $^-1$ **3.** $^-5$ **4.** 0

5. 6   **6.** 2   **7.** $^-3$ **8.** $^-6$

NOW USE APPLICATION A41. Graph the solution of the given equation. Use a number line.

9. $x + 3 = 4$ **10.** $x + 4 = 8$ **11.** $x + 1 = 6$

12. $x - 3 = 1$ **13.** $x - 4 = 2$ **14.** $x - 4 = 0$

15. $x + 4 = 3$ **16.** $x + 4 = 1$ **17.** $x + 4 = 2$

18. $x - 4 = ^-1$ **19.** $x - 2 = 2$ **20.** $x + 2 = 2$

21. $n - 3 = ^-2$ **22.** $n - 5 = ^-3$ **23.** $n - 5 = ^-5$

24. $y + 1 = ^-5$ **25.** $y + 3 = ^-3$ **26.** $y + 2 = ^-4$

27. $4 = x + 5$ **28.** $3 = x + 5$ **29.** $1 = x + 5$

30. $5 = y + 5$ **31.** $4 = y + 6$ **32.** $3 = y + 7$

33. $4 = x - 2$ **34.** $2 = x - 1$ **35.** $6 = x - 1$

36. $2x = ^-8$ **37.** $n - 6 = 3$ **38.** $n + 6 = 3$

39. $n - 6 = ^-3$ **40.** $x + 10 = 4$ **41.** $3n = 18$

42. $^-4y = 20$ **43.** $x + 20 = 30$ **44.** $x - 20 = ^-30$

For more practice: **Application Bank,** page 464.

Graphing the Solution of an Inequality

MODEL

Graph the solution.	Solve.	Graph.
$x > 3$	3 is not a solution. All numbers greater than 3 are solutions.	Use an open dot to show that 3 is not one of the solutions. Draw a heavy arrow pointing to the right.

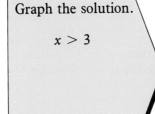

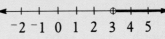

You can combine inequality signs with the equals sign.

$x \geq 3$ means "$x > 3$ or $x = 3$."
$x \leq 3$ means "$x < 3$ or $x = 3$."

MODEL

Graph the solution.	Solve.	Graph.
$x \leq {}^-3$	$^-3$ is a solution. All numbers less than $^-3$ are solutions.	Use a solid dot to show that $^-3$ is a solution. Draw a heavy arrow pointing to the left.

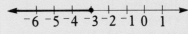

 NOW USE APPLICATION A42. Graph the solution on a number line.

1. $x > 2$ **2.** $x > 0$ **3.** $x > {}^-2$ **4.** $x > 4$

5. $n < 5$ **6.** $n < 2$ **7.** $n < {}^-1$ **8.** $n < 0$

9. $y > 1$ **10.** $y > {}^-1$ **11.** $y > 3$ **12.** $y > {}^-4$

13. $y \geq 2$ **14.** $y \geq 0$ **15.** $y \geq {}^-1$ **16.** $y \geq {}^-2$

17. $n \leq 5$ **18.** $n \leq 1$ **19.** $n \leq 0$ **20.** $n \leq {}^-3$

21. $y \geq 7$ **22.** $y \leq 9$ **23.** $x > {}^-7$ **24.** $x \leq 6$

For more practice: **Application Bank**, page 464.

Footloose

You are planning to take a backpacking trip during your summer vacation. To plan your route, you bought a trail guide of the area you will be visiting.

Guide to the Middle Range Trail

Points on the Route	Elev. (ft)	Dist. from pt. above (mi)
Golden Pass	10,900	0.0
Golden Pass Lake	10,350	0.6
North Fork of Lobo Creek	8,900	3.0
Lobo Pass Trail.	8,300	1.4
Lobo Creek Ford	7,700	1.6
Eagle Ridge.	9,950	4.6
Eagle Creek.	8,800	2.1
Rosemary Branch	9,250	2.3
Eagle Lakes Trail	9,500	1.2
Marie Lake	10,600	2.6
Belden Pass	10,872	1.3
Hart Lake	10,490	0.7
Onion Meadows	10,100	1.5
Singing Creek	9,700	1.7
Sawmill Meadows Trail	7,900	3.8
Modoc Pass Trail	8,000	1.7
Alpine Meadow.	8,300	1.5
East Fork Bridge.	8,350	2.1
Owens Meadow.	9,200	2.0
McGuire Meadow.	9,500	2.0
Canby Meadow.	9,900	1.0
Owens Lake	10,850	3.5
Rock Creek	9,200	1.3

You want to decide where you will stop to camp each night. Although you want to cover a reasonable distance, you also want to allow for a leisurely enough pace to enjoy the scenery, plants, and animals along the way. After discussing the trip with the other people who will be hiking with you, you all agree to some guidelines.

You will start at Golden Pass and finish at Rock Creek. You will camp each night at one of the places listed in the trail guide. You will hike as near to seven miles as possible each day, without going more than seven miles. On days when your elevation will rise by 2000 feet or more, you will hike as near to five miles as possible, without going more than five miles.

PROBLEM: Plan your trip. Decide where you will camp each night. Find out how far you will go each day, and how far you will hike on the entire trip.

Solution Plan: Pick your camping places by finding the total mileage between points. Choose the location that gives the greatest total for a day's hike without exceeding your guidelines.

WHAT DO YOU WANT TO KNOW? Use the information in the trail guide.

1. How far is it from Golden Pass to Golden Pass Lake?

2. How far is it from Golden Pass to the North Fork of Lobo Creek?

3. How far is it from Golden Pass to the Lobo Pass Trail?

4. How far is it from Golden Pass to Lobo Creek Ford?

5. How much does the elevation increase between Lobo Creek Ford and Eagle Ridge?

Use your hike guidelines to find out where you will spend each night and your mileage each day.

	6.	7.	8.	9.	10.	11.	12.
SAT	**SUN**	**MON**	**TUES**	**WED**	**THUR**	**FRI**	**SAT**
Lobo Creek Ford	?	?	?	?	?	?	Rock Creek
6.6 miles	? miles	? miles	? miles	? miles	? miles	? miles	? miles

13. How far do you hike on the whole trip, from Golden Pass to Rock Creek?

14. What is your average mileage per day? Round to the nearest tenth.

Skill 93 **Solve the equation.**

1. $y - 7 = 16$

2. $22 = m - 18$

Skill 94 **Solve the equation.**

3. $x + 28 = 35$

4. $39 = n + 11$

Application A41

Graph the solution on a number line.

5. $x + 2 = {}^-10$

6. $15 = y + 18$

Application A42

Graph the solution on a number line.

7. $x \leq {}^-1$

8. $c > 5$

Problem Solving

9. Your trail guide gives the distance of each point from the preceding point: Pine Lake 0 mi, Sink Hollow 2.3 mi, Green Summit 2.5 mi, Elk Meadow 1.6 mi. You want to hike no more than 6 miles. Where will you stop?

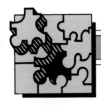

EXTRA!

Write a variable expression for each set of instructions. Evaluate for $n = 2, 4, 6$.

1. Pick a number.
Add 10.
Multiply by 2.
Divide by 4.
Subtract 5.
Multiply by 2.

2. Pick a number.
Add 7.
Multiply by 2.
Subtract 8.
Divide by 2.
Subtract the number picked.

Buying a House

There are many costs involved in buying a house. There are many questions to consider. For example, how large a down payment can you make? What interest rates do banks charge for a mortgage loan? How large a monthly mortgage payment can you afford?

Ken and Carol Rock agree to buy a house for $93,750. They can make a down payment of $18,750. What is the mortgage loan they need to finance the house?

$$\frac{\text{Selling}}{\text{price}} - \frac{\text{Down}}{\text{payment}} = \frac{\text{Mortgage}}{\text{loan}}$$

$$\$93,750 - \$18,750 = \$75,000$$

They shopped around for a bank with the best interest rate and loan conditions. They were offered two choices: a $75,000 mortgage loan at 16.75% interest for 25 years or at 17% interest for 20 years. Use the table to find out which mortgage results in a lower monthly payment.

MONTHLY PAYMENTS FOR $75,000 LOAN			
Annual Interest Rate	**Length of Loan (Years)**		
	20	**25**	**30**
16.75%	1085.87	1063.50	1054.05
17.00%	1100.11	1078.35	1069.26

The mortgage at 16.75% for 25 years has a lower monthly payment.

In figuring out the total amount paid over the life of the mortgage, which loan results in a lower total amount?

Monthly payment $\times$ Number of payments = Total amount paid

At 16.75%: $1063.50 $\times$ 300   = $319,050
At 17.00%: $1100.11 $\times$ 240   = $264,026

The mortgage at 17.00% for 20 years has the lower total amount paid.

1. House priced at $48,500.
   $9700 down payment.
   What is the mortgage loan?

2. $75,000 mortgage loan.
   17% interest rate for 30 years.
   What is the monthly payment?
   What is the total amount paid at the end of 30 years?

Solving Equations by Division

The equation $2x = 4$ may be called a multiplication equation because it is the same as $2 \cdot x = 4$.

Rule: To solve a multiplication equation, divide both sides by the same number.

MODELS

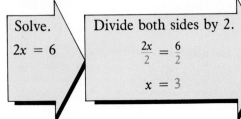

Solve.
$2x = 6$

Divide both sides by 2.
$\frac{2x}{2} = \frac{6}{2}$
$x = 3$

Check:
$2 \cdot 3 = 6$ True

Solution is 3.

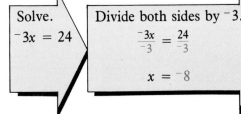

Solve.
$^-3x = 24$

Divide both sides by $^-3$.
$\frac{^-3x}{^-3} = \frac{24}{^-3}$
$x = {}^-8$

Check:
$^-3 \cdot {}^-8 = 24$ True

Solution is $^-8$.

GET READY. Simplify.

1. $\frac{3}{3}$

2. $\frac{^-4}{^-4}$

3. $\frac{6x}{6}$

4. $\frac{3a}{3}$

5. $1x$

6. $\frac{^-9x}{^-9}$

7. $\frac{8y}{8}$

8. $\frac{^-5m}{^-5}$

NOW USE SKILL 95. Solve the equation.

9. $5n = 15$ **10.** $4n = 16$ **11.** $7n = 28$ **12.** $9n = 27$

13. $8n = 72$ **14.** $6x = 54$ **15.** $11x = 121$ **16.** $12x = 144$

17. $4y = 52$ **18.** $8y = 128$ **19.** $5m = 125$ **20.** $7m = 567$

21. $15y = {}^-30$ **22.** $17x = {}^-51$ **23.** $20x = 100$ **24.** $^-5x = {}^-95$

25. $^-3x = 21$ **26.** $25y = 125$ **27.** $25y = {}^-125$ **28.** $14m = 42$

29. $5x = 40$ **30.** $4x = 40$ **31.** $^-4x = 16$ **32.** $4x = {}^-16$

33. $12y = 48$ **34.** $12y = {}^-48$ **35.** $10n = 200$ **36.** $3n = 0$

For more practice: **Skill Bank**, page 446.

Solving Equations by Multiplication

The equation $\frac{x}{3} = 2$ may be called a division equation because it is the same as $x \div 3 = 2$.

Rule: To solve a division equation, multiply both sides by the same number.

MODEL

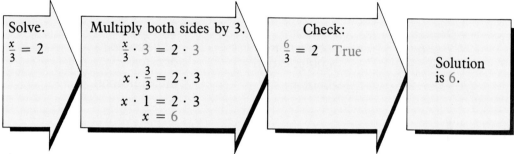

Solve.
$\frac{x}{3} = 2$

Multiply both sides by 3.
$$\frac{x}{3} \cdot 3 = 2 \cdot 3$$
$$x \cdot \frac{3}{3} = 2 \cdot 3$$
$$x \cdot 1 = 2 \cdot 3$$
$$x = 6$$

Check:
$\frac{6}{3} = 2$ True

Solution
is 6.

 GET READY. Simplify.

1. $4 \cdot \frac{1}{4}$ **2.** $7 \cdot \frac{1}{7}$ **3.** $\frac{4}{4}$ **4.** $\frac{^-4}{^-4}$

5. $\frac{8}{4}$ **6.** $^-5 \cdot \frac{a}{^-5}$ **7.** $3 \cdot \frac{x}{3}$ **8.** $10 \cdot \frac{x}{10}$

 NOW USE SKILL 96. Solve the equation.

9. $\frac{y}{3} = 3$ **10.** $\frac{x}{4} = 5$ **11.** $\frac{x}{2} = {}^-1$ **12.** $\frac{y}{2} = {}^-5$

13. $\frac{x}{4} = 3$ **14.** $\frac{x}{9} = 8$ **15.** $\frac{n}{8} = 8$ **16.** $\frac{n}{6} = 9$

17. $\frac{m}{5} = {}^-6$ **18.** $\frac{x}{^-1} = 11$ **19.** $\frac{q}{7} = 8$ **20.** $\frac{r}{9} = 7$

21. $\frac{c}{5} = 0$ **22.** $\frac{a}{9} = 0$ **23.** $\frac{y}{^-3} = {}^-4$ **24.** $\frac{y}{^-5} = {}^-7$

25. $\frac{y}{5} = 20$ **26.** $\frac{x}{4} = 30$ **27.** $\frac{n}{6} = 6$ **28.** $\frac{n}{25} = 5$

29. $\frac{n}{11} = 11$ **30.** $\frac{n}{8} = 12$ **31.** $\frac{x}{10} = 100$ **32.** $\frac{n}{100} = {}^-1$

33. $\frac{x}{16} = 5$ **34.** $\frac{x}{12} = {}^-12$ **35.** $\frac{y}{9} = 12$ **36.** $\frac{y}{9} = 9$

For more practice: **Skill Bank,** page 446.

Solving Equations

The first step when solving any equation is to decide which operation to use. Just remember that addition and subtraction are inverses; so are multiplication and division. If an equation shows one operation, use the inverse to solve.

Equation	Operation in equation	Operation to solve
$y + 9 = 10$	addition	subtraction
$x - 7 = 6$	subtraction	addition
$2x = 18$	multiplication	division
$\frac{y}{9} = 3$	division	multiplication

 GET READY. Name the operation to solve the equation.

1. $x + 2 = 3$

2. $y - 5 = 5$

3. $2 + n = 4$

4. $\frac{t}{5} = 6$

5. $9x = 36$

6. $\frac{x}{4} = -1$

 NOW USE SKILL 97. Solve the equation.

7. $n - 2 = 10$

8. $x + 3 = 12$

9. $5 + n = 15$

10. $10n = 100$

11. $4n = 32$

12. $5n = {}^-40$

13. $n + 7 = 35$

14. $x - 7 = 35$

15. $\frac{w}{7} = 35$

16. $5x = 35$

17. $\frac{x}{9} = 50$

18. $y - 4 = {}^-4$

19. $x + 4 = 4$

20. $\frac{v}{4} = {}^-1$

21. $4x = {}^-16$

22. $x + 17 = 52$

23. $n - 19 = 30$

24. $11x = 143$

25. $n + 11 = 9$

26. $n + 72 = 14$

27. $\frac{m}{8} = 14$

28. $50 + z = 49$

29. $z - 50 = 49$

30. $7y = 112$

31. $16n = {}^-64$

32. $15m = 615$

33. $14n = 210$

34. $x + 4 = 11.6$

35. $x - 2\frac{1}{2} = 8$

36. $\frac{n}{5} = 18.2$

For more practice: **Skill Bank,** page 447.

Using Two Operations

Sometimes you need to use two different operations to solve an
equation.

MODEL

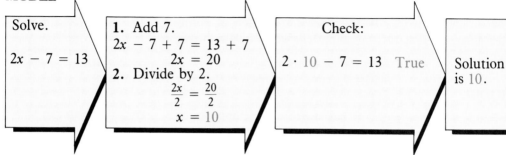

Solve.

$2x - 7 = 13$

1. Add 7.
$2x - 7 + 7 = 13 + 7$
$2x = 20$
2. Divide by 2.
$$\frac{2x}{2} = \frac{20}{2}$$
$x = 10$

Check:

$2 \cdot 10 - 7 = 13$ True

Solution
is 10.

 GET READY. Solve the equation.

1. $x - 6 = 16$ **2.** $2x = 22$ **3.** $n + 10 = 20$

4. $9 + y = {}^-4$ **5.** $7y = 21$ **6.** $x - 4 = 12$

7. $9n = {}^-27$ **8.** $4n = {}^-16$ **9.** $n + 4 = 0$

 NOW USE SKILL 98. Solve the equation.

10. $2x - 4 = 0$ **11.** $2x - 4 = 8$ **12.** $2x - 8 = 4$

13. $2x + 8 = 4$ **14.** $3x + 6 = 12$ **15.** $4x + 6 = 10$

16. $5n - 1 = 4$ **17.** $5n - 1 = 9$ **18.** $5n + 15 = 25$

19. $6 = 3n - 9$ **20.** $5 + 3n = 8$ **21.** $9x - 4 = 14$

22. $9x + 3 = 12$ **23.** $7y - 1 = 20$ **24.** $8y - 2 = 22$

25. $6n + 2 = 32$ **26.** $6n - 2 = 22$ **27.** $10x - 10 = 10$

28. $6y + 6 = 0$ **29.** $6y + 1 = 7$ **30.** $4n - 1 = {}^-21$

31. $5x - 6 = {}^-21$ **32.** $7n + 21 = 21$ **33.** $8x + 3 = 35$

34. $9y + 4 = 40$ **35.** $9y - 4 = {}^-40$ **36.** $8n - 5 = 75$

37. $11n - 7 = 70$ **38.** $11n + 7 = {}^-70$ **39.** $12x - 70 = 74$

For more practice: **Skill Bank,** page 447.

Graphs of Ordered Pairs

You can put two number lines together as shown. They cross at point zero on each line. You can use these lines, or **axes**, to graph number pairs. The horizontal axis is called the **x-axis**. The vertical axis is called the **y-axis**.

The ordered pair (3, ⁻4) describes the location of point *A*. To reach point *A* start from point 0, move 3 units to the right, then 4 units down.

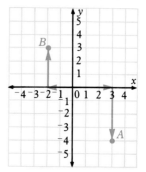

MODEL

Find the ordered pair that describes the location of point *B* in the graph above.

To reach *B* start at 0,
move 2 units
to the left ⟶ ⁻2
move 3 units up ⟶ 3

The ordered pair at point *B* is (⁻2, 3).

 NOW USE APPLICATION A43. Complete.

	Point	Moves from 0 Horizontal	Vertical	Ordered Pair
1.	C	2 right	3 up	(2, ?)
2.	D	3 right	0	(?, 0)
3.	E	? right	? up	(?, ?)
4.	F	? left	?	(⁻3, ?)
5.	G	?	? down	(?, ⁻3)
6.	H	?	?	(?, ?)
7.	I	?	?	(?, ?)
8.	J	?	?	(?, ?)
9.	K	0	?	(?, ?)

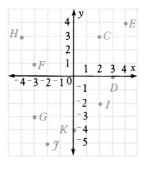

The numbers in an ordered pair that locates a point are called **coordinates.** The coordinates of point M are $(1, {}^-2)$.

Give the coordinates of the point.

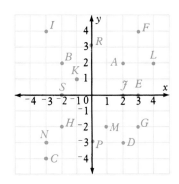

	Point	Coordinates
10.	A	$(?, ?)$
11.	B	$({}^-2, ?)$
12.	C	$(?, {}^-4)$
13.	D	$(?, ?)$
14.	E	$(?, 0)$

Give the letter of the point described.

15. $(3, 4)$ **16.** $(4, 2)$ **17.** $(1, {}^-2)$ **18.** $({}^-3, {}^-3)$

19. $(2, 0)$ **20.** $(3, {}^-2)$ **21.** $({}^-2, {}^-2)$ **22.** $({}^-3, 4)$

23. $({}^-1, 1)$ **24.** $(0, {}^-3)$ **25.** $(0, 3)$ **26.** $({}^-2, 0)$

Use graph paper to draw and number the axes. Locate the point described. Mark it with a dot and letter it on your grid.

	Point	Coordinates
27.	A	$(2, 3)$
28.	B	$({}^-2, 3)$
29.	C	$({}^-2, {}^-3)$
30.	D	$(2, {}^-3)$
31.	E	$(1, 0)$
32.	F	$(0, 1)$
33.	G	$(0, {}^-2)$
34.	H	$({}^-2, 0)$

35. Draw line segments to connect points A, B, C, and D. What have you drawn?

For more practice: **Application Bank,** page 465.

Rolling Along

You have been riding a ten-speed bike for a while. Now you are becoming a serious cyclist. You want to know more about the gear system of your bike.

A ten-speed bike has two sprockets, called chainwheels, in front. On your bike the small chainwheel has 40 teeth and the large chainwheel has 52 teeth. A ten-speed has five sprockets in the rear. Your rear sprockets have 14, 17, 20, 24, and 28 teeth. You have a 27-inch bike. This means that the diameter of your wheels is 27 inches.

You want to be able to compare the different gear combinations. You also want to know what are some reasonable speeds you could expect to achieve.

PROBLEM: For each of your ten gear combinations, how far does the bike go forward when you turn the pedals one complete turn? How fast can you expect to go, pedaling at a normal rate?

Solution Plan: Find the gear ratios. Multiply by the circumference of your wheel to find how far you go when the pedals are turned once. Then use your pedaling rate to find the speed in miles per hour.

WHAT DO YOU WANT TO KNOW? You want to know the gear ratio for each of the ten gear combinations.

In the first gear the chain is on the small chainwheel in front and the sprocket in the rear with 28 teeth. You figure that the gear ratio is 1.4 to the nearest tenth.

$$\text{Gear ratio in 1st gear} = \frac{40 \text{ chainwheel}}{28 \text{ rear sprocket}}$$

$$\text{Gear ratio} = 1.4$$

Complete the table of gear ratios. Round to the nearest tenth.

	Rear Sprocket	Chainwheel 40	Chainwheel 52
1.	14	5th: _?_	10th: _?_
2.	17	4th: _?_	9th: _?_
3.	20	3rd: _?_	8th: _?_
4.	24	2nd: _?_	7th: _?_
5.	28	1st: 1.4	6th: _?_

WHAT MORE DO YOU WANT TO KNOW?

6. What is the circumference of the wheel to the nearest inch?

7. What is the circumference of the wheel to the nearest foot?

NOW SOLVE THE PROBLEM. You want to know how far your bicycle will travel in each gear when the pedals are turned once.

The gear ratio is the number of times your wheel turns for one turn of the pedals. For example, when you turn the pedals once, in first gear, the wheels turn 1.4 times.

To find how far your bike travels when you turn the pedals once, multiply the gear ratio by the circumference of the wheel.

You figure that, in first gear, you move forward 9.8 feet every time you turn the pedals one full turn.

Gear ratio		Circumference of wheel	
1.4	×	7	= 9.8

Complete the table of distances traveled in each gear for one full turn of the pedals.

		8.	**9.**	**10.**	**11.**	**12.**	**13.**	**14.**	**15.**	**16.**
Gear	1st	2nd	3rd	4th	5th	6th	7th	8th	9th	10th
Distance (ft)	9.8	?	?	?	?	?	?	?	?	?

You know that you pedal at a rate of 60 turns per minute. This is 3600 turns per hour.

17. In 1st gear, how many feet do you travel an hour?

18. In 10th gear, how many feet do you travel in an hour?

Find your speed in miles per hour. There are 5280 feet in a mile. Round to the nearest tenth.

Miles per hour	=	Feet per hour	÷ 5280

19. 1st gear   **20.** 10th gear

Quiz

Skill 95

Solve the equation.

1. $^-15x = 75$ 2. $21y = 84$

Skill 96

Solve the equation.

3. $\frac{m}{9} = 18$ 4. $\frac{x}{5} = ^-21$

Skill 97

Solve the equation.

5. $n + 11 = 55$ 6. $4y = 56$

Skill 98

Solve the equation.

7. $10x + 7 = ^-23$ 8. $2x - 6 = 18$

Application A43

Use graph paper to draw and number the axes. Locate the point described. Mark it with a dot and letter it on your grid.

9. $P(^-6, ^-3)$ 10. $Q(4, ^-5)$ 11. $R(0, 7)$

Problem Solving

12. Your bike has 27-inch wheels. In 6th gear, the chainwheel has 50 teeth and the rear sprocket has 20 teeth. How far does your bike travel, to the nearest foot, when you turn the pedals once?

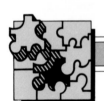

EXTRA!

Barry's little sister has a secret club. She and her friends meet every afternoon at a different location. They use a secret code to send messages about where to meet.

Use Code Delta to find the club's secret meeting places.

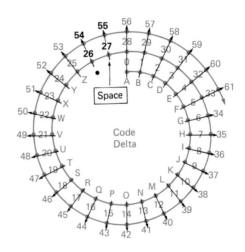

1. 56-2-17-42-46-18-27-19-7-4-27-17-14-18-4-27-1-20-18-7-26

2. 1-4-7-8-13-3-27-19-7-4-27-6-0-17-0-6-4-26

3. 19-22-4-13-19-24-27-15-0-2-4-18-27-13-14-17-19-7-27-14-5-27-19-7-4-27-0-15-15-11-4-27-19-17-4-4-26

Closing Costs

When a house is sold, the buyer and seller meet to transfer ownership. This meeting is also called a *closing*. At the closing the buyer pays the closing costs. The closing costs may include fees for the bank's lawyers, credit checks, loan origination, title search, documentation, and taxes. Some banks charge a flat fee for the closing costs, while others itemize the charges.

Many states have laws that set a maximum interest rate that banks and lending institutions can charge. When the trend in interest rates is above the maximum allowed by law, many institutions will add "points" to their basic mortgage charge. A point is equal to 1% of the mortgage loan and is included in the closing costs.

The Levitzes got a $55,000 mortgage loan from Fairfield City Bank. In addition to the charges shown, the bank charged 2 points. How much did the Levitzes pay for closing costs?

Title search:	$35
Appraisal fee:	$45
Credit report:	$10
Legal fee:	$400
Property tax:	$750

Sum of charges: $35 + $45 + $10 + $400 + $750 = $1240

2 points: 2% of $55,000 = $1100

Sum of charges + Points = Total closing cost
$1240   + $1100 = $2340

Find the total closing costs.

1. $83,000 mortgage loan
    bank fees:
    legal $470
    appraisal   75
    credit report   20
    3 points

2. $60,000 mortgage loan
    bank fees:
    title insurance $118
    recording   30
    plot plan   65
    property tax   600
    1 point

Equations with Two Variables

The equation $x + y = 6$ tells you that the sum of the two variables is 6. The value of the variables may be 1 and 5, or 4 and 2, or any other combination as long as the sum is 6. To see whether a pair of numbers is a solution, check by substituting the values in the equation.

For each value of x there is one and only one value of y. Write the values as an ordered pair (x, y). The ordered pair $(1, 5)$ means $x = 1$, $y = 5$; however, $(5, 1)$ means $x = 5$, $y = 1$.

MODEL

Equation: $x + y = 8$
Which of these ordered pairs are solutions?

$(3, 5)$
$(4, 5)$
$(11, {}^-3)$

Substitute.

$x + y = 8$

$3 + 5 = 8$ True
$4 + 5 = 8$ False
$11 + {}^-3 = 8$ True

The ordered pairs $(3, 5)$ and $(11, {}^-3)$ are solutions.

 GET READY. True or false?

1. $(x, y) = (4, {}^-5)$ means $x = 4$, $y = {}^-5$.

2. $(x, y) = ({}^-5, 4)$ means $x = 4$, $y = {}^-5$.

3. $(3, 4)$ is a solution of $x + y = {}^-7$ **4.** $({}^-1, {}^-2)$ is a solution of $x + y = {}^-3$.

 NOW USE SKILL 99. Is the given ordered pair a solution of the equation?

5. $x + y = 11$; $(8, 3)$ **6.** $x - y = 8$; $(16, 8)$

7. $x + y = 24$; $(12, 12)$ **8.** $x + y = 0$; $(4, {}^-4)$

9. $x + y = 12$; $(12, 0)$ **10.** $x - y = {}^-6$; $(6, 0)$

11. $y = 2x$; $(4, 8)$ **12.** $x = 2y$; $(10, 5)$

13. $x + 2y = 0$; $(4, {}^-2)$ **14.** $5x + y = 15$; $(3, 6)$

15. $y = 2x + 3$; $(0, 3)$ **16.** $y = 4x - 1$; $(3, 12)$

17. $y = 4 - 4x$; $(1, 0)$ **18.** $xy = 14$; $({}^-2, 7)$

19. $\frac{x}{7} = y$; $(21, 3)$ **20.** $3x + 2y = 10$; $(2, 2)$

               For more practice: **Skill Bank**, page 448.

Solving for One Variable

MODEL

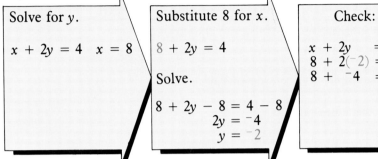

Solve for y.

$x + 2y = 4$ $x = 8$

Substitute 8 for x.

$8 + 2y = 4$

Solve.

$8 + 2y - 8 = 4 - 8$
$2y = {}^-4$
$y = {}^-2$

Check:

$x + 2y = 4$
$8 + 2({}^-2) = 4$
$8 + {}^-4 = 4$ True

$y = -2$

> **GET READY.** Solve the equation.

1. $2x - 8 = 0$ **2.** $3y = {}^-9$ **3.** $4y = 16$ **4.** $5 + 2y = 3$

5. $4 + 2x - 4 = 10$ **6.** $5 + 4y - 5 = 0$ **7.** $9 - 2x - 9 = 2$ **8.** $y + 1 = {}^-3$

> **NOW USE SKILL 100.** Find the value of y.

9. $y = x + 2; x = 2$ **10.** $y = x - 4; x = 4$ **11.** $y = 2x + 1; x = 1$

12. $y = 4x - 3; x = 3$ **13.** $y = 4x - 4; x = 1$ **14.** $y = 11x - 1; x = 0$

15. $x + y = 5; x = {}^-5$ **16.** $x - y = 16; x = 20$ **17.** $x + 3y = 6; x = 6$

18. $2x + y = 10; x = 2$ **19.** $8x + y = 8; x = 1$ **20.** $2x + 3y = 18; x = 3$

Find the missing number that will make the ordered pair a solution of the equation.

21. $y = 4 + x; (3, ?)$ **22.** $y = x - 8; (3, ?)$ **23.** $y = 10 - x; (3, ?)$

24. $x = y + 1; (?, 1)$ **25.** $x = y - 1; (?, {}^-1)$ **26.** $x = 12 + y; (?, {}^-2)$

27. $x + y = 9; (4, ?)$ **28.** $x + y = 15; (4, ?)$ **29.** $x - y = 4; (?, 4)$

30. $x + y = 16; (?, 9)$ **31.** $x + y = 16; (7, ?)$ **32.** $2x + y = 5; (?, 1)$

33. $2x + y = 11; (?, 3)$ **34.** $x + 3y = 8; (?, 2)$ **35.** $x + 3y = 8; ({}^-1, ?)$

36. $2x + 3y = 9; (?, 3)$ **37.** $2x + 3y = 4; (5, ?)$ **38.** $4x - 3y = 42; (15, ?)$

39. $y = 3x; (7, ?)$ **40.** $y = 3x; (?, 27)$ **41.** $y = 3x; (?, {}^-24)$

For more practice: **Skill Bank,** page 448.

Graphing Equations

To draw the graph of an equation such as $x + y = 1$, take these steps:

1: Substitute several values of x (at least three) in $x + y = 1$, and solve to get the y values. Show your values in a table.

$$x + y = 1$$

x	Substitute	y	Ordered Pair
$^-1$	$^-1 + y = 1$	2	$(^-1, 2)$
0	$0 + y = 1$	1	$(0, 1)$
1	$1 + y = 1$	0	$(1, 0)$
2	$2 + y = 1$	$^-1$	$(2, ^-1)$

2. Locate the points described by the ordered pairs. Mark a dot for each point.

3. Draw a line through the points you marked. This line is the graph of $x + y = 1$.

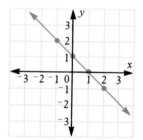

Note:

1. Any point on the line is a solution of $x + y = 1$. Test $(^-2, 3)$.
2. Any point not on the line is not a solution of $x + y = 1$. Test $(0, 2)$.

 GET READY. If $x = 2$, find the value of y.

1. $x + y = 4$ **2.** $x + y = 5$ **3.** $x + y = 10$

4. $2x + y = 1$ **5.** $2x - y = 3$ **6.** $2x - y = 1$

7. $y = x + 3$ **8.** $y = x - 3$ **9.** $y = x + 2$

10. $x + 2y = 4$ **11.** $x + 7y = 16$ **12.** $x - 2y = 4$

Is the given ordered pair a solution of the equation?

13. $x + y = 10$; $(3, 8)$ **14.** $x + y = 13$; $(8, 5)$

15. $y - x = 6$; $(9, 3)$ **16.** $y = 3x$; $(2, 6)$

17. $y = 2x - 5$; $(4, 3)$ **18.** $2x + 3y = 0$; $(^-3, ^-2)$

NOW USE APPLICATION A44. Complete the table. Mark the point for each ordered pair. Then draw a line through the points to graph the equation.

19. $y = x + 4$

x	y	(x, y)
-2	2	(-2, 2)
-1	3	?
0	?	?
1	?	?
2	?	?

20. $x - y = 1$

x	y	(x, y)
-1	?	?
0	?	?
1	0	?
2	?	?
3	?	?

21. $y = 2x$

x	y	(x, y)
-2	-4	?
-1	?	?
0	?	?
1	?	?
2	?	?

22. $x + y = 5$

x	y	(x, y)
-2	?	?
-1	?	?
0	?	?
1	?	?
2	?	?

23. $2x + y = 4$

x	y	(x, y)
-1	6	?
0	?	?
1	?	?
2	?	?
3	?	?

24. $3x - y = 4$

x	y	(x, y)
0	?	?
1	?	?
2	2	?
3	?	?
4	?	?

For more practice: **Application Bank,** page 465.

Pension Plan

You are thinking about retirement. You want to have some idea about the amount of pension payments you will receive when you retire. Your company has a handbook that describes the pension plan. You look up the information you need to figure your pension.

Your pension is based on your average annual earnings. This is figured by taking your earnings from the last 15 years, finding the five years during this period with the highest earnings, and taking the average of the earnings for those five years. For each year of service to the company you will receive $1\frac{2}{3}\%$ of your average annual earnings as your yearly pension payment. The normal retirement age is 65. You will be 65 in 1987.

The company gives you a record of your earnings for the last 15 years.

To play it safe, you assume that your annual earnings for 1985–1987 will be the same as in 1984.

PROBLEM: What will your pension be if you retire at 65 in 1987? What will your pension be if you take early retirement at age 62 in 1984?

Year	Earnings	Year	Earnings
1970	$14,206	1978	$18,491
1971	15,385	1979	19,520
1972	17,050	1980	20,748
1973	20,847	1981	26,625
1974	17,240	1982	23,274
1975	18,100	1983	23,560
1976	21,761	1984	24,080
1977	21,185		

Solution Plan: Figure your average annual earnings at age 62 and at age 65. Use the information from the handbook to find your pension.

WHAT DO YOU KNOW? You know in which years your earnings were highest.

1. If you retire at the end of 1984, in which 5 years did you have the highest earnings?

2. If you retire at the end of 1987, in which 5 years did you have the highest earnings?

WHAT DO YOU WANT TO KNOW? You want to know your average annual earnings. This is the average of the 5 years with the highest earnings.

3. If you retire at 65 in 1987, what are your average annual earnings?

4. If you retire at 62 in 1984, what are your average annual earnings?

WHAT MORE DO YOU WANT TO KNOW? You need to know your number of years of service. You started to work on January 1, 1964.

5. If you retire at the end of 1987, how many years of service do you have?

6. If you retire at the end of 1984, how many years of service do you have?

NOW SOLVE THE PROBLEM. You receive $1\frac{2}{3}\%$ of your average annual earnings for each year of service. You figure that you get 35% of average annual earnings if you retire early in 1984.

$$1\tfrac{2}{3}\% \times 21 = 35\%$$

7. What percent of average annual earnings do you get if you retire in 1987?

8. Find your yearly pension payment if you retire at 65 at the end of 1987.

9. Find your yearly pension payment if you retire at 62 at the end of 1984.

10. How much more is your yearly pension payment if you wait until age 65 to retire?

Quiz

Skill 99 **Is the given ordered pair a solution of the equation?**

1. $x + 8y = 19$; $(19, 0)$

2. $9y = 2x$; $(2, 9)$

Skill 100 **Find the number that will make the ordered pair a solution.**

3. $10x + y = 23$; $(0, ?)$

4. $3x - 2y = 25$; $(?, 7)$

Application A44

Problem Solving

Complete the table. Mark the point for each ordered pair. Then draw a line through the points to graph the equation.

5. $y = x - 7$

x	y	(x, y)
$^-3$	?	?
0	?	?
4	?	?

6. Your average annual earnings were $19,500. Your pension will be $1\frac{2}{3}\%$ of average annual earnings for each year of service. You began work on January 1, 1961, and you retire on December 31, 1984. How much is your pension?

CONSUMER NOTE

Nearly all homeowners have to pay real estate taxes. The tax money is used to support the operation of the town or county. The local government sets the tax rate and assesses the value of each home.

The tax rate may be expressed as mills per dollar of the assessed value. (A mill is one thousandth of a dollar, or $0.001.) A tax rate of 32 mills per dollar is the same as a tax rate of $32 per $1000.

The Renards' home has an assessed value of $54,000. The tax rate is 29 mills per dollar. Find their real estate tax.

$$29 \text{ mills per dollar} = \$29 \text{ per } \$1000$$
$$\$29 \times 54 = \$1566 \qquad \text{The Renards' tax is } \$1566.$$

Find the real estate tax.

1. Assessed value: $33,000   tax rate: 43 mills per dollar

2. Assessed value: $61,000   tax rate: $38 per $1000

3. Assessed value: $40,000   tax rate: 65 mills per dollar

4. Assessed value: $22,500   tax rate: $24 per $1000

Skills Review

Skill 93 **Solve the equation.** (pages 366–367)

1. $n - 5 = 9$ **2.** $8 - n = {}^-1$ **3.** $7 = n - 12$

4. $10 - n = 4$ **5.** $13 = y - 4$ **6.** $n - 8 = 17$

Skill 94 **Solve the equation.** (pages 368–369)

7. $x + 3 = 12$ **8.** $x + 11 = 11$ **9.** $x + 9 = 0$

10. $6 = x + 4$ **11.** ${}^-1 = x + 8$ **12.** $x + 10 = {}^-16$

Skill 95 **Solve the equation.** (page 376)

13. $6n = 18$ **14.** ${}^-4n = 36$ **15.** $12n = {}^-84$

16. ${}^-5x = 125$ **17.** $9n = 0$ **18.** $13n = {}^-13$

Skill 96 **Solve the equation.** (page 377)

19. $\frac{x}{6} = {}^-3$ **20.** $\frac{x}{3} = 45$ **21.** $\frac{x}{20} = {}^-1$

22. $\frac{x}{15} = 3$ **23.** $\frac{x}{100} = 0$ **24.** $\frac{x}{13} = {}^-13$

Skill 97 **Solve the equation.** (page 378)

25. $n - 4 = 10$ **26.** $5 + n = 16$ **27.** $6n = {}^-36$

28. $\frac{n}{14} = {}^-2$ **29.** ${}^-8n = 40$ **30.** $\frac{n}{9} = {}^-9$

Skill 98 **Solve the equation.** (page 379)

31. $6x - 2 = 16$ **32.** $13 + 4x = {}^-19$ **33.** $8x - 6 = 58$

34. $12x + 35 = 23$ **35.** $8x - 7 = 1$ **36.** $100 - 2x = {}^-2$

Skill 99 **Is the given ordered pair a solution of the equation?** (page 386)

37. $x = y + 3$; $(9, 6)$ **38.** $x - y = 6$; $(6, {}^-12)$ **39.** $x + y = 6$; $({}^-4, 9)$

40. $2x + 3y = 70$; $(8, 18)$ **41.** $\frac{x}{2} = y$; $(16, {}^-8)$ **42.** $y - x = 5$; $(0, 5)$

Skill 100 **Find the number that will make the ordered pair a solution.** (page 387)

43. $x + y = 13$; $(6, ?)$ **44.** $y - x = 1$; $(?, {}^-3)$ **45.** $2x + y = 0$; $(?, 2)$

46. $\frac{x}{y} = 6$; $({}^-36, ?)$ **47.** $xy = 0$; $(9, ?)$ **48.** $\frac{x}{7} = y$; $(63, ?)$

Application A41 Graph the solution of the given equation on a number line. (page 370)

1. $x - 4 = 3$

2. $2 = n - 6$

3. $y + 3 = {}^-2$

4. $^-4 = x + 4$

Application A42 Graph the solution of the given inequality on a number line. (page 371)

5. $n \leq 7$

6. $x > {}^-5$

7. $y \geq {}^-1$

8. $x < 2$

Application A43 Give the coordinates of the point. (pages 380–381)

Point	Coordinates
9. *A*	$(?, 4)$
10. *B*	$(4, ?)$
11. *C*	$(?, ?)$
12. *D*	$(?, ?)$

Locate each point on a grid. Use graph paper.

Point	Coordinates
13. *X*	$(0, 3)$
14. *Y*	$(^-2, ^-4)$
15. *Z*	$(2, ^-4)$

16. Draw line segments to connect points *X*, *Y*, and *Z* in order. What have you drawn?

Application A44 Complete the table. Then draw the graph using the ordered pairs. (pages 388–389)

17. $2x + y = 5$

x	*y*	*(x, y)*
$^-2$	9	?
0	?	?
3	?	?

Skill 93 Solve the equation.

1. $n - 18 = 26$ 2. $50 = x - 25$

Skill 94 Solve the equation.

3. $y + 16 = 43$ 4. $38 = m + 18$

Skill 95 Solve the equation.

5. $15x = 45$ 6. $6n = {}^-84$

Skill 96 Solve the equation.

7. $\frac{y}{10} = 15$ 8. $\frac{x}{12} = 11$

Skill 97 Solve the equation

9. $\frac{n}{15} = {}^-12$ 10. $y + 15 = 75$

Skill 98 Solve the equation.

11. $4x - 16 = 32$ 12. $6n - 45 = 45$

Skill 99 Is the given ordered pair a solution of the equation?

13. $2x + 4y = {}^-4$; $({}^-10, 4)$ 14. $\frac{x}{7} = 2y$; $(14, 1)$

Skill 100 Find the number that will make the ordered pair a solution.

15. $x + 5y = 20$; $(?, 4)$ 16. $2x - 3y = 7$; $(8, ?)$

Application A41

Graph the solution on a number line.

17. $n + 8 = 3$ 18. $-8 = t - 10$

Application A42

Graph the solution on a number line.

19. $p < 6$ 20. $q \leq {}^-2$

Application A43

Use graph paper to draw and number the axes. Locate the point described. Mark it with a dot and letter it on your grid.

21. $V(4, 5)$ 22. $W({}^-3, 6)$ 23. $R(2, {}^-3)$ 24. $S({}^-1, {}^-5)$

Application A44

Complete the table. Mark the point for each ordered pair. Then draw a line through the points to graph the equation.

25. $x + 2y = 5$

x	y	(x, y)
$^-1$	3	$(^-1, ?)$
3	1	?
5	?	?
7	?	?

Cumulative Review: Units 10–12

Skills

Skill 74 **Write a positive or negative number.** (page 302).

1. 325 ft above sea level
2. a loss of $90

Skill 75 **Write > or < in place of __?__.** (page 303)

3. 6 __?__ ⁻8
4. ⁻3 __?__ ⁻6
5. 7 __?__ ⁻7
6. ⁻2 __?__ ⁻1

Skill 76 (page 304)

7. ⁻3 + ⁻2
8. ⁻15 + ⁻4
9. ⁻12 + ⁻15
10. ⁻9 + ⁻19

Skill 77 (page 305)

11. 11 + ⁻3
12. ⁻24 + 12
13. ⁻2 + 10
14. 15 + ⁻36

Skill 78 (page 312)

15. 8 − 9
16. ⁻3 − 7
17. ⁻1 − 1
18. ⁻6 − 7

Skill 79 (page 313)

19. 13 − ⁻1
20. ⁻7 − ⁻2
21. 6 − ⁻6
22. ⁻15 − ⁻20

Skill 80 (page 314)

23. 3 × 4
24. ⁻8 × ⁻9
25. ⁻9 × ⁻3
26. ⁻16 ÷ ⁻4

Skill 81 (page 315)

27. 8 × ⁻4
28. ⁻2 × 2
29. ⁻18 ÷ 3
30. 25 ÷ ⁻5

Skill 82 **Simplify.** (page 322)

31. 5 + 7 × 4
32. 5 × 1 − 3
33. 4 − 4 × 4 + 3

Skill 83 **Simplify.** (page 323)

34. 6 ÷ (1 − 2)
35. (12 − 4) × 3
36. 5 − (3 × 3)

Skill 84 **State the meaning of the variable expression.** (page 334)

37. 22 − b
38. $\frac{z}{4}$
39. 2 × (c + d)

Skill 85 **State the meaning of the expression.** (page 335)

40. $10a$ **41.** $6 \cdot 9$ **42.** $10(x + y)$

Skill 86 **Find the value of the expression.**
Use $x = {-2}, y = 4, z = 1.$ (pages 336–337)

43. $x + 2z$ **44.** $x - y^2$ **45.** $\frac{yz}{x}$

Skill 87 **Simplify.** (page 344)

46. $a + 0$ **47.** $6 + x - 6$ **48.** $n \cdot 1$ **49.** $\frac{7}{7} \cdot c$

Skill 88 **Simplify by combining like terms.** (page 345)

50. $4a - 2a$ **51.** $8n + 3n$ **52.** $2b + 3b - b$ **53.** $2y + 3y - 5x$

Skill 89 **Write with exponents.** (page 346)

54. $6 \times 6 \times 6$ **55.** $a \cdot a \cdot a \cdot a$ **56.** $7 \cdot 7$ **57.** $(a - b)(a - b)$

Skill 90 **Find the value of the expression.**
Use $a = 2, b = 1, c = {-3}.$ (page 347)

58. $a^2 b$ **59.** $3b^2 c$ **60.** $(a - b)^2$ **61.** $4abc$

Skill 91 **Find the solution.** (page 354)

62. $\frac{m}{2} = 4$ **63.** $n + 9 = 18$ **64.** $\frac{y}{3} = 15$ **65.** $x - 3 = 7$

Skill 92 **Check these values for x: ${-1}, 0, 1, 2, 4.$ Which of these
values are solutions of the inequality?** (page 355)

66. $x > 2$ **67.** $x + 1 > 1$ **68.** $x + 6 < 7$ **69.** $\frac{x}{2} > 1$

Skill 93 **Solve the equation.** (pages 366–367)

70. $z - 4 = 16$ **71.** $n - 6 = {-5}$ **72.** $n - 25 = 5$

Skill 94 **Solve the equation.** (pages 368–369)

73. $a + 7 = 13$ **74.** $n + 28 = 11$ **75.** ${-15} = 7 + b$

Skill 95 **Solve the equation.** (page 376)   Skill 96 **Solve the equation.** (page 377)

76. $7n = 21$ **77.** $4n = {}^{-}12$ **78.** $\frac{w}{4} = 4$ **79.** $\frac{y}{7} = 5$

Skill 97 **Solve the equation.** (page 378)   Skill 98 **Solve the equation.** (page 379)

80. $f - 3 = 15$ **81.** $g + 6 = {}^{-}20$ **82.** $4x - 16 = 0$ **83.** $6n - 10 = 2$

Skill 99 **Is the given ordered pair a solution of the equation?** (page 386)

84. $x + y = 16$; $(8, 8)$ **85.** $y - 3 = x$; $(8, 5)$ **86.** $\frac{x}{3} = y$; $({}^{-}9, 3)$

Skill 100 **Find the number that will make the ordered pair a solution.** (page 387)

87. $x = 6 + y$; $(3, ?)$ **88.** $2x = 2y - 10$; $(2, ?)$ **89.** $y = 3x$; $(?, {}^{-}6)$

Applications

Application A35 **Write a sum of positive or negative numbers to represent each problem. Then add to solve the problem.** (pages 306–307)

1. Donald's savings account has a balance of $355. He withdraws $205 and $125, and then deposits $98. What is his new balance?

2. Ann climbed to an elevation of 3200 ft. She then descended 1000 ft and set up camp. The next day she ascended 500 ft to take photos. What was her elevation when she stopped to take photos?

Application A36 **Write a difference of positive or negative numbers to represent the situation. Then subtract to solve the problem.** (pages 316–317)

3. The daytime temperature is 5°F below zero. That night it falls to 20°F below zero. What is the change in the temprature?

4. In its first year your company's loss is $5000. In its second year its profit is $14,000. What is the change in your company's earnings?

Application A37 (pages 324–325)

5. Use the chart to find what time it is in each city when you make a 3:00 P.M. (EST) telephone call to

   (a) Denver

   (b) Los Angeles

   (c) New York.

Location	Time Difference U.S. Time Zone			
	EST	CST	MST	PST
Denver	⁻2	⁻1	0	1
Los Angeles	⁻3	⁻2	⁻1	0
New York	0	1	2	3

Application A38 **Write as a variable expression.** (pages 338–339)

6. 6 added to n **7.** 8 less than x **8.** the product of a and twice b
9. the ratio of m to n **10.** the sum of x and y **11.** 2 more than x

Application A39 **Use the formula and values to find the area.** (pages 348–349)

12. Square
$A = s^2$
$s = 3$ in.

13. Triangle
$A = \frac{1}{2} bh$
$b = 4$ ft, $h = 9$ ft

14. Rectangle
$A = lw$
$l = 4$ cm, $w = 7$ cm

Application A40 **Select the answer that best fits the problem.** (pages 356–357)

15. Juice costs 50¢. Juice and toast cost 85¢.

  a. $n + 50n = 85$ **b.** $50 + n = 85$ **c.** $2n = 85$ **d.** $n - 50 = 85$

Application A41 **Graph the solution of the given equation. Use a number line.** (page 370)

16. $x + 5 = 6$ **17.** $x + 2 = {}^{-}4$ **18.** $n - 10 = {}^{-}6$

Application A42 **Graph the solution on a number line.** (page 371)

19. $x > 5$ **20.** $n \leq 7$ **21.** $y > {}^{-}1$

Application A43 **Use graph paper to draw and number the axes. Locate the point described. Mark it with a dot and letter it on your grid.** (pages 380–381)

22. $A(1, 4)$ **23.** $B({}^{-}3, 2)$ **24.** $C({}^{-}2, {}^{-}5)$

Application A44 **Complete the table. Mark the point for each ordered pair. Then draw a line through the points to graph the equation.** (pages 388–389)

25. $y = x + 6$

x	y	(x, y)
$^{-}1$	5	$(^{-}1, 5)$
0	?	?
1	?	?
2	?	?

26. $y = 3x$

x	y	(x, y)
$^{-}2$	$^{-}6$	$(^{-}2, ^{-}6)$
$^{-}1$	?	?
0	?	?
1	?	?

Sequential Skills Outline

Sequential Skills Outline

Sequential Skills Outline

Sequential Applications Outline

Sequential Applications Outline

Skill 1
(pages 2–3)

462.59

The digit 6 is in the tens' place. Its value is 60.

The digit 5 is in the tenths' place. Its value is 0.5, or $\frac{5}{10}$.

Give the place and value of the underlined digit.

1. 4̲6   **2.** 30̲9   **3.** 9̲0   **4.** 5̲28

5. 4̲09   **6.** 12̲,306   **7.** 98,0̲06   **8.** 2̲4,970

9. 23.17̲   **10.** 6̲.05   **11.** 142.9̲6   **12.** 300.5̲

13. 3.009̲   **14.** 267̲.591   **15.** 7̲2.098   **16.** 100̲0.59

17. 6.05̲   **18.** 142.9̲6   **19.** 600.5̲   **20.** 300.95̲

21. 72.098̲   **22.** 1254.00̲5   **23.** 690.0̲31   **24.** 2̲,321,625.3   **25.** 325,4̲27.031

Skill 2
(pages 4–5)

16,872.06 is 16,900 to the nearest hundred.

42.2346 is 42.23 to the nearest hundredth.

Round each number to the nearest hundred.

26. 306   **27.** 1496   **28.** 4960   **29.** 5009

30. 142.6   **31.** 6908.5   **32.** 989.5   **33.** 4360.2

Round each number to the nearest ten.

34. 78   **35.** 62   **36.** 75   **37.** 89.4

38. 467   **39.** 382.7   **40.** 497.9   **41.** 161.43   **42.** 677.211

Round each number to the nearest thousand.

43. 17,900   **44.** 24,390   **45.** 10,986.2   **46.** 2099   **47.** 3500

Round each number to the nearest tenth.

48. 6.29   **49.** 10.39   **50.** 29.08   **51.** 39.96   **52.** 34.24

53. 0.469   **54.** 190.85   **55.** 0.082   **56.** 0.925   **57.** 3.45

Round each number to the nearest hundredth.

58. 9.091   **59.** 27.354   **60.** 103.096   **61.** 5640.063   **62.** 3.284

63. 0.527   **64.** 0.908   **65.** 10.998   **66.** 207.836   **67.** 82.483

Skill 3
(page 12)

$$\begin{array}{r} 325 \\ + 142 \\ \hline 467 \end{array}$$

1. $\begin{array}{r} 93 \\ + 2 \\ \hline \end{array}$ **2.** $\begin{array}{r} 67 \\ + 1 \\ \hline \end{array}$ **3.** $\begin{array}{r} 34 \\ + 5 \\ \hline \end{array}$ **4.** $\begin{array}{r} 21 \\ + 7 \\ \hline \end{array}$

5. $\begin{array}{r} 85 \\ + 3 \\ \hline \end{array}$ **6.** $\begin{array}{r} 62 \\ + 7 \\ \hline \end{array}$ **7.** $\begin{array}{r} 75 \\ + 4 \\ \hline \end{array}$ **8.** $\begin{array}{r} 97 \\ + 2 \\ \hline \end{array}$

9. $\begin{array}{r} 23 \\ + 46 \\ \hline \end{array}$ **10.** $\begin{array}{r} 17 \\ + 81 \\ \hline \end{array}$ **11.** $\begin{array}{r} 42 \\ + 37 \\ \hline \end{array}$ **12.** $\begin{array}{r} 25 \\ + 43 \\ \hline \end{array}$ **13.** $\begin{array}{r} 32 \\ + 51 \\ \hline \end{array}$

14. $\begin{array}{r} 526 \\ + 41 \\ \hline \end{array}$ **15.** $\begin{array}{r} 109 \\ + 70 \\ \hline \end{array}$ **16.** $\begin{array}{r} 420 \\ + 63 \\ \hline \end{array}$ **17.** $\begin{array}{r} 256 \\ + 32 \\ \hline \end{array}$ **18.** $\begin{array}{r} 720 \\ + 79 \\ \hline \end{array}$

19. $\begin{array}{r} 523 \\ + 161 \\ \hline \end{array}$ **20.** $\begin{array}{r} 408 \\ + 391 \\ \hline \end{array}$ **21.** $\begin{array}{r} 624 \\ + 134 \\ \hline \end{array}$ **22.** $\begin{array}{r} 270 \\ + 126 \\ \hline \end{array}$ **23.** $\begin{array}{r} 635 \\ + 254 \\ \hline \end{array}$

24. $\begin{array}{r} 6104 \\ + 2352 \\ \hline \end{array}$ **25.** $\begin{array}{r} 4860 \\ + 2019 \\ \hline \end{array}$ **26.** $\begin{array}{r} 2096 \\ + 1502 \\ \hline \end{array}$ **27.** $\begin{array}{r} 3061 \\ + 1814 \\ \hline \end{array}$ **28.** $\begin{array}{r} 5186 \\ + 3210 \\ \hline \end{array}$

Skill 4
(page 13)

$$\begin{array}{r} \overset{11}{4698} \\ + 8164 \\ \hline 12{,}862 \end{array}$$

29. $\begin{array}{r} 26 \\ + 47 \\ \hline \end{array}$ **30.** $\begin{array}{r} 88 \\ + 34 \\ \hline \end{array}$ **31.** $\begin{array}{r} 35 \\ + 75 \\ \hline \end{array}$ **32.** $\begin{array}{r} 19 \\ + 89 \\ \hline \end{array}$

33. $\begin{array}{r} 73 \\ + 68 \\ \hline \end{array}$ **34.** $\begin{array}{r} 68 \\ + 49 \\ \hline \end{array}$ **35.** $\begin{array}{r} 94 \\ + 37 \\ \hline \end{array}$ **36.** $\begin{array}{r} 86 \\ + 76 \\ \hline \end{array}$

37. $\begin{array}{r} 623 \\ + 94 \\ \hline \end{array}$ **38.** $\begin{array}{r} 406 \\ + 97 \\ \hline \end{array}$ **39.** $\begin{array}{r} 295 \\ + 59 \\ \hline \end{array}$ **40.** $\begin{array}{r} 752 \\ + 83 \\ \hline \end{array}$ **41.** $\begin{array}{r} 376 \\ + 48 \\ \hline \end{array}$

42. $\begin{array}{r} 428 \\ + 196 \\ \hline \end{array}$ **43.** $\begin{array}{r} 196 \\ + 505 \\ \hline \end{array}$ **44.** $\begin{array}{r} 763 \\ + 168 \\ \hline \end{array}$ **45.** $\begin{array}{r} 507 \\ + 483 \\ \hline \end{array}$ **46.** $\begin{array}{r} 624 \\ + 199 \\ \hline \end{array}$

47. $\begin{array}{r} 586 \\ + 679 \\ \hline \end{array}$ **48.** $\begin{array}{r} 493 \\ + 978 \\ \hline \end{array}$ **49.** $\begin{array}{r} 392 \\ + 608 \\ \hline \end{array}$ **50.** $\begin{array}{r} 229 \\ + 785 \\ \hline \end{array}$ **51.** $\begin{array}{r} 686 \\ + 867 \\ \hline \end{array}$

52. $\begin{array}{r} 7680 \\ + 1979 \\ \hline \end{array}$ **53.** $\begin{array}{r} 2398 \\ + 9746 \\ \hline \end{array}$ **54.** $\begin{array}{r} 5050 \\ + 5060 \\ \hline \end{array}$ **55.** $\begin{array}{r} 6273 \\ + 3108 \\ \hline \end{array}$ **56.** $\begin{array}{r} 3986 \\ + 8085 \\ \hline \end{array}$

Skill 5

(page 14)

4726
− 3415
1311

1.   86
  − 4

2.   42
  − 1

3.   59
  − 6

4.   39
  − 4

5.   46
  − 24

6.   78
  − 30

7.   92
  − 32

8.   54
  − 33

9.   820
  − 520

10.   359
  − 23

11.   597
  − 65

12.   850
  − 20

13.   635
  − 30

14.   238
  − 15

15.   965
  − 304

16.   453
  − 253

17.   590
  − 480

18.   856
  − 342

19.   6892
  − 1571

20.   8936
  − 4214

21.   4099
  − 3097

22.   5508
  − 3205

23.   2875
  − 1523

24.   4829
  − 2512

25.   6457
  − 1316

26.   6572
  − 1161

27.   7736
  − 2424

28.   8089
  − 4038

Skill 6

(page 15)

2 14 12 10 15
3 5 , 3 1 5
− 1 5 , 8 4 7
1 9 , 4 6 8

29.   80
  − 15

30.   47
  − 19

31.   93
  − 78

32.   32
  − 17

33.   263
  − 17

34.   476
  − 29

35.   508
  − 32

36.   625
  − 83

37.   786
  − 298

38.   403
  − 186

39.   510
  − 295

40.   801
  − 799

41.   532
  − 495

42.   4073
  − 2981

43.   5000
  − 2509

44.   2001
  − 1906

45.   6983
  − 4997

46.   8090
  − 5989

47.   30,179
  − 26,189

48.   89,041
  − 56,958

49.   76,504
  − 60,876

50.   80,005
  − 39,487

51.   84,013
  − 47,219

52.   52,161
  − 13,974

53.   86,413
  − 18,847

54.   57,317
  − 18,498

Skill 7
(page 22)

```
  1 21
 11.097
230.28
+ 1.973
───────
243.350
```

1.   31.49
   + 67.83

2.   1.098
   + 6.995

3.   271.5
   + 729.6

4.   46.09
   + 3.725

5.   439.8
   + 2.785

6.   5.097
   + 123.4

7.   858.5
   72.49
   + 106.08

8.   12.83
   1.509
   + 304.89

9.   7201.5
   26.003
   + 849.85

10.   86.28
   100.5
   + 1.097

11. 28.3 + 4.061 + 92.8

12. 1986.21 + 8.005 + 497.6

13. 1.007 + 46.93 + 90.889

14. 260.504 + 86.067 + 5.90

15. $26.30 + $198.00 + $79.95

16. $10.98 + $39.77 + $105.49

17. $156.25 + $230.95 + $16.50

18. $5096.60 + $26.93 + $792.89

Skill 8
(page 23)

```
2 18 8  11 14
3 8 9 . 2 4
− 1 9 8 . 5 6
─────────────
  1 9 0 . 6 8
```

19.   46.37
   − 23.28

20.   7.534
   − 1.282

21.   690.5
   − 358.3

22.   89.04
   − 26.98

23.   30.04
   − 15.96

24.   625.7
   − 489.8

25.   39.46
   − 8.984

26.   31.05
   − 0.982

27.   56.07
   − 9.806

28. 100.2
   − 8.923

29. 4936.2 − 27.381

30. 599.2 − 38.99

31. 60.09 − 8.597

32. 26.3 − 18.009

33. $16.75 − $2.89

34. $357.02 − $199.76

35. $1000 − $19.65

36. $650 − $327.98

37. $1905 − $27.82

38. $32 − $6.59

Skill 9
(page 34)

1. 23
    $\times$ 2

2. 42
    $\times$ 2

3. 31
    $\times$ 3

4. 24
    $\times$ 1

313
$\times$ 2
626

5. 203
    $\times$ 2

6. 130
    $\times$ 3

7. 112
    $\times$ 3

8. 302
    $\times$ 2

9. 1042
    $\times$ 2

10. 2231
    $\times$ 3

11. 3605
    $\times$ 0

12. 2112
    $\times$ 3

13. 1003
    $\times$ 3

Skill 10
(page 35)

14. 18
    $\times$ 2

15. 42
    $\times$ 3

16. 83
    $\times$ 2

17. 26
    $\times$ 3

7
649
$\times$ 8
5192

18. 502
    $\times$ 4

19. 391
    $\times$ 3

20. 217
    $\times$ 3

21. 193
    $\times$ 2

22. 4806
    $\times$ 4

23. 7839
    $\times$ 6

24. 2075
    $\times$ 5

25. 6937
    $\times$ 8

26. 4890
    $\times$ 7

Skill 11
(pages 36–37)

27. 27
    $\times$ 30

28. 85
    $\times$ 70

29. 48
    $\times$ 60

30. 32
    $\times$ 90

1
836
$\times$ 200
167,200

31. 938
    $\times$ 40

32. 407
    $\times$ 80

33. 216
    $\times$ 50

34. 693
    $\times$ 20

2452
$\times$ 1000
2,452,000

35. 708
    $\times$ 300

36. 560
    $\times$ 800

37. 679
    $\times$ 500

38. 456
    $\times$ 700

39. 5845
    $\times$ 2000

40. 7090
    $\times$ 6000

41. 8132
    $\times$ 9000

42. 2600
    $\times$ 4000

43. 6005
    $\times$ 7000

44. 4296
    $\times$ 2000

45. 5369
    $\times$ 4000

46. 2875
    $\times$ 8000

47. 7957
    $\times$ 5000

48. 4844
    $\times$ 6000

Skill Bank

Skill 12
(page 44)

$$648 \\ \times\ 26 \\ \overline{3888} \\ 12960 \\ \overline{16,848}$$

1. 56×38

2. 79×81

3. 32×87

4. 61×42

5. 683×72

6. 509×35

7. 780×67

8. 493×48

9. 380×56

10. 710×27

11. 907×77

12. 416×64

13. 847×75

14. 6859×39

15. 4076×42

16. 2895×65

17. 5900×18

18. 9789×23

19. 2102×83

20. 6009×96

21. 7146×53

22. 5020×68

23. 8637×75

24. 4319×56

25. 2674×72

26. 8662×85

27. 9174×68

28. 8036×77

Skill 13
(page 45)

$$387 \\ \times\ 423 \\ \overline{1161} \\ 7740 \\ 154800 \\ \overline{163,701}$$

29. 792×241

30. 583×327

31. 448×336

32. 813×318

33. 245×337

34. 772×509

35. 890×706

36. 369×248

37. 308×280

38. 460×504

39. 328×207

40. 945×210

41. 7012×638

42. 8196×273

43. 4380×529

44. 3088×431

45. 6992×827

46. 4051×3107

47. 2709×1340

48. 7266×2407

49. 4905×6203

50. 7048×5104

51. 7923×2536

52. 2648×5927

53. 8158×4932

54. 4692×9543

55. 5872×6928

Skill 14
(pages 46–47)

$$\begin{array}{r} 690 \\ \times\ 407 \\ \hline 4830 \\ 0000 \\ 276000 \\ \hline 280{,}830 \end{array}$$

1. $\begin{array}{r} 207 \\ \times\ 60 \\ \hline \end{array}$

2. $\begin{array}{r} 580 \\ \times\ 30 \\ \hline \end{array}$

3. $\begin{array}{r} 402 \\ \times\ 80 \\ \hline \end{array}$

4. $\begin{array}{r} 630 \\ \times\ 90 \\ \hline \end{array}$

5. $\begin{array}{r} 806 \\ \times\ 350 \\ \hline \end{array}$

6. $\begin{array}{r} 120 \\ \times\ 705 \\ \hline \end{array}$

7. $\begin{array}{r} 430 \\ \times\ 680 \\ \hline \end{array}$

8. $\begin{array}{r} 290 \\ \times\ 409 \\ \hline \end{array}$

9. $\begin{array}{r} 505 \\ \times\ 606 \\ \hline \end{array}$

10. $\begin{array}{r} 460 \\ \times\ 504 \\ \hline \end{array}$

11. $\begin{array}{r} 290 \\ \times\ 150 \\ \hline \end{array}$

12. $\begin{array}{r} 610 \\ \times\ 203 \\ \hline \end{array}$

13. $\begin{array}{r} 4106 \\ \times\ 390 \\ \hline \end{array}$

14. $\begin{array}{r} 8096 \\ \times\ 402 \\ \hline \end{array}$

15. $\begin{array}{r} 9560 \\ \times\ 809 \\ \hline \end{array}$

16. $\begin{array}{r} 6089 \\ \times\ 720 \\ \hline \end{array}$

17. $\begin{array}{r} 3045 \\ \times\ 280 \\ \hline \end{array}$

18. $\begin{array}{r} 2009 \\ \times\ 420 \\ \hline \end{array}$

19. $\begin{array}{r} 5900 \\ \times\ 590 \\ \hline \end{array}$

20. $\begin{array}{r} 2630 \\ \times\ 400 \\ \hline \end{array}$

21. $\begin{array}{r} 8001 \\ \times\ 606 \\ \hline \end{array}$

22. $\begin{array}{r} 4800 \\ \times\ 902 \\ \hline \end{array}$

23. $\begin{array}{r} 4064 \\ \times\ 1009 \\ \hline \end{array}$

24. $\begin{array}{r} 5280 \\ \times\ 6900 \\ \hline \end{array}$

25. $\begin{array}{r} 2708 \\ \times\ 4900 \\ \hline \end{array}$

26. $\begin{array}{r} 7090 \\ \times\ 5020 \\ \hline \end{array}$

27. $\begin{array}{r} 3070 \\ \times\ 7300 \\ \hline \end{array}$

Skill 15
(pages 54–55)

$$\begin{array}{r} 16.48 \\ \times\ 5.7 \\ \hline 11536 \\ 82400 \\ \hline 93.936 \end{array}$$

28. $\begin{array}{r} 3.64 \\ \times\ 13 \\ \hline \end{array}$

29. $\begin{array}{r} 24.06 \\ \times\ 31 \\ \hline \end{array}$

30. $\begin{array}{r} 4.908 \\ \times\ 24 \\ \hline \end{array}$

31. $\begin{array}{r} 7.042 \\ \times\ 61 \\ \hline \end{array}$

32. $\begin{array}{r} 62.8 \\ \times\ 3.17 \\ \hline \end{array}$

33. $\begin{array}{r} 9.03 \\ \times\ 8.69 \\ \hline \end{array}$

34. $\begin{array}{r} 21.5 \\ \times\ 5.95 \\ \hline \end{array}$

35. $\begin{array}{r} 8.71 \\ \times\ 28.6 \\ \hline \end{array}$

36. $\begin{array}{r} 0.652 \\ \times\ 0.07 \\ \hline \end{array}$

37. $\begin{array}{r} 38.2 \\ \times\ 41.7 \\ \hline \end{array}$

38. $\begin{array}{r} 12.2 \\ \times\ 6.21 \\ \hline \end{array}$

39. $\begin{array}{r} 4.69 \\ \times\ 25.1 \\ \hline \end{array}$

40. $\begin{array}{r} 41.1 \\ \times\ 33.2 \\ \hline \end{array}$

41. $\begin{array}{r} 7.18 \\ \times\ 6.82 \\ \hline \end{array}$

42. $\begin{array}{r} 4.8 \\ \times\ 0.15 \\ \hline \end{array}$

43. $\begin{array}{r} 0.39 \\ \times\ 0.12 \\ \hline \end{array}$

44. $\begin{array}{r} 1.6 \\ \times\ 0.89 \\ \hline \end{array}$

45. $\begin{array}{r} 0.52 \\ \times\ 0.16 \\ \hline \end{array}$

46. $\begin{array}{r} 0.41 \\ \times\ 0.25 \\ \hline \end{array}$

47. $\begin{array}{r} 0.1002 \\ \times\ 0.4 \\ \hline \end{array}$

48. $\begin{array}{r} 5.216 \\ \times\ 0.05 \\ \hline \end{array}$

49. $\begin{array}{r} 63.01 \\ \times\ 0.003 \\ \hline \end{array}$

50. $\begin{array}{r} 0.4906 \\ \times\ 0.78 \\ \hline \end{array}$

51. $\begin{array}{r} 0.6107 \\ \times\ 0.5003 \\ \hline \end{array}$

52. $\begin{array}{r} 0.7001 \\ \times\ 0.01 \\ \hline \end{array}$

53. $\begin{array}{r} 0.2510 \\ \times\ 0.002 \\ \hline \end{array}$

54. $\begin{array}{r} 0.0002 \\ \times\ 0.002 \\ \hline \end{array}$

55. $\begin{array}{r} 0.0202 \\ \times\ 0.13 \\ \hline \end{array}$

Skill Bank

1. $5\overline{)75}$ **2.** $4\overline{)76}$ **3.** $6\overline{)90}$ **4.** $8\overline{)96}$

$$\begin{array}{r} 73 \\ 4\overline{)292} \\ -28 \\ \hline 12 \\ -12 \\ \hline 0 \end{array}$$

5. $5\overline{)285}$ **6.** $6\overline{)426}$ **7.** $4\overline{)116}$ **8.** $8\overline{)376}$

9. $6\overline{)912}$ **10.** $2\overline{)816}$ **11.** $5\overline{)925}$ **12.** $3\overline{)831}$

13. $7\overline{)5285}$ **14.** $5\overline{)4285}$ **15.** $4\overline{)2792}$ **16.** $8\overline{)5232}$

17. $4\overline{)3216}$ **18.** $6\overline{)3624}$ **19.** $5\overline{)4235}$ **20.** $7\overline{)6454}$ **21.** $6\overline{)3768}$

22. $9\overline{)5688}$ **23.** $9\overline{)6597}$ **24.** $8\overline{)1264}$ **25.** $8\overline{)1064}$ **26.** $5\overline{)4225}$

27. $2\overline{)1090}$ **28.** $3\overline{)1020}$ **29.** $6\overline{)1218}$ **30.** $7\overline{)1421}$ **31.** $6\overline{)1086}$

Show remainders as R.

32. $4\overline{)39}$ **33.** $9\overline{)84}$ **34.** $6\overline{)71}$ **35.** $5\overline{)92}$

$$\begin{array}{r} 95R1 \\ 5\overline{)476} \\ -45 \\ \hline 26 \\ -25 \\ \hline 1 \end{array}$$

36. $4\overline{)206}$ **37.** $7\overline{)594}$ **38.** $5\overline{)156}$ **39.** $6\overline{)568}$

40. $8\overline{)249}$ **41.** $5\overline{)428}$ **42.** $3\overline{)235}$ **43.** $8\overline{)380}$

44. $3\overline{)1102}$ **45.** $9\overline{)4216}$ **46.** $8\overline{)5692}$ **47.** $9\overline{)5077}$

48. $7\overline{)2430}$ **49.** $8\overline{)5235}$ **50.** $6\overline{)3769}$ **51.** $5\overline{)4239}$ **52.** $7\overline{)6455}$

53. $6\overline{)1420}$ **54.** $4\overline{)1335}$ **55.** $6\overline{)5048}$ **56.** $2\overline{)1011}$ **57.** $3\overline{)1030}$

58. $6\overline{)1220}$ **59.** $7\overline{)1422}$ **60.** $6\overline{)1096}$ **61.** $5\overline{)2296}$ **62.** $6\overline{)3628}$

Skill 18
(pages 68–69)

$$\begin{array}{r} 21\text{R}6 \\ 59)\overline{1245} \\ -\ 118 \\ \hline 65 \\ -\ 59 \\ \hline 6 \end{array}$$

$$\begin{array}{r} 15\text{R}386 \\ 421)\overline{6701} \\ -\ 421 \\ \hline 2491 \\ -\ 2105 \\ \hline 386 \end{array}$$

Show remainders as R.

1. $47)\overline{57}$ 2. $72)\overline{93}$ 3. $85)\overline{91}$ 4. $64)\overline{73}$

5. $36)\overline{870}$ 6. $15)\overline{783}$ 7. $62)\overline{402}$ 8. $47)\overline{393}$

9. $68)\overline{4763}$ 10. $82)\overline{9461}$ 11. $37)\overline{4792}$ 12. $73)\overline{6371}$

13. $86)\overline{9696}$ 14. $35)\overline{6165}$ 15. $42)\overline{3107}$ 16. $83)\overline{8797}$

17. $305)\overline{760}$ 18. $248)\overline{804}$ 19. $610)\overline{932}$ 20. $157)\overline{698}$

21. $191)\overline{3846}$ 22. $273)\overline{8080}$ 23. $309)\overline{4860}$ 24. $420)\overline{6459}$

25. $434)\overline{29,603}$ 26. $706)\overline{81,940}$ 27. $226)\overline{26,003}$ 28. $378)\overline{20,493}$

29. $689)\overline{250,047}$ 30. $505)\overline{492,169}$ 31. $243)\overline{789,000}$ 32. $802)\overline{390,580}$

Skill 19
(pages 76–77)

$$\begin{array}{r} 59\frac{12}{23} \\ 23)\overline{1369} \\ -\ 115 \\ \hline 219 \\ -\ 207 \\ \hline 12 \end{array}$$

$$\begin{array}{r} 12\frac{87}{735} \\ 735)\overline{8907} \\ -\ 735 \\ \hline 1557 \\ -\ 1470 \\ \hline 87 \end{array}$$

Show remainders as fractions.

33. $49)\overline{76}$ 34. $32)\overline{90}$ 35. $91)\overline{94}$ 36. $69)\overline{99}$

37. $82)\overline{708}$ 38. $47)\overline{349}$ 39. $22)\overline{840}$ 40. $16)\overline{509}$

41. $47)\overline{4721}$ 42. $32)\overline{6934}$ 43. $61)\overline{8712}$ 44. $89)\overline{9443}$

45. $26)\overline{4194}$ 46. $19)\overline{2499}$ 47. $52)\overline{6125}$ 48. $63)\overline{9027}$

49. $426)\overline{970}$ 50. $403)\overline{870}$ 51. $290)\overline{688}$ 52. $183)\overline{692}$

53. $416)\overline{8067}$ 54. $527)\overline{6193}$ 55. $319)\overline{5282}$

56. $704)\overline{8970}$ 57. $532)\overline{8767}$ 58. $181)\overline{4900}$

59. $620)\overline{70,480}$ 60. $498)\overline{26,304}$ 61. $381)\overline{65,583}$ 62. $270)\overline{49,007}$

413

Skill Bank

Skill 20
(pages 78–79)

$$6.48, \text{ or } 6.5$$
$$43\overline{)279.00}$$
$$-\ 258$$
$$\overline{\quad 210}$$
$$-\ 172$$
$$\overline{\quad 380}$$
$$-\ 344$$
$$\overline{\quad 36}$$

$$1.208, \text{ or } 1.21$$
$$302\overline{)365.000}$$
$$-\ 302$$
$$\overline{\quad 630}$$
$$-\ 604$$
$$\overline{\quad 2600}$$
$$-\ 2416$$
$$\overline{\quad 184}$$

Divide to hundredths. Round to the nearest tenth.

1. $5\overline{)17}$ **2.** $6\overline{)37}$ **3.** $8\overline{)70}$ **4.** $7\overline{)91}$

5. $3\overline{)782}$ **6.** $6\overline{)253}$ **7.** $7\overline{)188}$ **8.** $9\overline{)236}$

9. $29\overline{)341}$ **10.** $43\overline{)666}$ **11.** $61\overline{)702}$ **12.** $74\overline{)267}$

13. $71\overline{)8012}$ **14.** $82\overline{)9603}$ **15.** $26\overline{)4211}$ **16.** $37\overline{)8946}$

Divide to thousandths. Round to the nearest hundredth.

17. $7\overline{)20}$ **18.** $8\overline{)49}$ **19.** $5\overline{)74}$ **20.** $3\overline{)92}$

21. $43\overline{)904}$ **22.** $34\overline{)686}$ **23.** $29\overline{)372}$ **24.** $16\overline{)403}$

25. $238\overline{)870}$ **26.** $109\overline{)476}$ **27.** $389\overline{)902}$ **28.** $416\overline{)644}$

29. $226\overline{)6204}$ **30.** $349\overline{)7081}$ **31.** $870\overline{)6902}$ **32.** $530\overline{)1325}$

Skill 21
(pages 86–87)

$$4.841, \text{ or } 4.84$$
$$0.48.\overline{)2.32.400}$$
$$-\ 192$$
$$\overline{\quad 404}$$
$$-\ 384$$
$$\overline{\quad 200}$$
$$-\ 192$$
$$\overline{\quad 80}$$
$$-\ 48$$
$$\overline{\quad 32}$$

Divide to hundredths. Round to the nearest tenth.

33. $3.2\overline{)5.69}$ **34.** $4.3\overline{)7.82}$ **35.** $6.9\overline{)8.92}$ **36.** $5.1\overline{)9.89}$

37. $6.91\overline{)7.08}$ **38.** $1.63\overline{)4.26}$ **39.** $8.91\overline{)9.02}$ **40.** $3.12\overline{)7.89}$

41. $4.02\overline{)47.9}$ **42.** $2.76\overline{)18.6}$ **43.** $7.80\overline{)14.4}$ **44.** $6.23\overline{)38.7}$

Divide to thousandths. Round to the nearest hundredth.

45. $0.45\overline{)1.27}$ **46.** $0.19\overline{)3.94}$ **47.** $0.52\overline{)5.68}$ **48.** $0.94\overline{)2.07}$

49. $0.17\overline{)2.301}$ **50.** $0.46\overline{)5.307}$ **51.** $0.87\overline{)2.742}$

52. $0.92\overline{)1.034}$ **53.** $0.12\overline{)1.192}$ **54.** $0.13\overline{)2.911}$ **55.** $0.32\overline{)7.890}$

56. $0.24\overline{)0.977}$ **57.** $0.76\overline{)16.325}$ **58.** $0.16\overline{)11.111}$ **59.** $0.10\overline{)80.123}$

Skill 22
(page 102)

1. Complete the table. List the factors of 10.

Factors of 8				
Trial factors	1	2	3	4
Other factors	8	4	no	2

Trial factors	1	2	3	4	5
Other factors	10	?	?	?	?

Factors of 8: 1, 2, 4, 8

List the factors of each number.

2. 2	**3.** 5	**4.** 6	**5.** 8	**6.** 9	**7.** 10
8. 11	**9.** 14	**10.** 15	**11.** 18	**12.** 24	**13.** 28
14. 32	**15.** 36	**16.** 42	**17.** 45	**18.** 56	**19.** 58
20. 60	**21.** 64	**22.** 110	**23.** 73	**24.** 75	**25.** 76

Skill 23
(page 103)

Factors of 10: 1, 2, 5, 10
Factors of 15: 1, 3, 5, 15
Common factors: 1, 5
Greatest common factor (GCF): 5

Complete.

26. Factors of 4: 1, 2, 4
Factors of 8: 1, 2, 4, 8
Common factors: __?__, __?__, __?__
GCF: __?__

Write the GCF.

27. 3 and 4	**28.** 4 and 6	**29.** 3 and 9	**30.** 8 and 4
31. 6 and 8	**32.** 7 and 6	**33.** 10 and 5	**34.** 3 and 12
35. 16 and 4	**36.** 8 and 20	**37.** 24 and 9	**38.** 36 and 4
39. 56 and 8	**40.** 6 and 42	**41.** 9 and 21	**42.** 25 and 5
43. 64 and 8	**44.** 60 and 9	**45.** 24 and 16	**46.** 36 and 12
47. 42 and 14	**48.** 64 and 16	**49.** 72 and 36	**50.** 50 and 80
51. 4, 6, and 12	**52.** 6, 19, and 15	**53.** 10, 5, and 15	**54.** 10, 20, and 25

Skill Bank

Skill 24
(page 104)

The GCF of 8 and 12 is 4.

$$\frac{8}{12} = \frac{8 \div 4}{12 \div 4} = \frac{2}{3}$$

In lowest terms, $\frac{8}{12} = \frac{2}{3}$.

Write in lowest terms.

1. $\frac{2}{8} = \frac{2 \div 2}{8 \div 2} = ?$ **2.** $\frac{4}{6} = \frac{4 \div 2}{6 \div 2} = ?$

3. $\frac{6}{12} = \frac{6 \div 6}{12 \div 6} = ?$ **4.** $\frac{10}{12} = \frac{10 \div 2}{12 \div 2} = ?$

Use the GCF to write in lowest terms.

5. $\frac{2}{6}$ **6.** $\frac{4}{8}$ **7.** $\frac{3}{9}$

8. $\frac{4}{12}$ **9.** $\frac{6}{12}$ **10.** $\frac{4}{16}$ **11.** $\frac{9}{18}$ **12.** $\frac{8}{20}$ **13.** $\frac{6}{21}$

14. $\frac{14}{21}$ **15.** $\frac{24}{32}$ **16.** $\frac{40}{64}$ **17.** $\frac{16}{64}$ **18.** $\frac{25}{200}$ **19.** $\frac{75}{100}$

20. $7\frac{2}{4}$ **21.** $8\frac{3}{4}$ **22.** $9\frac{4}{8}$ **23.** $8\frac{6}{9}$ **24.** $4\frac{5}{10}$

25. $2\frac{4}{36}$ **26.** $10\frac{10}{12}$ **27.** $12\frac{12}{36}$ **28.** $14\frac{15}{45}$ **29.** $20\frac{20}{24}$

Skill 25
(page 105)

$$\frac{2}{3} = \frac{?}{12}$$

$\llcorner \times 4 \lrcorner$

$$\frac{2}{3} = \frac{2 \times 4}{3 \times 4} = \frac{8}{12}$$

In higher terms, $\frac{2}{3} = \frac{8}{12}$.

Write the fraction in higher terms.

30. $\frac{1}{4} = \frac{1 \times 2}{4 \times 2} = ?$ **31.** $\frac{2}{3} = \frac{2 \times 2}{3 \times 2} = ?$

32. $\frac{2}{5} = \frac{2 \times 3}{5 \times 3} = ?$ **33.** $\frac{1}{3} = \frac{1 \times 4}{3 \times 4} = ?$

34. $\frac{1}{2} = \frac{?}{8}$ **35.** $\frac{1}{2} = \frac{?}{10}$ **36.** $\frac{1}{2} = \frac{6}{?}$

37. $\frac{1}{3} = \frac{?}{9}$ **38.** $\frac{1}{5} = \frac{?}{25}$ **39.** $\frac{1}{8} = \frac{3}{?}$

40. $\frac{1}{6} = \frac{6}{?}$ **41.** $\frac{3}{6} = \frac{?}{18}$ **42.** $\frac{4}{8} = \frac{?}{16}$ **43.** $\frac{5}{9} = \frac{25}{?}$

44. $\frac{6}{8} = \frac{?}{40}$ **45.** $\frac{7}{9} = \frac{?}{36}$ **46.** $\frac{10}{12} = \frac{?}{36}$ **47.** $\frac{15}{24} = \frac{?}{48}$

48. $\frac{15}{20} = \frac{?}{100}$ **49.** $\frac{24}{36} = \frac{?}{72}$ **50.** $\frac{18}{40} = \frac{?}{80}$ **51.** $5\frac{2}{3} = 5\frac{?}{6}$

52. $7\frac{4}{5} = 7\frac{?}{15}$ **53.** $8\frac{3}{7} = 8\frac{9}{?}$ **54.** $9\frac{4}{9} = 9\frac{?}{27}$ **55.** $6\frac{6}{8} = 6\frac{18}{?}$

Skill 26
(page 112)

Multiples of 3			
3×1	3×2	3×3	3×4
3	6	9	12

Complete the list of multiples.

1. 4: 4, 8, 12, ?, ?, ?

2. 5: 5, 10, 15, ?, ?, ?

3. 10: 10, 20, 30, ?, ?, ?

4. 25: 25, 50, 75, ?, ?, ?

Write the first four multiples of the number.

5. 2   **6.** 6   **7.** 8   **8.** 9   **9.** 11

10. 12   **11.** 14   **12.** 15   **13.** 18   **14.** 19

15. 21   **16.** 24   **17.** 26   **18.** 32   **19.** 40

20. 42   **21.** 45   **22.** 52   **23.** 56   **24.** 60

Skill 27
(page 113)

Multiples of 2: 2, 4, 6, 8, . . .
Multiples of 4: 4, 8, 12, 16, . . .
Multiples of 8: 8, 16, 24, 32, . . .

The least common
multiple (LCM) is 8.

Write the least common multiple (LCM).

25. 2 and 3   **26.** 4 and 5

27. 3 and 4   **28.** 9 and 3

29. 10 and 3   **30.** 20 and 30

31. 5 and 12   **32.** 5 and 10

33. 6 and 15   **34.** 4 and 18   **35.** 45 and 30   **36.** 15 and 12

37. 24 and 30   **38.** 12 and 3   **39.** 16 and 8   **40.** 42 and 4

41. 60 and 15   **42.** 15 and 10   **43.** 16 and 12   **44.** 18 and 12

45. 3, 2, and 6   **46.** 4, 8, and 6   **47.** 9, 3, and 2

48. 6, 10, and 5   **49.** 9, 12, and 4   **50.** 14, 6, and 7

51. 10, 15, and 6   **52.** 5, 10, and 12   **53.** 24, 6, and 12

(page 114)

The least common denominator (LCD) of

$$\frac{1}{3} \text{ and } \frac{1}{4}$$

is the LCM of 3 and 4.
LCM = 12   LCD = 12

Write the LCD of the fractions.

1. $\frac{1}{2}, \frac{1}{3}$

2. $\frac{1}{4}, \frac{1}{5}$

3. $\frac{1}{6}, \frac{1}{7}$

4. $\frac{1}{8}, \frac{1}{5}$

5. $\frac{1}{3}, \frac{5}{6}$

6. $\frac{3}{8}, \frac{1}{2}$

7. $\frac{3}{8}, \frac{1}{4}$

8. $\frac{4}{9}, \frac{2}{3}$

9. $\frac{1}{4}, \frac{2}{3}$

10. $\frac{7}{10}, \frac{2}{5}$

11. $\frac{5}{6}, \frac{7}{12}$

12. $\frac{3}{4}, \frac{1}{3}$

13. $\frac{4}{5}, \frac{5}{6}$

14. $\frac{2}{7}, \frac{1}{2}$

15. $\frac{7}{8}, \frac{2}{3}$

16. $\frac{5}{8}, \frac{1}{4}$

17. $\frac{7}{10}, \frac{5}{6}$

18. $\frac{5}{12}, \frac{4}{5}$

19. $\frac{5}{9}, \frac{5}{12}$

20. $\frac{1}{6}, \frac{7}{15}$

21. $\frac{2}{5}, \frac{2}{3}$

22. $\frac{11}{12}, \frac{2}{3}, \frac{1}{6}$

23. $\frac{5}{8}, \frac{5}{12}, \frac{5}{6}$

24. $\frac{5}{9}, \frac{11}{36}, \frac{7}{18}$

25. $\frac{1}{2}, \frac{2}{5}, \frac{1}{6}$

(page 115)

The LCD of $\frac{2}{3}$ and $\frac{3}{4}$ is 12.

$$\frac{2}{3} = \frac{2 \times 4}{3 \times 4} = \frac{8}{12}$$

$$\frac{3}{4} = \frac{3 \times 3}{4 \times 3} = \frac{9}{12}$$

Complete. Write as fractions with the LCD.

26. $\frac{1}{3} = \frac{1 \times 2}{3 \times 2} = \frac{?}{6}$

$\frac{1}{2} = \frac{1 \times 3}{2 \times 3} = \frac{?}{6}$

27. $\frac{1}{4} = \frac{1 \times 5}{4 \times 5} = \frac{?}{20}$

$\frac{1}{5} = \frac{1 \times 4}{5 \times 4} = \frac{?}{20}$

28. $\frac{1}{6} = \frac{1 \times 1}{6 \times 1} = ?$

$\frac{1}{3} = \frac{1 \times 2}{3 \times 2} = ?$

29. $\frac{3}{8} = \frac{3 \times 1}{8 \times 1} = ?$

$\frac{1}{2} = \frac{1 \times 4}{2 \times 4} = ?$

Write as fractions with the LCD.

30. $\frac{1}{2}, \frac{3}{8}$

31. $\frac{2}{3}, \frac{5}{6}$

32. $\frac{2}{3}, \frac{4}{9}$

33. $\frac{1}{2}, \frac{5}{6}$

34. $\frac{1}{2}, \frac{3}{5}$

35. $\frac{2}{3}, \frac{2}{7}$

36. $\frac{3}{5}, \frac{5}{6}$

37. $\frac{4}{9}, \frac{3}{4}$

38. $\frac{1}{3}, \frac{5}{12}$

39. $\frac{5}{18}, \frac{2}{9}$

40. $\frac{1}{4}, \frac{4}{5}$

41. $\frac{7}{32}, \frac{5}{8}$

42. $\frac{2}{5}, \frac{5}{8}$

43. $\frac{1}{3}, \frac{3}{5}$

44. $\frac{4}{15}, \frac{5}{6}$

45. $\frac{5}{6}, \frac{3}{4}, \frac{5}{18}$

46. $\frac{1}{10}, \frac{2}{5}, \frac{1}{2}$

47. $\frac{3}{20}, \frac{2}{5}, \frac{1}{2}$

48. $\frac{5}{24}, \frac{1}{8}, \frac{5}{12}$

Skill 30
(page 122)

Write as a whole number or a mixed number in lowest terms.

$$1\frac{2}{6}$$
$$\frac{8}{6} \rightarrow 6\overline{)8}$$

$$\frac{8}{6} = 1\frac{2}{6} = 1\frac{1}{3}$$

1. $\frac{6}{2}$ 2. $\frac{8}{3}$ 3. $\frac{9}{4}$

4. $\frac{5}{1}$ 5. $\frac{4}{3}$ 6. $\frac{6}{4}$

7. $\frac{10}{3}$ 8. $\frac{12}{4}$ 9. $\frac{16}{5}$

10. $\frac{18}{3}$ 11. $\frac{20}{9}$ 12. $\frac{26}{7}$

13. $\frac{24}{6}$ 14. $\frac{30}{16}$ 15. $\frac{32}{18}$ 16. $\frac{34}{20}$ 17. $\frac{12}{5}$ 18. $\frac{13}{7}$

19. $\frac{14}{9}$ 20. $\frac{15}{5}$ 21. $\frac{15}{6}$ 22. $\frac{16}{3}$ 23. $\frac{21}{2}$ 24. $\frac{7}{4}$

25. $\frac{8}{5}$ 26. $\frac{8}{4}$ 27. $\frac{10}{3}$ 28. $\frac{9}{2}$ 29. $\frac{18}{9}$ 30. $\frac{18}{5}$

31. $\frac{63}{63}$ 32. $\frac{40}{32}$ 33. $\frac{63}{40}$ 34. $\frac{74}{13}$ 35. $\frac{402}{20}$ 36. $\frac{205}{48}$

Skill 31
(page 123)

Write as a fraction.

$$3\frac{2}{3} = \frac{?}{?}$$
$$3 \times 3 = 9$$
$$9 + 2 = 11$$
$$3\frac{2}{3} = \frac{11}{3}$$

37. $3\frac{1}{3}$ 38. $4\frac{1}{6}$ 39. $5\frac{1}{8}$

40. $9\frac{1}{4}$ 41. $8\frac{1}{7}$ 42. $2\frac{1}{2}$

43. $6\frac{2}{3}$ 44. $4\frac{3}{4}$ 45. $7\frac{1}{6}$

46. $8\frac{3}{5}$ 47. $9\frac{1}{8}$ 48. $6\frac{4}{5}$

49. $12\frac{4}{5}$ 50. $14\frac{2}{3}$ 51. $15\frac{3}{4}$ 52. $20\frac{5}{6}$ 53. $24\frac{7}{8}$ 54. $31\frac{2}{3}$

55. $5\frac{1}{10}$ 56. $4\frac{1}{12}$ 57. $3\frac{1}{16}$ 58. $2\frac{1}{18}$ 59. $4\frac{1}{20}$ 60. $1\frac{23}{40}$

61. $10\frac{5}{12}$ 62. $14\frac{3}{16}$ 63. $12\frac{5}{18}$ 64. $20\frac{1}{24}$ 65. $25\frac{5}{8}$ 66. $16\frac{5}{18}$

Skill 32
(page 134)

Write the answer in lowest terms.

$\dfrac{3}{8}$
$+\dfrac{7}{8}$
$\overline{\dfrac{10}{8}} = 1\dfrac{2}{8} = 1\dfrac{1}{4}$

1. $\dfrac{3}{5}$
$+\dfrac{1}{5}$

2. $\dfrac{1}{6}$
$+\dfrac{5}{6}$

3. $\dfrac{2}{7}$
$+\dfrac{4}{7}$

4. $\dfrac{3}{8}$
$+\dfrac{3}{8}$

5. $\dfrac{3}{5}$
$+\dfrac{3}{5}$

6. $\dfrac{5}{8}$
$+\dfrac{1}{8}$

7. $\dfrac{3}{7}$
$+\dfrac{2}{7}$

8. $\dfrac{1}{4}$
$+\dfrac{1}{4}$

9. $\dfrac{7}{8}$
$+\dfrac{7}{8}$

10. $\dfrac{4}{7}$
$+\dfrac{6}{7}$

11. $\dfrac{5}{6}$
$+\dfrac{5}{6}$

12. $\dfrac{2}{3}$
$+\dfrac{2}{3}$

13. $\dfrac{7}{12}$
$+\dfrac{1}{12}$

14. $\dfrac{9}{16}$
$+\dfrac{7}{16}$

15. $\dfrac{1}{4} + \dfrac{1}{4} + \dfrac{1}{4}$

16. $\dfrac{2}{9} + \dfrac{4}{9} + \dfrac{1}{9}$

17. $\dfrac{4}{15} + \dfrac{7}{15} + \dfrac{8}{15}$

Skill 33
(page 135)

Write the answer in lowest terms.

$2\dfrac{3}{4}$
$+1\dfrac{3}{4}$
$\overline{3\dfrac{6}{4}} = 4\dfrac{2}{4} = 4\dfrac{1}{2}$

18. $1\dfrac{1}{4}$
$+3\dfrac{3}{4}$

19. $5\dfrac{4}{7}$
$+3\dfrac{3}{7}$

20. $4\dfrac{5}{6}$
$+2\dfrac{5}{6}$

21. $7\dfrac{5}{8}$
$+2\dfrac{7}{8}$

22. $6\dfrac{1}{6}$
$+4\dfrac{5}{6}$

23. $3\dfrac{4}{9}$
$+2\dfrac{7}{9}$

24. $10\dfrac{9}{10}$
$+3\dfrac{3}{10}$

25. $4\dfrac{9}{14}$
$+12\dfrac{5}{14}$

26. $15\dfrac{13}{16}$
$+5\dfrac{5}{16}$

27. $20\dfrac{15}{16}$
$+28\dfrac{5}{16}$

28. $41\dfrac{20}{21}$
$+17\dfrac{17}{21}$

29. $4\dfrac{1}{4} + 1\dfrac{1}{4} + 3\dfrac{3}{4}$

30. $5\dfrac{4}{9} + 2\dfrac{4}{9} + 4\dfrac{4}{9}$

31. $2\dfrac{5}{12} + 10\dfrac{5}{12} + 3\dfrac{5}{12}$

32. $1\dfrac{1}{8} + 3\dfrac{5}{8} + 2\dfrac{3}{8}$

33. $4\dfrac{1}{6} + 2\dfrac{5}{6} + 1\dfrac{1}{6}$

34. $3\dfrac{2}{7} + 4\dfrac{4}{7} + 2\dfrac{1}{7}$

Skill 34
(page 136)

$$\frac{3}{4} = \frac{15}{20} \leftarrow$$
$$+\frac{2}{5} = \frac{8}{20} \leftarrow \text{LCD}$$
$$\frac{23}{20} = 1\frac{3}{20}$$

Write the answer in lowest terms.

1. $\frac{1}{2}$
$+\frac{1}{3}$

2. $\frac{2}{3}$
$+\frac{1}{4}$

3. $\frac{3}{5}$
$+\frac{2}{3}$

4. $\frac{5}{8}$
$+\frac{1}{4}$

5. $\frac{5}{6}$
$+\frac{2}{3}$

6. $\frac{1}{2}$
$+\frac{3}{8}$

7. $\frac{1}{4}$
$+\frac{1}{10}$

8. $\frac{3}{4}$
$+\frac{4}{15}$

9. $\frac{5}{12}$
$+\frac{3}{5}$

10. $\frac{4}{11}$
$+\frac{1}{6}$

11. $\frac{10}{13}$
$+\frac{1}{3}$

12. $\frac{1}{2} + \frac{2}{5} + \frac{3}{10}$

13. $\frac{4}{9} + \frac{2}{3} + \frac{1}{2}$

14. $\frac{3}{4} + \frac{1}{2} + \frac{3}{10}$

Skill 35
(page 137)

$$2\frac{1}{4} = 2\frac{3}{12} \leftarrow$$
$$+ 3\frac{1}{3} = 3\frac{4}{12} \leftarrow \text{LCD}$$
$$5\frac{7}{12}$$

Write the answer in lowest terms.

15. $1\frac{1}{2}$
$+ 4\frac{1}{4}$

16. $5\frac{1}{3}$
$+ 7\frac{1}{5}$

17. $1\frac{1}{2}$
$+ 6\frac{3}{8}$

18. $4\frac{1}{4}$
$+ 5\frac{3}{8}$

19. $3\frac{3}{4}$
$+ 2\frac{1}{5}$

20. $7\frac{1}{6}$
$+ 1\frac{2}{3}$

21. $28\frac{3}{8}$
$+ 11\frac{1}{12}$

22. $45\frac{1}{14}$
$+ 44\frac{1}{6}$

23. $61\frac{2}{9}$
$+ 28\frac{11}{24}$

24. $54\frac{1}{12}$
$+ 33\frac{3}{5}$

25. $41\frac{1}{2}$
$+ 35\frac{4}{11}$

26. $3\frac{1}{2} + 1\frac{1}{8} + 4\frac{3}{4}$

27. $12\frac{1}{3} + 10\frac{2}{15} + 8\frac{3}{5}$

Skill Bank

Skill 36
(page 144)

Write the answer in lowest terms.

$$\frac{5}{8}$$
$$-\frac{3}{8}$$
$$\overline{\frac{2}{8}} = \frac{1}{4}$$

1. $\frac{2}{3}$
$-\frac{1}{3}$

2. $\frac{3}{4}$
$-\frac{1}{4}$

3. $\frac{5}{6}$
$-\frac{1}{6}$

4. $\frac{7}{8}$
$-\frac{1}{8}$

5. $\frac{3}{5}$
$-\frac{1}{5}$

6. $\frac{4}{5}$
$-\frac{4}{5}$

7. $\frac{15}{16}$
$-\frac{11}{16}$

8. $\frac{11}{12}$
$-\frac{7}{12}$

9. $\frac{9}{10}$
$-\frac{3}{10}$

10. $\frac{14}{15}$
$-\frac{11}{15}$

11. $\frac{19}{20}$
$-\frac{13}{20}$

12. $\frac{7}{10} - \frac{3}{10}$

13. $\frac{7}{8} - \frac{5}{8}$

14. $\frac{7}{12} - \frac{5}{12}$

15. $\frac{17}{24} - \frac{11}{24}$

Skill 37
(page 145)

Write the answer in lowest terms.

$$6\frac{5}{8}$$
$$-3\frac{1}{8}$$
$$\overline{3\frac{4}{8}} = 3\frac{1}{2}$$

16. $2\frac{3}{4}$
$-1\frac{1}{4}$

17. $4\frac{5}{6}$
$-2\frac{1}{6}$

18. $5\frac{4}{5}$
$-3\frac{3}{5}$

19. $7\frac{8}{9}$
$-3\frac{5}{9}$

20. $6\frac{2}{3}$
$-2\frac{2}{3}$

21. $9\frac{5}{6}$
$-6\frac{1}{6}$

22. $18\frac{4}{5}$
$-7\frac{2}{5}$

23. $20\frac{2}{3}$
$-6\frac{1}{3}$

24. $5\frac{11}{12}$
$-1\frac{7}{12}$

25. $6\frac{9}{14}$
$-3\frac{3}{14}$

26. $26\frac{13}{15}$
$-18\frac{13}{15}$

27. $7\frac{5}{8} - 5\frac{3}{8}$

28. $10\frac{7}{9} - 5\frac{4}{9}$

29. $7\frac{4}{15} - 6\frac{1}{15}$

Skill 38
(page 146)

Write the answer in lowest terms.

$$6 = 5\frac{6}{6}$$
$$-3\frac{1}{6} = 3\frac{1}{6}$$
$$\overline{\phantom{-3\frac{1}{6}}}$$
$$2\frac{5}{6}$$

1. $\begin{array}{r} 6 \\ -\frac{3}{4} \\ \hline \end{array}$ **2.** $\begin{array}{r} 7 \\ -\frac{7}{8} \\ \hline \end{array}$ **3.** $\begin{array}{r} 8 \\ -\frac{1}{6} \\ \hline \end{array}$

4. $\begin{array}{r} 10 \\ -\frac{4}{5} \\ \hline \end{array}$ **5.** $\begin{array}{r} 11 \\ -\frac{1}{3} \\ \hline \end{array}$ **6.** $\begin{array}{r} 18 \\ -\frac{5}{8} \\ \hline \end{array}$

7. $\begin{array}{r} 4 \\ -1\frac{3}{4} \\ \hline \end{array}$ **8.** $\begin{array}{r} 5 \\ -3\frac{5}{7} \\ \hline \end{array}$ **9.** $\begin{array}{r} 7 \\ -4\frac{3}{8} \\ \hline \end{array}$ **10.** $\begin{array}{r} 6 \\ -2\frac{2}{5} \\ \hline \end{array}$ **11.** $\begin{array}{r} 9 \\ -6\frac{7}{9} \\ \hline \end{array}$

12. $6 - \frac{5}{6}$ **13.** $11 - \frac{5}{9}$ **14.** $20 - \frac{7}{12}$ **15.** $4 - \frac{7}{12}$

Skill 39
(page 147)

Write the answer in lowest terms.

$$7\frac{1}{8} = 6\frac{9}{8}$$
$$-4\frac{3}{8} = 4\frac{3}{8}$$
$$\overline{\phantom{-4\frac{3}{8}}}$$
$$2\frac{6}{8} = 2\frac{3}{4}$$

16. $\begin{array}{r} 7\frac{1}{3} \\ -2\frac{2}{3} \\ \hline \end{array}$ **17.** $\begin{array}{r} 6\frac{2}{5} \\ -4\frac{4}{5} \\ \hline \end{array}$ **18.** $\begin{array}{r} 3\frac{5}{8} \\ -1\frac{7}{8} \\ \hline \end{array}$

19. $\begin{array}{r} 5\frac{5}{9} \\ -3\frac{8}{9} \\ \hline \end{array}$ **20.** $\begin{array}{r} 7\frac{3}{7} \\ -2\frac{5}{7} \\ \hline \end{array}$ **21.** $\begin{array}{r} 4\frac{3}{5} \\ -1\frac{4}{5} \\ \hline \end{array}$

22. $\begin{array}{r} 18\frac{7}{9} \\ -6\frac{8}{9} \\ \hline \end{array}$ **23.** $\begin{array}{r} 15\frac{4}{7} \\ -3\frac{6}{7} \\ \hline \end{array}$ **24.** $\begin{array}{r} 12\frac{1}{4} \\ -1\frac{3}{4} \\ \hline \end{array}$ **25.** $\begin{array}{r} 20\frac{5}{16} \\ -12\frac{9}{16} \\ \hline \end{array}$ **26.** $\begin{array}{r} 28\frac{1}{18} \\ -19\frac{5}{18} \\ \hline \end{array}$

27. $8\frac{5}{9} - 6\frac{8}{9}$ **28.** $15\frac{1}{6} - 9\frac{5}{6}$ **29.** $24\frac{3}{20} - 18\frac{13}{20}$

Skill 40
(page 154)

$$\frac{3}{4} = \frac{15}{20}$$

LCD

$$-\frac{2}{5} = \frac{8}{20}$$

$$\frac{7}{20}$$

Write the answer in lowest terms.

1. $\frac{1}{2}$
    $-\frac{1}{4}$

2. $\frac{1}{2}$
    $-\frac{3}{8}$

3. $\frac{2}{3}$
    $-\frac{5}{9}$

4. $\frac{5}{7}$
    $-\frac{1}{4}$

5. $\frac{8}{9}$
    $-\frac{1}{6}$

6. $\frac{4}{5}$
    $-\frac{3}{7}$

7. $\frac{7}{12}$
    $-\frac{1}{4}$

8. $\frac{2}{3}$
    $-\frac{4}{15}$

9. $\frac{7}{9}$
    $-\frac{1}{6}$

10. $\frac{13}{14}$
    $-\frac{1}{4}$

11. $\frac{14}{15}$
    $-\frac{1}{5}$

12. $\frac{4}{7} - \frac{5}{14}$

13. $\frac{8}{11} - \frac{1}{3}$

14. $\frac{7}{12} - \frac{4}{9}$

15. $\frac{5}{8} - \frac{2}{5}$

Skill 41
(page 155)

$$9\frac{1}{8} = 9\frac{1}{8} = 8\frac{9}{8}$$

$$-5\frac{1}{2} = 5\frac{4}{8} = 5\frac{4}{8}$$

$$3\frac{5}{8}$$

Write the answer in lowest terms.

16. $5\frac{4}{9}$
    $-4\frac{2}{3}$

17. $6\frac{3}{8}$
    $-5\frac{1}{2}$

18. $7\frac{1}{2}$
    $-3\frac{5}{6}$

19. $5\frac{2}{9}$
    $-2\frac{2}{3}$

20. $9\frac{1}{8}$
    $-5\frac{1}{2}$

21. $5\frac{1}{2}$
    $-1\frac{3}{4}$

22. $18\frac{2}{7}$
    $-6\frac{2}{3}$

23. $16\frac{3}{8}$
    $-5\frac{2}{3}$

24. $18\frac{1}{3}$
    $-9\frac{1}{2}$

25. $16\frac{1}{18}$
    $-7\frac{1}{4}$

26. $26\frac{5}{12}$
    $-9\frac{4}{9}$

27. $5\frac{2}{7} - 4\frac{3}{14}$

28. $20\frac{4}{11} - 6\frac{8}{9}$

29. $8\frac{1}{4} - 4\frac{3}{8}$

Skill 42
(page 166)

$$\frac{3}{8} \times \frac{6}{7}$$
$$\downarrow$$
$$\frac{3 \times 6}{8 \times 7} = \frac{18}{56} = \frac{9}{28}$$

Write the answer in lowest terms.

1. $\frac{1}{3} \times \frac{1}{3}$ **2.** $\frac{1}{4} \times \frac{1}{2}$ **3.** $\frac{1}{5} \times \frac{1}{6}$

4. $\frac{3}{5} \times \frac{1}{8}$ **5.** $\frac{1}{7} \times \frac{4}{9}$ **6.** $\frac{1}{9} \times \frac{3}{7}$

7. $\frac{4}{5} \times \frac{7}{8}$ **8.** $\frac{3}{4} \times \frac{7}{9}$ **9.** $\frac{5}{6} \times \frac{7}{8}$

10. $\frac{5}{9} \times \frac{8}{9}$ **11.** $\frac{5}{8} \times \frac{2}{3}$ **12.** $\frac{8}{9} \times \frac{4}{5}$ **13.** $\frac{4}{5} \times \frac{5}{6}$

14. $\frac{4}{5} \times \frac{3}{4}$ **15.** $\frac{4}{7} \times \frac{3}{8}$ **16.** $\frac{5}{7} \times \frac{2}{3}$ **17.** $\frac{7}{9} \times \frac{4}{5}$

18. $\frac{2}{5} \times \frac{1}{10}$ **19.** $\frac{1}{12} \times \frac{3}{7}$ **20.** $\frac{5}{6} \times \frac{7}{12}$ **21.** $\frac{5}{16} \times \frac{2}{3}$

22. $\frac{3}{4} \times \frac{13}{16}$ **23.** $\frac{11}{12} \times \frac{3}{5}$ **24.** $\frac{13}{16} \times \frac{2}{5}$ **25.** $\frac{9}{11} \times \frac{5}{6}$

Skill 43
(page 167)

$$6 \times \frac{3}{4}$$
$$\downarrow$$
$$\frac{6}{1} \times \frac{3}{4} = \frac{18}{4} = 4\frac{2}{4} = 4\frac{1}{2}$$

Write the answer in lowest terms.

26. $6 \times \frac{1}{3}$ **27.** $\frac{1}{5} \times 7$

28. $5 \times \frac{4}{9}$ **29.** $\frac{5}{6} \times 4$

30. $5 \times \frac{2}{5}$ **31.** $9 \times \frac{1}{2}$

32. $10 \times \frac{4}{5}$ **33.** $\frac{5}{6} \times 12$ **34.** $\frac{5}{9} \times 11$ **35.** $\frac{5}{8} \times 15$

36. $9 \times \frac{2}{3}$ **37.** $8 \times \frac{4}{5}$ **38.** $\frac{5}{6} \times 3$ **39.** $2 \times \frac{7}{8}$

40. $5 \times \frac{1}{10}$ **41.** $3 \times \frac{7}{12}$ **42.** $6 \times \frac{5}{16}$ **43.** $9 \times \frac{5}{18}$

44. $6 \times \frac{11}{12}$ **45.** $\frac{11}{24} \times 4$ **46.** $2 \times \frac{13}{20}$ **47.** $3 \times \frac{13}{16}$

Skill 44
(page 168)

$$\frac{3}{4} \times 1\frac{2}{3}$$
$$\downarrow$$
$$\frac{3}{4} \times \frac{5}{3} = \frac{15}{12} = 1\frac{1}{4}$$

Write the answer in lowest terms.

1. $\frac{1}{2} \times 3\frac{1}{4}$　　　　2. $\frac{1}{5} \times 2\frac{1}{6}$

3. $4\frac{5}{8} \times \frac{1}{7}$　　　　4. $2\frac{1}{9} \times \frac{3}{4}$

5. $\frac{1}{6} \times 1\frac{3}{4}$　　　　6. $1\frac{4}{5} \times \frac{1}{4}$

7. $\frac{2}{3} \times 1\frac{3}{4}$　　8. $\frac{5}{8} \times 1\frac{2}{3}$　　9. $2\frac{2}{9} \times \frac{3}{5}$　　10. $\frac{3}{8} \times 2\frac{2}{3}$

11. $3\frac{2}{5} \times \frac{3}{4}$　　12. $\frac{2}{3} \times 3\frac{5}{6}$　　13. $\frac{2}{5} \times 4\frac{3}{7}$　　14. $\frac{1}{6} \times 2\frac{3}{8}$

15. $\frac{7}{9} \times 1\frac{2}{7}$　　16. $1\frac{4}{5} \times \frac{5}{6}$　　17. $\frac{2}{7} \times 1\frac{3}{4}$　　18. $2\frac{2}{5} \times \frac{5}{6}$

19. $1\frac{1}{16} \times \frac{4}{5}$　　20. $1\frac{3}{4} \times \frac{7}{10}$　　21. $1\frac{5}{12} \times \frac{2}{11}$　　22. $\frac{1}{5} \times 2\frac{3}{10}$

Skill 45
(page 169)

$$2\frac{2}{3} \times 1\frac{4}{5}$$
$$\downarrow$$
$$\frac{8}{3} \times \frac{9}{5} = \frac{72}{15} = 4\frac{4}{5}$$

Write the answer in lowest terms.

23. $2\frac{1}{3} \times 1\frac{1}{2}$　　　　24. $2\frac{1}{3} \times 1\frac{1}{8}$

25. $1\frac{5}{6} \times 4\frac{1}{7}$　　　　26. $3\frac{1}{6} \times 2\frac{3}{4}$

27. $1\frac{3}{7} \times 1\frac{4}{5}$　　　　28. $2\frac{5}{6} \times 1\frac{4}{5}$

29. $2\frac{6}{7} \times 2\frac{4}{5}$　　30. $3\frac{3}{5} \times 6\frac{2}{3}$　　31. $2\frac{2}{9} \times 1\frac{4}{5}$　　32. $2\frac{2}{5} \times 1\frac{7}{8}$

33. $1\frac{1}{12} \times 2\frac{1}{2}$　　34. $1\frac{3}{4} \times 1\frac{1}{16}$　　35. $2\frac{3}{10} \times 1\frac{1}{5}$　　36. $1\frac{2}{11} \times 2\frac{3}{11}$

37. $3 \times 3\frac{1}{5}$　　38. $2\frac{1}{8} \times 2$　　39. $3 \times 4\frac{1}{3}$　　40. $5 \times 2\frac{1}{4}$

41. $1\frac{3}{8} \times 4$　　42. $2 \times 2\frac{3}{8}$　　43. $1\frac{3}{8} \times 3$　　44. $5 \times 1\frac{3}{5}$

Skill 46
(page 176)

Find the reciprocal.

1. 3 **2.** 8 **3.** 15 **4.** 20

$\frac{7}{9} \times ? = 1$

$\frac{7}{9} \times \frac{9}{7} = 1$

5. $\frac{1}{2}$ **6.** $\frac{1}{7}$ **7.** $\frac{4}{5}$ **8.** $\frac{5}{8}$

The reciprocal of $\frac{7}{9}$ is $\frac{9}{7}$.

9. $\frac{7}{3}$ **10.** $\frac{8}{5}$ **11.** $\frac{6}{5}$ **12.** $\frac{9}{7}$

13. $1\frac{1}{3}$ **14.** $2\frac{1}{5}$ **15.** $3\frac{5}{6}$ **16.** $4\frac{7}{10}$ **17.** $2\frac{11}{16}$ **18.** $1\frac{5}{18}$

Skill 47
(page 177)

Write the answer in lowest terms.

19. $\frac{1}{4} \div \frac{1}{2}$ **20.** $\frac{1}{9} \div \frac{1}{5}$ **21.** $\frac{3}{5} \div \frac{3}{4}$

$\frac{4}{5} \div \frac{1}{6}$

$\downarrow$

$\frac{4}{5} \times \frac{6}{1} = \frac{24}{5} = 4\frac{4}{5}$

22. $\frac{2}{3} \div \frac{1}{3}$ **23.** $\frac{3}{5} \div \frac{1}{4}$ **24.** $\frac{5}{9} \div \frac{2}{9}$

25. $\frac{5}{2} \div \frac{5}{6}$ **26.** $\frac{1}{4} \div \frac{7}{6}$ **27.** $\frac{5}{6} \div \frac{8}{5}$

28. $\frac{1}{9} \div \frac{11}{3}$ **29.** $\frac{3}{7} \div \frac{10}{3}$ **30.** $\frac{5}{8} \div \frac{13}{8}$ **31.** $\frac{5}{6} \div \frac{11}{6}$

Skill 48
(page 178)

Write the answer in lowest terms.

32. $8 \div \frac{1}{7}$ **33.** $10 \div \frac{1}{5}$ **34.** $4 \div \frac{1}{8}$

$6 \div \frac{1}{3}$

$\downarrow$

$\frac{6}{1} \times \frac{3}{1} = \frac{18}{1} = 18$

35. $3 \div \frac{1}{12}$ **36.** $5 \div \frac{7}{12}$ **37.** $3 \div \frac{2}{3}$

38. $5 \div \frac{5}{6}$ **39.** $9 \div \frac{6}{7}$ **40.** $12 \div \frac{3}{4}$

41. $3 \div \frac{7}{4}$ **42.** $4 \div \frac{8}{5}$ **43.** $5 \div \frac{5}{3}$ **44.** $15 \div \frac{3}{2}$

45. $\frac{1}{3} \div 2$ **46.** $\frac{1}{4} \div 5$ **47.** $\frac{2}{3} \div 2$ **48.** $\frac{4}{5} \div 4$

Skill 49
(page 179)

$$2\frac{3}{4} \div \frac{2}{3}$$
$$\downarrow$$
$$\frac{11}{4} \times \frac{3}{2} = \frac{33}{8} = 4\frac{1}{8}$$

Write the answer in lowest terms.

1. $3\frac{1}{7} \div \frac{1}{4}$ 2. $4\frac{1}{6} \div \frac{1}{3}$

3. $3\frac{3}{5} \div \frac{1}{6}$ 4. $4\frac{2}{3} \div \frac{1}{8}$

5. $5\frac{2}{3} \div \frac{9}{4}$ 6. $2\frac{4}{7} \div \frac{5}{3}$

7. $\frac{1}{3} \div 4\frac{3}{4}$ 8. $\frac{4}{5} \div 5\frac{2}{3}$ 9. $\frac{6}{7} \div 7\frac{4}{5}$ 10. $\frac{7}{9} \div 3\frac{5}{6}$

Skill 50
(page 186)

$$2\frac{4}{5} \div 4$$
$$\downarrow$$
$$\frac{14}{5} \times \frac{1}{4} = \frac{14}{20} = \frac{7}{10}$$

Write the answer in lowest terms.

11. $5\frac{1}{3} \div 2$ 12. $6\frac{1}{4} \div 3$

13. $2\frac{5}{18} \div 5$ 14. $1\frac{7}{20} \div 2$

15. $5 \div 3\frac{4}{7}$ 16. $6 \div 3\frac{5}{8}$

17. $3 \div 2\frac{4}{9}$ 18. $2 \div 1\frac{5}{6}$ 19. $1 \div 2\frac{2}{7}$ 20. $9 \div 3\frac{5}{6}$

Skill 51
(page 187)

$$2\frac{2}{3} \div 1\frac{4}{5}$$
$$\downarrow$$
$$\frac{8}{3} \div \frac{9}{5} = \frac{8}{3} \times \frac{5}{9} = \frac{40}{27} = 1\frac{13}{27}$$

Write the answer in lowest terms.

21. $1\frac{1}{4} \div 1\frac{1}{4}$ 22. $4\frac{1}{5} \div 1\frac{1}{3}$

23. $3\frac{1}{8} \div 7\frac{1}{2}$ 24. $7\frac{1}{2} \div 3\frac{1}{8}$

25. $5\frac{1}{4} \div 8\frac{1}{6}$ 26. $4\frac{3}{8} \div 1\frac{3}{4}$

27. $3\frac{3}{5} \div 2\frac{3}{4}$ 28. $4\frac{2}{3} \div 2\frac{4}{5}$ 29. $2\frac{3}{7} \div 2\frac{4}{5}$ 30. $1\frac{3}{4} \div 3\frac{2}{5}$

Skill 52
(page 202)

53 students
and
2 teachers

The ratio of students
to teachers is
53:2, or $\frac{53}{2}$.

Write the ratio using a colon (:) and as a fraction.

1. 12 wins to 5 losses

2. 3 balls and 2 strikes

3. 30 out of 33 newspapers delivered

4. $24.00 saved out of $93.00 earned

Write the ratio as a fraction in lowest terms.

5. 64 of the 128 plants   **6.** $10.00 out of $30.00

7. 6 boys in a class of 24   **8.** 5 out of 15 girls   **9.** 20 of the 24 pages

---

Skill 53
(page 203)

$$\frac{4}{9} \overset{?}{=} \frac{8}{18}$$

$$\frac{4}{9} \overset{?}{\times} \frac{8}{18}$$

$$4 \times 18 \overset{?}{=} 9 \times 8$$
$$72 \overset{?}{=} 72 \ \text{Yes}$$

Are the ratios equal? Write yes or no.

10. $\frac{4}{8}, \frac{2}{4}$ **11.** $\frac{2}{3}, \frac{4}{5}$ **12.** $\frac{1}{3}, \frac{4}{6}$

13. $\frac{3}{4}, \frac{5}{8}$ **14.** $\frac{3}{9}, \frac{1}{3}$ **15.** $\frac{4}{7}, \frac{5}{9}$

16. $\frac{4}{6}, \frac{2}{3}$ **17.** $\frac{1}{2}, \frac{3}{6}$ **18.** $\frac{8}{9}, \frac{4}{5}$

19. $\frac{2}{3}, \frac{7}{12}$ **20.** $\frac{4}{5}, \frac{8}{10}$ **21.** $\frac{3}{9}, \frac{9}{27}$

22. $\frac{5}{8}, \frac{9}{12}$ **23.** $\frac{4}{6}, \frac{12}{18}$ **24.** $\frac{6}{9}, \frac{24}{35}$

Skill 54
(pages 204–205)

Solve for a.
$$\frac{5}{7} = \frac{13}{a}$$
$$5 \times a = 7 \times 13$$
$$5 \times a = 91$$
$$a = 91 \div 5$$
$$a = 18\frac{1}{5}$$

Solve the proportion.

25. $\frac{15}{2} = \frac{3}{a}$ **26.** $\frac{m}{6} = \frac{4}{1}$ **27.** $\frac{36}{9} = \frac{y}{1}$

28. $\frac{2}{4} = \frac{x}{8}$ **29.** $\frac{1}{6} = \frac{6}{c}$ **30.** $\frac{b}{7} = \frac{9}{21}$

31. $\frac{s}{15} = \frac{36}{45}$ **32.** $\frac{20}{n} = \frac{40}{80}$ **33.** $\frac{45}{5} = \frac{9}{x}$

34. $\frac{4}{100} = \frac{m}{50}$ **35.** $\frac{6}{100} = \frac{3}{s}$ **36.** $\frac{t}{100} = \frac{6}{50}$

37. $\frac{n}{60} = \frac{75}{100}$ **38.** $\frac{t}{450} = \frac{12}{100}$ **39.** $\frac{x}{6} = \frac{50}{100}$

Skill Bank

120 points in 8 games

Rate

Write the rate as a fraction.

1. 72 sales in 16 h
2. 200 m in 30 s
3. 3 games in 2 days
4. 16 letters in 4 min

6 buses in 12 hours =
x buses in 36 hours

buses → $\frac{6}{12} = \frac{x}{36}$ ← buses
hours →   ← hours

Write the equal rates as a proportion.

5. 3 windows in 5 rooms = x windows in 10 rooms
6. $32 for 4 books = $128 for b books
7. 10 apples for 4 people = 40 apples for m people
8. 30 houses on 6 blocks = a houses on 2 blocks

$10 saved in 3 days =
t saved in 15 days

$\frac{10}{3} = \frac{t}{15}$
$150 = 3 \times t$
$50 = t$

Write the equal rates as a proportion. Solve the proportion.

9. 3 TV sets fixed in 8 h = b sets fixed in 32 h
10. 4 problems per page = y problems in 144 pages
11. 120 people in 4 rooms = m people in 10 rooms
12. 3 loaves of bread in 9 days = 27 loaves in x days
13. 15 books per shelf = d books on 9 shelves
14. 7 letters in 3 days = 35 letters in y days

$60 for 5 shirts
↓
d for 1 shirt
$\frac{60}{5} = \frac{d}{1}$
$60 \times 1 = 5 \times d$
$60 = 5 \times d$
$60 \div 5 = d$
$12 = d$

Unit rate: $12 per shirt.

Find the unit rate.

15. 18 repairs in 3 days
16. 630 people in 7 rows
17. 25 hits in 20 games
18. $48 for 6 dinners

Compare the unit rates to answer the question.

19. A: $224 in 40 h
B: $147 in 25 h
Which is the higher rate of pay?

20. A: $12.50 in 8 days
B: $460 in 40 days
Who had the lower daily expenses?

Skill 58
(page 234)

$$\frac{3}{7} \rightarrow 7\overline{)3.000}^{\,0.428}$$

To the nearest hundredth:

$$\frac{3}{7} = 0.43$$

Write as a decimal. Round to the nearest hundredth when necessary.

1. $\frac{1}{3}$ **2.** $\frac{1}{2}$ **3.** $\frac{1}{6}$ **4.** $\frac{3}{5}$

5. $\frac{1}{5}$ **6.** $\frac{1}{4}$ **7.** $\frac{1}{9}$ **8.** $\frac{2}{3}$

9. $\frac{1}{10}$ **10.** $\frac{3}{10}$ **11.** $\frac{8}{25}$ **12.** $\frac{9}{29}$

13. $\frac{7}{10}$ **14.** $\frac{3}{20}$ **15.** $\frac{7}{25}$ **16.** $\frac{8}{50}$

Skill 59
(page 235)

$$0.150 = \frac{150}{1000}$$

In lowest terms:

$$0.150 = \frac{3}{20}$$

Write as a fraction in lowest terms.

17. 0.2   **18.** 0.3   **19.** 0.5   **20.** 0.6

21. 0.07   **22.** 0.04   **23.** 0.06   **24.** 0.09

25. 0.25   **26.** 0.33   **27.** 0.45   **28.** 0.50

29. 0.375   **30.** 0.448   **31.** 0.775   **32.** 0.432

Skill 60
(page 236)

$$2\frac{3}{8} = 2.375$$

To the nearest hundredth:

$$2\frac{3}{8} = 2.38$$

Write as a decimal. Round to the nearest hundredth when necessary.

33. $1\frac{1}{2}$ **34.** $2\frac{1}{3}$ **35.** $5\frac{1}{6}$ **36.** $2\frac{4}{9}$

37. $4\frac{1}{5}$ **38.** $7\frac{1}{4}$ **39.** $8\frac{1}{9}$ **40.** $4\frac{1}{6}$

41. $12\frac{1}{4}$ **42.** $15\frac{1}{8}$ **43.** $17\frac{1}{9}$ **44.** $13\frac{1}{5}$

45. $2\frac{1}{12}$ **46.** $4\frac{1}{10}$ **47.** $6\frac{1}{20}$ **48.** $8\frac{1}{25}$

Skill 61
(page 237)

$$6.25 = 6\frac{25}{100}$$

In lowest terms:

$$6.25 = 6\frac{1}{4}$$

Write as a mixed number in lowest terms.

49. 9.1   **50.** 7.4   **51.** 6.3   **52.** 8.5

53. 6.32   **54.** 7.45   **55.** 8.56   **56.** 4.01

57. 30.35   **58.** 42.25   **59.** 9.300   **60.** 7.500

Skill 62
(page 244)

Write as a fraction with a denominator of 100.

$$\frac{2}{20} = \frac{?}{100}$$

$$\frac{2 \times 5}{20 \times 5} = \frac{10}{100}$$

$$\frac{2}{20} = \frac{10}{100}$$

1. $\frac{3}{4}$ 2. $\frac{2}{5}$ 3. $\frac{3}{5}$ 4. $\frac{2}{4}$

5. $\frac{1}{5}$ 6. $\frac{1}{2}$ 7. $\frac{1}{4}$ 8. $\frac{4}{5}$

9. $\frac{3}{10}$ 10. $\frac{7}{10}$ 11. $\frac{7}{20}$ 12. $\frac{9}{20}$

13. $\frac{14}{25}$ 14. $\frac{17}{20}$ 15. $\frac{13}{50}$ 16. $\frac{13}{25}$

17. $\frac{41}{10}$ 18. $\frac{46}{25}$ 19. $\frac{57}{20}$ 20. $\frac{89}{50}$ 21. $\frac{63}{25}$

Skill 63
(page 245)

Write as a percent.

$$\frac{3}{20} = ?\%$$

$$\frac{3}{20} = \frac{15}{100}$$

$$\frac{3}{20} = 15\%$$

22. $\frac{4}{100}$ 23. $\frac{6}{100}$ 24. $\frac{2}{100}$ 25. $\frac{5}{100}$

26. $\frac{7}{100}$ 27. $\frac{9}{100}$ 28. $\frac{5}{100}$ 29. $\frac{8}{100}$

30. $\frac{10}{100}$ 31. $\frac{15}{100}$ 32. $\frac{12}{100}$ 33. $\frac{18}{100}$

34. $\frac{1}{4}$ 35. $\frac{3}{4}$ 36. $\frac{4}{5}$ 37. $\frac{2}{5}$ 38. $\frac{1}{2}$

39. $\frac{24}{25}$ 40. $\frac{18}{50}$ 41. $\frac{14}{25}$ 42. $\frac{25}{25}$ 43. $\frac{10}{10}$

Skill 64
(page 246)

Write as a percent.

$$0.54 = ?\%$$

$$0.54 \rightarrow 54\%$$

44. 0.02   45. 0.04   46. 0.07   47. 0.05

48. 0.46   49. 0.57   50. 0.63   51. 0.48

52. 0.15   53. 0.68   54. 0.92   55. 0.10   56. 0.99

57. 5.03   58. 7.06   59. 4.25   60. 9.36   61. 5.42

62. 9.17   63. 8.04   64. 1.75   65. 3.86   66. 5.10

Skill 65
(page 247)

$2\frac{1}{3} = ?\%$

$2\frac{1}{3} = 2.333$

$2.\overset{\curvearrowright}{3}33 \to 233.3\%$

Write as a percent.

1. $\frac{1}{4}$ 2. $\frac{1}{8}$ 3. $\frac{1}{2}$

4. $\frac{2}{5}$ 5. $\frac{4}{5}$ 6. $\frac{7}{8}$

7. $\frac{7}{20}$ 8. $\frac{3}{10}$ 9. $\frac{8}{25}$

10. $3\frac{1}{2}$ 11. $4\frac{1}{4}$ 12. $7\frac{2}{5}$

13. $8\frac{1}{5}$ 14. $2\frac{3}{4}$ 15. $9\frac{1}{8}$

Skill 66
(page 254)

$1\overset{\curvearrowright}{2}0\% \to 1.20$

$00\overset{\curvearrowright}{2}.3\% \to 0.023$

Write as a decimal.

16. 2% 17. 5% 18. 7%

19. 10% 20. 12% 21. 17%

22. 32% 23. 45% 24. 50%

25. 110% 26. 125% 27. 130% 28. 142% 29. 180%

30. 2.0% 31. 3.4% 32. 2.7% 33. 1.8% 34. 6.3%

Skill 67
(page 255)

1.5%
↓
$0.015 = \frac{15}{1000}$
In lowest terms:
$1.5\% = \frac{3}{200}$

125%
↓
$1.25 = \frac{125}{100}$
As a mixed number
in lowest terms:
$125\% = 1\frac{1}{4}$

Write as a fraction or mixed number in lowest terms.

35. 8% 36. 3% 37. 5%

38. 4% 39. 7% 40. 1%

41. 3.8% 42. 1.5% 43. 6.9%

44. 12.7% 45. 37.5% 46. 85.6%

47. 12% 48. 15% 49. 18%

50. 23% 51. 25% 52. 46%

53. 110% 54. 104% 55. 105%

56. 180% 57. 130% 58. 142%

59. 137% 60. 156% 61. 175%

15% of 24 is 3.6.
↓   ↓
15% × 24 = 3.6

What is 25% of 60?
↓   ↓   ↓
N = 25% × 60

Write as a number sentence.

1. 10% of 40 is 4.

2. 12% of 50 is 6.

3. 20% of 60 is 12.

4. 30% of 100 is 30.

5. 4% of 40 is 1.6.

6. 8% of 80 is 6.4.

7. 5% of 400 is 20.

8. 40% of 60 is 24.

9. 70% of 80 is 56.

10. 50% of 160 is 80.

11. 85% of 40 is 34.

12. 25% of 44 is 11.

13. 75% of 400 is 300.

14. What is 5% of 80?

15. What is 5% of 600?

16. What is 3% of 900?

17. What is 20% of 100?

18. What is 40% of 125?

19. What is 65% of 600?

20. What is 65% of 600?

21. What is 55% of 180?

22. What is 70% of 300?

Percent   Base   Percentage
↓   ↓   ↓
80%   ×   75   =   60

Percentage   Percent   Base
↓   ↓   ↓
N =   25%   ×   336

Identify the percent.

23. 15% of 800 is 120.

24. 30% of 500 is 150.

25. What is 50% of 450?

26. What is 25% of 144?

27. What percent of 400 is 240?

28. What is 18% of 950?

29. What percent of 75 is 60?

Identify the base.

30. 200% of 30 is 60.

31. 120% of 300 is 360.

32. 80% of 75 is 60.

33. What percent of 50 is 38?

Identify the percentage.

34. 76% of 50 is 38.

35. What is 30% of 50?

36. 9 is what percent of 60?

37. What is 8% of 650?

Identify the percent, base, and percentage.

38. 10% of 40 is 4.

39. 25% of 200 is 50.

40. What is 80% of 45?

Skill 70
(pages 268–269)

What is 50% of 54?
$$N = 50\% \times 54$$
$$N = \frac{50}{100} \times 54$$
$$N = 27$$

18% of 35 is
what number?
$$18\% \times 35 = N$$
$$0.18 \times 35 = N$$
$$6.3 = N$$

Write the answer as a mixed number in lowest terms or a whole number.

1. What is 5% of 40?

2. What is 2% of 100?

3. What is 10% of 130?

4. What is 20% of 460?

5. 5% of 60 is what number?

6. 9% of 200 is what number?

7. 12% of 50 is what number?

8. 10% of 40 is what number?

9. What is 52% of 70?

10. 13% of 260 is what number?

Write the answer as a decimal.

11. 20% of 40 is what number?

12. 60% of 50 is what number?

13. 30% of 90 is what number?

14. What is 110% of 60?

Skill 71
(pages 276–277)

20% of what number is 8?
$$20\% \times N = 8$$
$$\frac{20}{100} \times N = 8$$
$$\frac{1}{5} \times N = 8$$
$$N = 8 \div \frac{1}{5}$$
$$N = 40$$

7 is 14% of what number?
$$7 = 14\% \times N$$
$$7 = 0.14 \times N$$
$$7 \div 0.14 = N$$
$$50 = N$$

Write the answer as a mixed number in lowest terms or a whole number.

15. 3% of what number is 12?

16. 10% of what number is 15?

17. 8% of what number is 64?

18. 9% of what number is 54?

19. 25% of what number is 100?

20. 50% of what number is 400?

21. 49 is 7% of what number?

22. 28 is 4% of what number?

23. 6 is 18% of what number?

24. 2 is 24% of what number?

Round to the nearest tenth when necessary.

25. 24 is 25% of what number?

26. 42 is 75% of what number?

27. 77 is 100% of what number?

28. 14 is 25% of what number?

29. 16 is 50% of what number?

30. 13 is 20% of what number?

31. 66% of what number is 21?

32. 60% of what number is 92?

Skill Bank

Skill 72
(pages 278–279)

3 is what
percent of 6?

$$3 = N \times 6$$
$$3 \div 6 = N$$
$$0.50 = N$$
$$50\% = N$$

What percent of
15 is 3?

$$N \times 15 = 3$$
$$N = 3 \div 15$$
$$N = 0.20$$
$$N = 20\%$$

1. 12 is what percent of 30?

2. What percent of 70 is 7?

3. 60 is what percent of 80?

4. What percent of 50 is 30?

5. What percent of 55 is 44?

6. 48 is what percent of 64?

7. What percent of 6 is 6?

8. 77 is what percent of 77?

9. 16 is what percent of 32?

10. What percent of 65 is 13?

11. What percent of 56 is 42?

12. 72 is what percent of 96?

13. 40 is what percent of 92?

14. What percent of 75 is 25?

15. What percent of 92 is 60?

16. 40 is what percent of 320?

Skill 73
(pages 286–287)

What percent of
64 is 48?

$$\frac{N}{100} = \frac{48}{64}$$
$$N \times 64 = 100 \times 48$$
$$N = 75$$
75% of 64
is 48.

Write as a proportion.

17. What percent of 40 is 20?

18. What percent of 80 is 16?

19. 60 is what percent of 80?

20. 9 is what percent of 90?

Answer the question.

21. What percent of 32 is 16?

22. What percent of 50 is 30?

23. 20 is what percent of 40?

24. 200 is what percent of 800?

15% of 60
is what number?

$$\frac{15}{100} = \frac{N}{60}$$
$$15 \times 60 = 100 \times N$$
$$9 = N$$
15% of 60
is 9.

Write as a proportion.

25. 80% of 20 is what number?

26. 60% of 50 is what number?

27. What is 60% of 35?

28. What is 4% of 200?

Answer the question.

29. What is 100% of 40?

30. 40% of 125 is what number?

31. 90% of what number is 72?

32. 28 is 28% of what number?

Skill 74
(page 302)

4° above zero → $^+4$
8° below zero → $^-8$
$30 earned → $^+30$
$15 spent → $^-15$
10 steps up → $^+10$
3 steps down → $^-3$

Write a positive or negative number.

1. 30 m above sea level
2. 95 m below sea level
3. 15 kg lost
4. 38 kg gained
5. 10° increase
6. 20° decrease
7. $20 loss
8. $50 gain
9. 10 games won
10. 3 games lost
11. 800 m up
12. 16 m below water level
13. $80 profit
14. Loss of 18 kg
15. Down 15 steps
16. A mine 600 m deep
17. Jet 20 km up
18. Water level 2 m above normal
19. Submarine 80 m below sea level
20. 6 more guests
21. 17% increase
22. pool 30 ft deep
23. 15 new students
24. 10 fewer chairs
25. 2 more hours

Skill 75
(page 303)

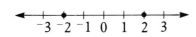

$^-2$ is less than 2.
$^-2 < 2$

$^-1$ is greater than $^-2$.
$^-1 > ^-2$

The opposite of $^-2$ is 2.

Write > or < in place of __?__.

26. 1 __?__ 3
27. $^-2$ __?__ $^-3$
28. $^-2$ __?__ $^-1$
29. 1 __?__ 2
30. 3 __?__ 2
31. $^-3$ __?__ 0
32. 1 __?__ $^-1$
33. 2 __?__ $^-3$
34. $^-3$ __?__ 3
35. 2 __?__ $^-2$

Write the opposite.

36. 3
37. 1
38. $^-1$
39. $^-4$
40. 18
41. $^-56$
42. 88
43. 25
44. $^-86$
45. 96
46. $^-26$
47. 16

Skill 76
(page 304)

Use a number line when necessary.

Add.

$2 + 5 = 7$

$^-3 + ^-1 = ^-4$

1. $2 + 8$ **2.** $6 + 3$ **3.** $7 + 4$

4. $^-4 + ^-5$ **5.** $^-6 + ^-1$ **6.** $^-3 + ^-2$

7. $7 + 2$ **8.** $8 + 8$ **9.** $6 + 9$

10. $^-3 + ^-8$ **11.** $^-7 + ^-4$ **12.** $^-2 + ^-10$

13. $10 + 6$ **14.** $^-6 + ^-6$ **15.** $9 + 8$ **16.** $^-7 + ^-5$

17. $^-10 + ^-5$ **18.** $^-8 + ^-8$ **19.** $5 + 6$ **20.** $6 + 9$

21. $^-10 + ^-10$ **22.** $^-1 + ^-7$ **23.** $10 + 9$ **24.** $7 + 8$

25. $19 + 19$ **26.** $^-29 + ^-18$ **27.** $25 + 17$ **28.** $^-50 + ^-25$

29. $^-32 + ^-12$ **30.** $36 + 24$ **31.** $^-25 + ^-25$ **32.** $75 + 16$

Skill 77
(page 305)

Use a number line when necessary.

Add.

$^-2 + 5 = 3$

$3 + ^-7 = ^-4$

33. $^-6 + 2$ **34.** $5 + ^-1$ **35.** $9 + ^-3$

36. $2 + ^-3$ **37.** $5 + ^-7$ **38.** $6 + ^-8$

39. $10 + ^-2$ **40.** $6 + ^-7$ **41.** $15 + ^-7$

42. $^-6 + 3$ **43.** $5 + ^-5$ **44.** $3 + ^-18$

45. $^-7 + 7$ **46.** $^-15 + 7$ **47.** $14 + ^-4$ **48.** $^-14 + 4$

49. $13 + ^-7$ **50.** $^-13 + 13$ **51.** $^-13 + 7$ **52.** $^-7 + 13$

53. $15 + ^-20$ **54.** $35 + ^-36$ **55.** $50 + ^-100$ **56.** $100 + ^-50$

57. $^-34 + 35$ **58.** $^-35 + 34$ **59.** $^-34 + 34$ **60.** $35 + ^-35$

61. $18 + ^-17$ **62.** $^-17 + 13$ **63.** $13 + ^-10$ **64.** $^-10 + 20$

Skill 78
(page 312)

To subtract,
add the opposite.

$8 - 3 \rightarrow 8 + {}^-3 = 5$

${}^-9 - 4 \rightarrow {}^-9 + {}^-4 = {}^-13$

$2 - 7 \rightarrow 2 + {}^-7 = {}^-5$

1. $7 - 6$ **2.** $3 - 1$

3. $8 - 3$ **4.** $6 - 4$

5. $10 - 2$ **6.** $5 - 4$

7. ${}^-8 - 5$ **8.** ${}^-12 - 7$

9. ${}^-11 - 9$ **10.** ${}^-10 - 8$

11. ${}^-6 - 3$ **12.** ${}^-4 - 8$ **13.** ${}^-5 - 4$ **14.** ${}^-7 - 2$

15. $2 - 9$ **16.** $3 - 8$ **17.** $4 - 7$ **18.** $5 - 6$

19. $12 - 27$ **20.** ${}^-20 - 8$ **21.** $15 - 30$ **22.** ${}^-21 - 7$

23. $10 - 15$ **24.** $21 - 19$ **25.** ${}^-35 - 12$ **26.** ${}^-27 - 17$

27. $19 - 14$ **28.** $37 - 48$ **29.** ${}^-40 - 44$ **30.** ${}^-15 - 31$

Skill 79
(page 313)

To subtract,
add the opposite.

$6 - {}^-4 \rightarrow 6 + 4 = 10$

${}^-5 - {}^-3 \rightarrow {}^-5 + 3 = {}^-2$

${}^-2 - {}^-7 \rightarrow {}^-2 + 7 = 5$

31. $7 - {}^-5$ **32.** $6 - {}^-3$

33. $5 - {}^-3$ **34.** $4 - {}^-2$

35. $3 - {}^-2$ **36.** $2 - {}^-1$

37. ${}^-9 - {}^-6$ **38.** ${}^-8 - {}^-5$

39. ${}^-7 - {}^-4$ **40.** ${}^-6 - {}^-3$

41. ${}^-10 - {}^-8$ **42.** ${}^-11 - {}^-9$ **43.** ${}^-12 - {}^-11$ **44.** ${}^-13 - {}^-12$

45. ${}^-3 - {}^-7$ **46.** ${}^-6 - {}^-8$ **47.** ${}^-2 - {}^-9$ **48.** ${}^-4 - {}^-6$

49. ${}^-8 - {}^-9$ **50.** ${}^-5 - {}^-8$ **51.** ${}^-10 - {}^-12$ **52.** ${}^-12 - {}^-15$

53. ${}^-27 - {}^-35$ **54.** ${}^-36 - {}^-14$ **55.** $22 - {}^-18$ **56.** $28 - {}^-16$

57. ${}^-43 - {}^-59$ **58.** ${}^-27 - {}^-60$ **59.** $100 - {}^-74$ **60.** $85 - {}^-96$

Skill 80
(page 314)

Multiply.

$5 \times 2 = 10$
Same sign
$^-5 \times ^-2 = 10$

The product
is positive.

Divide.

$6 \div 2 = 3$
Same sign
$^-6 \div ^-2 = 3$

The quotient
is positive.

1. 2×6
2. $^-8 \times ^-5$
3. $^-1 \times ^-7$
4. $^-9 \times ^-3$
5. $^-3 \times ^-8$
6. 8×2
7. 4×3
8. $^-1 \times ^-5$
9. $^-6 \times ^-4$
10. $^-7 \times ^-4$
11. $^-3 \times ^-7$
12. 2×7
13. 3×3
14. $^-3 \times ^-4$
15. $^-3 \times ^-1$
16. $^-2 \times ^-15$
17. $^-14 \times ^-1$
18. $^-10 \times ^-3$
19. $6 \div 3$
20. $^-4 \div ^-2$
21. $^-5 \div ^-1$
22. $^-25 \div ^-5$
23. $^-24 \div ^-6$
24. $18 \div 2$
25. $64 \div 8$
26. $^-81 \div ^-9$
27. $56 \div 7$
28. $^-24 \div ^-12$
29. $^-62 \div ^-2$
30. $^-50 \div ^-25$

Skill 81
(page 315)

Multiply.

$5 \times ^-2 = ^-10$
Different signs
$^-5 \times 2 = ^-10$

The product
is negative.

Divide.

$6 \div ^-2 = ^-3$
Different signs
$^-6 \div 2 = ^-3$

The quotient
is negative.

31. $^-2 \times 5$
32. $3 \times ^-6$
33. $7 \times ^-4$
34. $11 \times ^-2$
35. $^-3 \times 10$
36. $4 \times ^-12$
37. $^-50 \times 4$
38. $22 \times ^-3$
39. $^-33 \times 3$
40. $6 \times ^-8$
41. $7 \times ^-9$
42. $9 \times ^-9$
43. $^-20 \times 7$
44. $^-30 \times 8$
45. $^-40 \times 4$
46. $43 \times ^-3$
47. $42 \times ^-4$
48. $50 \times ^-1$
49. $42 \div ^-6$
50. $^-44 \div 4$
51. $^-35 \div 35$
52. $55 \div ^-5$
53. $^-84 \div 4$
54. $50 \div ^-50$
55. $30 \div ^-15$
56. $^-26 \div 2$
57. $^-75 \div 3$
58. $^-64 \div 8$
59. $99 \div ^-3$
60. $^-88 \div 4$

Skill 82
(page 322)

Simplify.

$6 + 2 \times 4$

$6 + \quad 8$

14

$6 + 2 \times 4 = 14$

Simplify.

$8 - 6 \div 2 + {}^-2$

$8 - \quad 3 \quad + {}^-2$

$5 + {}^-2$

3

$8 - 6 \div 2 + {}^-2 = 3$

Simplify.

1. $4 + 5 \times 3$

2. $6 \times 3 - 5$

3. $8 - 4 \times 5$

4. $8 - 6 \div 2$

5. $6 + {}^-3 \times 3$

6. $10 \div 2 - 4$

7. $14 \div 2 - 3$

8. ${}^-9 \times 2 + 4$

9. $4 \times 2 + 2 \div 2$

10. $6 - 6 \div 2 + {}^-2$

11. $7 \times 2 + 4 \div 2$

12. $5 + 7 - 8 \div 2$

13. $7 \div 7 - 18 \div 9$

14. $8 \times 2 - 4 \div 2$

15. $8 + 8 \div 2 - 1 \times {}^-1$

16. $2 + 2 \div 2 - 2$

17. $6 + 4 - 8 \div 2$

18. $6 - 6 \div 3 + {}^-2$

19. $4 \times 2 + 4 \div 2$

20. $3 + 3 \times 3 \div 3$

Skill 83
(page 323)

Simplify.

$5 \times (3 + 4)$

$5 \times \quad 7$

35

$5 \times (3 + 4) = 35$

Simplify.

$[6 \times (3 + 2)] - 4$

$[6 \times \quad 5] - 4$

$30 - 4$

26

$[6 \times (3 + 2)] - 4 = 26$

Simplify.

21. $7 - (4 \times 1)$

22. $(8 - 2) \times 6$

23. $(8 + 8) \div 8$

24. $7 \times (5 + 2)$

25. $9 - (8 - 8)$

26. $16 \div (4 - 2)$

27. $6 \times (3 + 4)$

28. $7 - (2 \times 3)$

29. $[5 \times (4 + 1)] - 5$

30. $[6 \times (7 - 5)] - 6$

31. $[7 - (3 - 2)] + 6$

32. $[7 \times (9 - 4)] + 5$

33. $[11 \times (6 + 3)] + 1$

34. $6 - [(4 \times 4) + 3]$

35. $5 - [5 \times (4 + 2)]$

36. $4 + (5 + 1) \times 2$

37. $[7 \times (5 - 3)] + 14$

38. $[18 \div (6 - 3)] - 3$

39. $6 \times [(14 \div 2) - 4]$

40. $7 - [(4 \times 4) + 1]$

Skill 84
(page 334)

State the meaning of the expression.

1. $7 + n$ **2.** $y + b$ **3.** $5 \times m$

$(a + b)2$
Add b to a.
Multiply by 2.

4. $(t \times 4) + 3$ **5.** $\dfrac{12}{w}$ **6.** $\dfrac{11 - a}{y}$

7. $3 \times (b - 1)$ **8.** $\dfrac{k + s}{4}$ **9.** $16 \times y$

10. $5 \times (b + 9)$ **11.** $(n - r) + 7$ **12.** $(g - 18) \times 4$ **13.** $(17 \times d) - c$

14. $\left(\dfrac{k + 6}{2}\right) \times 4$ **15.** $(w + 6) \times 8$ **16.** $(a + b) \times c$ **17.** $a + (b \times c)$

Skill 85
(page 335)

State the meaning of the expression.

18. $7 \cdot 3$ **19.** $a \times 6$ **20.** cd

$2(d + e)$
Add e to d.
Multiply by 2.

21. $4m$ **22.** $9k + 2$ **23.** $s - 3d$

24. $a(4)$ **25.** $(1 + m)(2)$ **26.** $4 \cdot (t - 8)$

27. $(10m) \cdot c$ **28.** $8bq$ **29.** $(7n - 8)a$ **30.** $9(ab)$

31. $\left(\dfrac{c}{d}\right)4$ **32.** $\dfrac{2m}{v}$ **33.** $\left(\dfrac{f + i}{2}\right)a$ **34.** qrs

---

Skill 86
(pages 336–337)

Find the value of the expression. Use $x = 3$, $y = 7$, $z = {}^{-}2$.

35. $9x$ **36.** $4y$ **37.** $3z$

$\dfrac{x + y}{z} = \dfrac{3 + 7}{{}^{-}2}$
$= {}^{-}5$

38. $8 - x$ **39.** $y + x$ **40.** $z + x$

41. $12 - y + z$ **42.** $y + x - 8$ **43.** $y(7 + x)$

44. $\dfrac{x}{y}$ **45.** $x - 9 + y$ **46.** $\dfrac{3x}{15}$ **47.** $\dfrac{49}{y}$

48. $(x + 1)z$ **49.** $\dfrac{y - x}{z}$ **50.** $x - \dfrac{14}{y}$ **51.** $\dfrac{4(y - x)}{16}$

Skill 87
(page 344)

$$x(9 - 8)$$
$$x(1)$$
$$x$$

Simplify.

1. $x + 0$ **2.** $2x + 0$ **3.** $x - x$

4. $n \cdot 0$ **5.** $3n \cdot 1$ **6.** $b \cdot 1$

7. $\frac{x}{x}$ **8.** $\frac{4x}{4}$ **9.** $\frac{8x}{x}$

10. $17 \cdot 0 + x$ **11.** $6 + y - y$ **12.** $0 \cdot 4n$ **13.** $12 + b - 12$

14. $\frac{0}{n}$ **15.** $\frac{5 \cdot 0}{n}$ **16.** $(x - 4)\frac{0}{x}$ **17.** $(n + 1)0$

18. $3 \cdot \frac{n}{n}$ **19.** $(x + 0) \cdot \frac{1}{x}$ **20.** $4 \cdot c \cdot \frac{c}{c}$ **21.** $(9 + b + x)\frac{0}{x}$

Skill 88
(page 345)

$$7c + 2c + 8$$
$$9c + 8$$

Simplify by combining like terms.

22. $4y - 3y$ **23.** $7n + 8n$ **24.** $a + 2a$

25. $8y + 2y$ **26.** $5c + c$ **27.** $18x - 2x$

28. $x + 3x + x$ **29.** $9m + m + 3m$ **30.** $7p - 4p + 8$

31. $5x + 2x + y$ **32.** $9m + 8 + 7m$ **33.** $6x - 4x + x$ **34.** $4x + 8x - 11x$

35. $6 + 8z - 9z$ **36.** $8y - 2y + t$ **37.** $3r - 6r + 3r$ **38.** $8y - 2y + 9z$

39. $4 + 2z + z$ **40.** $7m + 2t - 3m$ **41.** $5n + 9n - 6n$ **42.** $13x - 8x - 5x$

43. $13m - 9 + 6$ **44.** $2t + 4t - 6t$ **45.** $11x - 3 + 8x$ **46.** $30 - 4x - 9$

Skill 89
(page 346)

$$3b \cdot b \cdot a$$
$$3b^2 \cdot a$$
$$3b^2 a$$

Write with exponents.

47. $6 \times 6 \times 6$ **48.** $3 \cdot 2 \cdot 3 \cdot 2 \cdot 3$ **49.** $(5 \cdot 5)a$

50. $(ac)(ac)$ **51.** $x \cdot x \cdot y$ **52.** $4f \cdot g \cdot f$

53. $7x \cdot x \cdot y$ **54.** $(6 \times z)(6 \times z)$ **55.** $9x \cdot x \cdot y$

56. $(2m - 7)(2m - 7)$ **57.** $9x \cdot x \cdot y \cdot y$ **58.** $(x - y)(x - y)$

59. $5a \cdot a \cdot b \cdot b \cdot b$ **60.** $(a + c)(a + c)(a + c)$ **61.** $6(a + x)(a + x)$

Skill 90
(page 347)

Find the value. Use $x = 3$, $y = 4$, $z = {}^-2$.

$2 + x - y$
$\downarrow$
$2 + 3 - 4 = 1$

1. x^2 **2.** y^3 **3.** z^2

4. xy **5.** $8x$ **6.** $(x - y)(z)$

7. $(y + 1)^2$ **8.** $6y^2$ **9.** $(xz)^2$

10. z^3 **11.** $3x^2z$ **12.** $(5y)^2$ **13.** $(x + 1)^2$

14. $(x + y)^2$ **15.** $(x + z)^2$ **16.** $(x - y)^3$ **17.** $(x^2 - 9)^2$

18. $(x + 7)^2 + y$ **19.** $(4 + z + x)^3$ **20.** $(x + y + z)^2$ **21.** $(x - y + z)^2$

Skill 91
(page 354)

Find the solution.

$-3m = {}^-27$
Try $m = 9$.
${}^-3(9) = {}^-27$
      True

The solution
is 9.

22. $2x = 14$ **23.** $4n = 56$ **24.** $8n = 0$

25. $15x = 30$ **26.** $x + 1 = 6$ **27.** $x - 1 = 4$

28. $1 + a = 15$ **29.** $3 + d = 6$ **30.** $m + 8 = 7$

31. $x + 4 = 0$ **32.** $\frac{n}{5} = 3$ **33.** $x + 2 = 2$

34. $7x = {}^-49$ **35.** $3x - x = {}^-2$ **36.** $3r - 0 = {}^-18$

Skill 92
(page 355)

Check these values for x: ${}^-3$, ${}^-2$, ${}^-1$, 0, 1, 2. Which of these values are solutions of the inequality?

$x + {}^-1 > 0$

Is 1 a solution?
$1 + {}^-1 > 0$ False

Is 2 a solution?
$2 + {}^-1 > 0$ True

Is 3 a solution?
$3 + {}^-1 > 0$ True

2 and 3 are
solutions of
$x + {}^-1 > 0$.

37. $x > 0$ **38.** $x < {}^-1$ **39.** $x > 1$

40. $x + 1 > {}^-1$ **41.** $x + 2 < 0$ **42.** $x + 2 > 1$

43. $2 - x > 0$ **44.** $x - 1 < 0$ **45.** $x + 3 > 0$

46. $\frac{x}{2} > {}^-1$ **47.** $6x < 2$ **48.** $x - 1 > 1$

49. $x + {}^-3 < 0$ **50.** ${}^-2x > 0$ **51.** $2x > 0$

52. $2x > {}^-1$ **53.** $2x < {}^-1$ **54.** $3x > 0$

Skill 93
(pages 366–367)

$$n - 8 = 12$$
$$n - 8 + 8 = 12 + 8$$
$$n = 20$$

Solve the equation.

1. $n - 7 = 12$

2. $x - 6 = 13$

3. $y - 10 = 28$

4. $z - 13 = 15$

5. $m - 8 = 10$

6. $n - 14 = 16$

7. $x - 11 = 36$

8. $y - 19 = 20$

9. $a - 5 = 34$

10. $6 = m - 2$

11. $8 = x - 10$

12. $14 = a - 4$

13. $b - 18 = 6$

14. $x - 12 = 12$

15. $y - 24 = 0$

16. $16 = n - 8$

17. $15 = x - 30$

18. $25 = n - 25$

19. $0 = n - 11$

20. $40 = n - 40$

21. $0 = x - 42$

22. $9 = y - 3$

23. $10 = y - 7$

24. $18 = y - 4$

25. $56 = y - 24$

26. $63 = y - 72$

27. $54 = n - 90$

Skill 94
(pages 368–369)

$$x + 3 = 10$$
$$x + 3 - 3 = 10 - 3$$
$$x = 7$$

Solve the equation.

28. $x + 8 = 20$

29. $y + 18 = 24$

30. $a + 12 = 20$

31. $n + 10 = 12$

32. $s + 18 = 33$

33. $m + 27 = 48$

34. $y + 9 = 24$

35. $t + 33 = 36$

36. $x + 28 = 60$

37. $z + 28 = 64$

38. $n + 56 = 64$

39. $m + 27 = 32$

40. $n + 85 = 85$

41. $x + 45 = 40$

42. $m + 30 = 70$

43. $p + 35 = 70$

44. $w + 18 = 12$

45. $n + 24 = 6$

46. $x + {}^-8 = {}^-8$

47. ${}^-24 = n + 12$

48. ${}^-48 = n + 42$

49. $y + {}^-42 = 48$

50. $x + 77 = {}^-119$

51. $n + {}^-56 = 28$

Skill Bank

$$8x = 24$$
$$\frac{8x}{8} = \frac{24}{8}$$
$$x = 3$$

Solve the equation.

1. $6x = 48$ **2.** $8y = 32$ **3.** $5m = 60$

4. $12n = 96$ **5.** $18m = 72$ **6.** $22y = 88$

7. $24x = 72$ **8.** $10n = 60$ **9.** $14n = 56$

10. $6p = {}^-48$ **11.** $4m = {}^-28$ **12.** $8x = 64$

13. $^-3x = 27$ **14.** $^-9y = {}^-63$ **15.** $^-12m = 60$ **16.** $^-15p = {}^-45$

17. $6n = {}^-18$ **18.** $8x = 40$ **19.** $^-16y = 16$ **20.** $20s = {}^-100$

21. $7n = 0$ **22.** $13p = 52$ **23.** $3y = {}^-48$ **24.** $^-6m = 240$

25. $12y = 3$ **26.** $45x = 5$ **27.** $^-72m = {}^-4$ **28.** $56t = 8$

29. $75n = 5$ **30.** $36z = 4$ **31.** $60y = 12$ **32.** $75m = 3$

$$\frac{x}{9} = 6$$
$$\frac{x}{9} \cdot 9 = 6 \cdot 9$$
$$x = 54$$

Solve the equation.

33. $\frac{y}{3} = 9$ **34.** $\frac{x}{2} = 8$ **35.** $\frac{m}{4} = {}^-1$

36. $\frac{n}{4} = {}^-5$ **37.** $\frac{a}{6} = 7$ **38.** $\frac{t}{5} = 7$

39. $\frac{x}{9} = 4$ **40.** $\frac{s}{7} = 3$ **41.** $\frac{n}{-2} = 14$

42. $\frac{v}{-1} = 70$ **43.** $\frac{x}{5} = 60$ **44.** $\frac{r}{6} = 72$

45. $\frac{n}{8} = 0$ **46.** $\frac{m}{12} = 5$ **47.** $\frac{x}{21} = {}^-3$ **48.** $\frac{y}{-7} = 7$

49. $\frac{a}{20} = 5$ **50.** $\frac{y}{32} = {}^-2$ **51.** $\frac{m}{15} = 4$ **52.** $\frac{n}{17} = {}^-6$

53. $\frac{x}{18} = {}^-3$ **54.** $\frac{m}{27} = {}^-7$ **55.** $\frac{a}{-14} = 5$ **56.** $\frac{d}{13} = {}^-10$

57. $\frac{t}{24} = 5$ **58.** $\frac{i}{-16} = 0$ **59.** $\frac{x}{10} = 10$ **60.** $\frac{m}{21} = {}^-8$

61. $\frac{n}{31} = 4$ **62.** $\frac{p}{45} = 2$ **63.** $\frac{m}{75} = 0$ **64.** $\frac{x}{84} = {}^-1$

Skill 97
(page 378)

$$x + 2 = 6$$
$$x + 2 - 2 = 6 - 2$$
$$x = 4$$

$$y - 2 = 6$$
$$y - 2 + 2 = 6 + 2$$
$$y = 8$$

$$2m = 6$$
$$\frac{2m}{2} = \frac{6}{2}$$
$$m = 3$$

$$\frac{n}{2} = 6$$
$$\frac{n}{2} \cdot 2 = 6 \cdot 2$$
$$n = 12$$

Solve the equation.

1. $n - 3 = 11$

2. $x + 6 = 12$

3. $4 + p = 14$

4. $9y = 90$

5. $8t = 24$

6. $6n = {}^{-}36$

7. $n + 8 = 30$

8. $x - 35 = 7$

9. $\frac{m}{4} = 36$

10. $x + 2 = {}^{-}2$

11. $\frac{m}{6} = {}^{-}1$

12. $n - 2 = {}^{-}4$

13. $p + 3 = {}^{-}3$

14. $\frac{w}{7} = {}^{-}7$

15. $8x = {}^{-}16$

16. $y + 15 = 48$

17. $n + 30 = {}^{-}19$

18. $2x = 140$

19. $n + 9 = {}^{-}11$

20. $m + {}^{-}47 = 12$

21. $\frac{x}{{}^{-}7} = 77$

22. $n - 5 = 40$

Skill 98
(page 379)

$$9x + 3 = 39$$
$$9x + 3 - 3 = 39 - 3$$
$$9x = 36$$
$$\frac{9x}{9} = \frac{36}{9}$$
$$x = 4$$

Solve the equation.

23. $4n - 8 = 0$

24. $4x - 8 = 16$

25. $4y - 16 = 8$

26. $6p + 24 = 12$

27. $3m + 6 = 15$

28. $5x + 6 = 21$

29. $7n - 1 = 13$

30. $6x - 1 = 11$

31. $12 = 9a - 15$

32. $6 = 50 - 4y$

33. $8c - 7 = 33$

34. $34 = 28 - 3d$

35. $9n + 7 = 52$

36. $4p - 8 = 36$

37. $10 = 20 - 5x$

38. $7y + 7 = 0$

39. $3y - 7 = 2$

40. $8n - 2 = {}^{-}42$

41. $10x - 12 = {}^{-}42$

42. $4m + 32 = 32$

43. $8n + 4 = 76$

44. $40 = 4 - 9y$

45. $12y - 3 = {}^{-}39$

Skill 99
(page 386)

Is the given ordered pair a solution of the equation?

Is ($^-$7, 2) a
solution of
$x + y = 5$?
$^-7 + 2 = 5$ True
($^-$7, 2) is a
solution.

1. $x + y = 13$; (6, 7)

2. $x - y = 2$; (7, 5)

3. $x - y = {}^-7$; (2, $^-$8)

4. $y = 4x$; (2, 8)

5. $2x - y = 6$; (8, 10)

6. $x + 3y = 2$; (4, $^-$2)

7. $y = x - 15$; (11, $^-$3)

8. $xy = {}^-28$; ($^-$7, $^-$4)

9. $\frac{x}{4} = y$; (28, 7)

10. $3x + 2y = 11$; (2, 3)

11. $4x - 2y = 16$; (3, $^-$2)

12. $3xy = {}^-90$; (6, $^-$5)

13. $xy = {}^-24$; (8, $^-$3)

14. $\frac{x}{7} = y$; (3, 21)

15. $y = 6x - 1$; ($^-$2, 13)

16. $^-5y = x$; ($^-$8, 40)

17. $9x + 9y = 0$; ($^-$1, 0)

18. $3y - 4x = {}^-6$; (6, 6)

19. $y = 4x - 1$; ($^-$3, $^-$13)

20. $5x + y = 0$; (2, $^-$10)

21. $xy = 72$; ($^-$8, $^-$9)

22. $y = 4x - 4$; ($^-$1, 0)

Skill 100
(page 387)

Find the missing number that will make the ordered pair a solution.

$y = 4 + x$; (?, 12)
Substitute 12 for y.
$12 = 4 + x$
$8 = x$
(8, 12) is a
solution.

23. $y = 3 + x$; (7, ?)

24. $x = y - 8$; (?, 11)

25. $y = 10 + x$; ($^-$7, ?)

26. $x = y + 7$; (?, $^-$4)

27. $x = y - 3$; (?, $^-$3)

28. $y = 15 + x$; (0, ?)

29. $x + y = 8$; ($^-$8, ?)

30. $x + y = 15$; (7, ?)

31. $x - y = 5$; (?, 6)

32. $x + y = 11$; ($^-$4, ?)

33. $2x + y = 6$; (8, ?)

34. $x + 4y = 10$; (2, ?)

35. $7x + 2y = 15$; (?, 4)

36. $7 = x + 3y$; (?, 6)

37. $4x + 2y = 16$; ($^-$5, ?)

38. $x = {}^-2y$; (8, ?)

39. $20 = 4x - 8y$; (3, ?)

40. $17x - 4y = 0$; (?, 0)

41. $y = 4x$; (6, ?)

42. $12x = 12 + 3y$; (7, ?)

43. $3x + 2y = 1$; (5, ?)

44. $y = (2 + x)x$; (9, ?)

Application A1
(pages 6–7)

Which is larger?

689.5
or
692.3

692.3 is larger.

Compare the two numbers. Which is larger?

1. 47.56 or 52.036   **2.** 332 or 391.4

3. 9875 or 9089   **4.** $30.75 or $28.99

5. $249.98 or $250   **6.** $800.10 or $790.25

7. $1000 or $998   **8.** $562.39 or $559.90

9. $7625 or $8050   **10.** 6.937 or 6.978

11. 50,680 or 49,975 **12.** 0.0839 or 0.0903

Application A2
(pages 16–17)

Estimate the
difference to the
nearest tenth.

45.389 ⟶ 45.4
− 27.307 ⟶ − 27.3
_____
18.1

Estimate the sum to the nearest thousand.

13. 6093 + 8502   **14.** 3461 + 15,630

15. 24,980 + 38,600   **16.** 4588 + 9200 + 18,750

Estimate the difference to the nearest hundred.

17. 809 − 680   **18.** 949 − 320

19. 5670 − 958   **20.** 9312 − 4871

21. 10,220 − 7580   **22.** 13,476 − 13,105

Application A3
(pages 24–25)

Estimate the sum to
the nearest thousand.

25,750 ⟶ 26,000
42,300 ⟶ 42,000
+ 70,985 ⟶ + 71,000
_____
139,000

Estimate the sum to the nearest whole number.

23. 72.36 + 84.53   **24.** 0.89 + 5.29

25. $32.95 + $17.50   **26.** $588.50 + $791.29

Estimate the difference to the nearest tenth.

27. 51.723 − 29.508   **28.** 85 − 71.82

29. 125.55 − 107.91   **30.** $532.48 − $67.25

Application A4
(pages 38–39)

9 yd 2 ft = 9 yd + 2 ft
= 27 ft + 2 ft
= 29 ft

Complete the statement.

1. 12 yd 2 ft = 12 yd + 2 ft
= __?__ ft + 2 ft
= __?__ ft

2. 7 ft 2 in.
= __?__ in. + 2 in.
= __?__ in.

3. 9 yd 6 in. = 9 yd + 6 in.
= __?__ in. + 6 in.
= __?__ in.

4. 2 yd 2 ft 8 in.
= 2 yd + 2 ft + 8 in.
= __?__ in.

5. You have one ladder that measures 4 yd 2 ft. Another ladder measures 10 ft. How much shorter is the second ladder?

Application A5
(pages 48–49)

6 lb 10 oz = 6 lb + 10 oz
= 96 oz + 10 oz
= 106 oz

Complete the statement.

6. 7 lb 12 oz = 7 lb + 12 oz
= __?__ oz + 12 oz
= __?__ oz

7. 12 lb 5 oz
= __?__ oz + 5 oz
= __?__ oz

Which weighs more?

8. 5 lb package of soil or 70 oz package of soil?

9. 7 lb 8 oz box of paper or 130 oz box of paper?

10. You can buy nuts in 2-pound cans for $5.79 each or in 8-ounce cans for $1.50 each. Which is the better buy?

Application A6
(pages 56–57)

5 gal 2 qt = 5 gal + 2 qt
= 20 qt + 2 qt
= 22 qt

Complete the statement.

11. 6 qt 1 pt = 6 qt + 1 pt
= __?__ pt + 1 pt
= __?__ pt

12. 8 gal 3 qt
= __?__ qt + 3 qt
= __?__ qt

Find the amount.

13. 3 pints of yogurt.
79¢ per pint.
Total cost?

14. 4 pints water.
Double the amount.
How many quarts?

Application A7
(pages 70–71)

9.6 km = _?_ m
9.6 × 1000 = 9600
9.6 km = 9600 m

15.3 m = _?_ km
15.3 × 0.001 = 0.0153
15.3 m = 0.0153 km

Complete the table.

	1.	2.	3.	4.	5.
km	5.8	23	?	?	10.4
m	?	?	1600	42,500	?

Arrange in order of length, starting with the shortest.

6. 300 m   18 km   5500 m   0.6 km

Application A8
(pages 80–81)

7.24 m = _?_ cm
7.24 × 100 = 724
7.24 m = 724 cm

820 mm = _?_ cm
820 × 0.1 = 82
820 mm = 82 cm

Complete the table.

	7.	8.	9.	10.	11.
m	3	?	?	?	6.1
cm	300	12	5	?	?
mm	?	?	50	8500	?

Which length is more reasonable?

12. Pocket calculator: 14 cm or 14 mm?

Application A9
(page 88)

62 kg = _?_ g
62 × 1000 = 62,000
62 kg = 62,000 g

5420 g = _?_ kg
5420 × 0.001 = 5.42
5420 g = 5.42 kg

Complete the statement.

13. 94 kg = _?_ g

14. 7.08 kg = _?_ g

15. 10,320 g = _?_ kg

16. 652 g = _?_ kg

Which is more?

17. a 350 g box or a 0.6 kg box?

18. a 15 kg dog or a 12,500 g dog?

Application A10
(page 89)

9.3 L = _?_ mL
9.3 × 1000 = 9300
9.3 L = 9300 mL

7045 mL = _?_ L
7045 × 0.001 = 7.045
7045 mL = 7.045 L

Complete the statement.

19. 6500 mL = _?_ L

20. 8.25 L = _?_ mL

21. 46 L = _?_ mL

22. 14,250 mL = _?_ L

Arrange in order, starting with the smallest amount.

23. 62.1 mL   962 mL   0.9 L   0.075 L

24. 12,500 mL   21 L   861 mL   10.3 L

451

Application Bank

Application A11
(page 106)

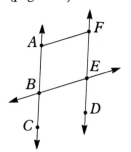

Use the diagram to name the following.

1. two parallel lines

2. two rays intersecting at a point

3. two intersecting lines

4. corners of the polygon

5. the point where $\overrightarrow{EF}$ and $\overrightarrow{EB}$ intersect each other

6. six line segments

Line: $\overleftrightarrow{AC}$

Line segment: $\overline{BE}$

Ray: $\overrightarrow{ED}$

Polygon: $ABEF$

Draw each of the following.

7. $\overleftrightarrow{RS}$

8. $\overleftrightarrow{XY}$

9. two parallel lines

10. a 3-sided polygon

11. $\overline{PQ}$

12. three lines intersecting at a point

Application A12
(page 107)

Right angle

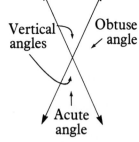
Vertical angles
Obtuse angle
Acute angle

Is the angle right, acute, or obtuse?

13.

14.

15.

16.

Draw each of the following.

17. obtuse angle

18. acute angle

19. right angle

20. vertical angles

452

Application A13
(pages 116–117)

In triangle *ABC*, what is the measure of ∠*C*?

1. ∠*A* = 50°
∠*B* = 30°
∠*C* = __?__

2. ∠*A* = 90°
∠*B* = 35°
∠*C* = __?__

3. ∠*A* = 130°
∠*B* = 25°
∠*C* = __?__

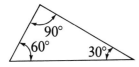

The sum of the
measures of the
angles of any
triangle is 180°.

Is the triangle equilateral, isosceles, or right?

4.

5.

6.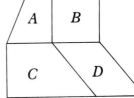

Answer the question.

7. Can an isosceles triangle be a right triangle?

8. Can a right triangle be equilateral?

9. In an equilateral triangle, what is the measure of each angle?

Application A14
(pages 124–125)

Types of quadrilaterals

Trapezoid   Rectangle

Parallelogram   Square

Circle

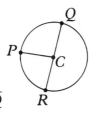

Center: *C*
Radius: *CP*
Diameter: *RQ*

Name each type of quadrilateral in the diagram.

10. *A*

11. *B*

12. *C*

13. *D*

14. *E*

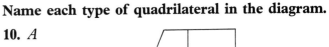

Answer the question.

15. If the radius of a circle is 4.2 cm long, how long is the diameter?

16. If the diameter of a circle is 18 cm long, how long is the radius?

Application A15
(page 138)

$$9^2 = 9 \times 9 = 81$$
$$14^2 = 14 \times 14 = 196$$
$$(5.6)^2 = 5.6 \times 5.6 = 31.36$$

$$\sqrt{81} = 9 \text{ because } 81 = 9^2$$
$$\sqrt{400} = 20 \text{ because } 400 = 20^2$$

Complete the statement.

1. $7^2 = $ __?__

2. $13^2 = $ __?__

3. $25^2 = $ __?__

4. $(1.9)^2 = $ __?__

5. $(0.04)^2 = $ __?__

6. $(3.2)^2 = $ __?__

7. $50^2 = $ __?__

8. $81^2 = $ __?__

9. $100^2 = $ __?__

10. $\sqrt{9} = $ __?__

11. $\sqrt{1} = $ __?__

12. $\sqrt{25} = $ __?__

13. $\sqrt{169} = $ __?__

14. $\sqrt{289} = $ __?__

15. $\sqrt{625} = $ __?__

16. $\sqrt{361} = $ __?__

17. $\sqrt{441} = $ __?__

18. $\sqrt{324} = $ __?__

Application A16
(page 139)

Pythagorean Theorem

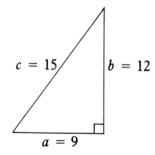

$$a = 9$$
$$a^2 + b^2 = c^2$$
$$9^2 + 12^2 = 15^2$$
$$81 + 144 = 225$$

The lengths of the sides of a triangle are given. Is the triangle a right triangle?

19. 4 cm, 9 cm, 12 cm

20. 5 yd, 12 yd, 13 yd

21. 7 in., 24 in., 25 in.

22. 5 cm, 8 cm, 10 cm

23. 8 m, 10 m, 13 m

24. 14 in., 48 in., 50 in.

25. 18 ft, 24 ft, 28 ft

26. 10 cm, 8 cm, 6 cm

27. 14 in., 12 in., 9 in.

28. 25 ft, 20 ft, 15 ft

Application A17
(pages 148–149)

Find the perimeter.

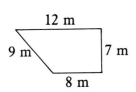

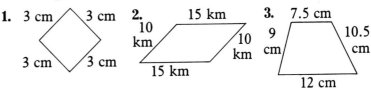

1. 3 cm / 3 cm / 3 cm / 3 cm

2. 15 km, 10 km, 10 km, 15 km

3. 7.5 cm, 9 cm, 10.5 cm, 12 cm

Perimeter:
12 + 7 + 8 + 9 = 36

Perimeter is 36 m.

4. 9 in., 7.2 in., 9 in.

5. 12 m, 5 m

6. 6 ft, 6 ft, 6 ft, 4.5 ft, 4.5 ft

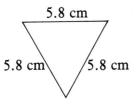

5.8 cm / 5.8 cm / 5.8 cm

Perimeter:
3 × 5.8 = 17.4

Perimeter is 17.4 cm.

Complete the table.

	7.	**8.**	**9.**	**10.**	**11.**
Number of equal sides	4	10	6	12	9
Length of one side	12 cm	7.5 cm	4.8 ft	9 in.	14.6 m
Perimeter	?	?	?	?	?

Application A18
(pages 156–157)

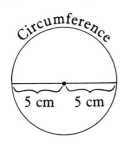

Circumference

5 cm   5 cm

Diameter = 2 × radius

Circumference = π × diameter
C = 3.14 × 10 = 31.4

Circumference is 31.4 cm.

Complete the table.

	12.	**13.**	**14.**	**15.**
Radius	6 cm	?	12 m	?
Diameter	12 cm	20 ft	?	8.4 km
Circumference	?	62.8 ft	?	?

16. What is the circumference of a circular swimming pool with a diameter of 14 m?

17. What is the circumference of a wheel with a radius of 36 cm?

18. What is the circumference of a circular flower garden with a radius of 4 m?

Application A19
(page 170)

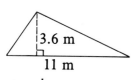

4.2 km

8 km

Area: $8 \times 4.2 = 33.6$

Area is 33.6 km^2.

Complete the table.

	1.	2.	3.	4.	5.
Length	15 cm	28 ft	18.4 m	9.3 km	$10\frac{1}{2}$ in.
Width	12 cm	14 ft	13.6 m	6 km	$7\frac{1}{4}$ in.
Area	?	?	?	?	?

6. What is the area of a garden which is 18 m long and 15 m wide?

7. What is the area of a carpet 9 ft wide and 12 ft long?

Application A20
(page 171)

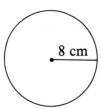

3.6 m

11 m

$A = \frac{1}{2} \times b \times h$

$A = 19.8$

Area is 19.8 m^2.

Complete the table.

	8.	9.	10.	11.	12.
Base	20 cm	4.8 m	8 in.	5.2 ft	25 cm
Height	12 cm	2.4 m	$9\frac{1}{2}$ in.	10 ft	8 cm
Area	?	?	?	?	?

13. What is the area in square feet of a triangle with base 18 ft and height 8 ft?

14. Find the area of a triangle with base 4 m and height 7 m.

Application A21
(page 180)

8 cm

$A = \pi \times r^2$

$A = 200.96$

Area is 200.96 cm^2.

Complete the table. Round to the nearest tenth.

	15.	16.	17.	18.	19.
Radius	9 cm	14 in.	3.5 m	10 ft	15 m
Area	?	?	?	?	?

20. Which area is larger? Square cake with side 8 in. or round cake with 6 in. radius?

21. Which area is larger? Circle with diameter 1 meter or square with side 1 meter.

22. What is the area of a circular table top with a radius of 100 cm?

Application A22
(page 181)
semicircle

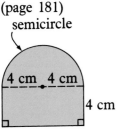

Area of semicircle
$A = 25.12$ cm
Area of rectangle
$A = 32$ cm^2
Total Area $= 57.12$ cm^2

Find the area of the figure.

1.

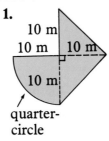

quarter-circle

2.

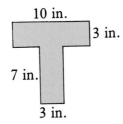

3.

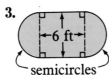

semicircles

4.

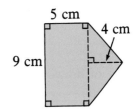

Application A23
(page 188)

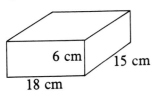

Volume of
rectangular prism

$V = l \times w \times h$

$V = 1620$ cm^3

Complete the table.

	5.	**6.**	**7.**	**8.**	**9.**
Length	10 cm	2 m	10 ft	3 cm	20 in.
Width	8 cm	1.8 m	$8\frac{1}{2}$ ft	1.5 cm	8 in.
Height	7 cm	1.2 m	6 ft	10 cm	4.5 in.
Volume	?	?	?	?	?

10. Which has the greater volume? A carton that is 4 ft by 6 ft by 3 ft or a carton that has length, width, and height each 4.2 ft.

Application A24
(page 189)

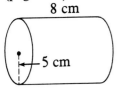

Volume of cylinder

$V = \pi \times r^2 \times h$

$V = 628$ cm^3

Complete the table.

	11.	**12.**	**13.**	**14.**	**15.**
Radius of base	4 in.	10 m	6 cm	3 in.	5 cm
Height	8 in.	10 m	9.5 cm	1 in.	7 cm
Volume	?	?	?	?	?

16. A pipe is 5 m long. The radius is 1 m. Find the volume.

17. A soup can is 10 cm high and has a base with radius 4 cm. What is the volume?

457

Application A25
(pages 206–207)

There are 50 tickets in a box: 30 game tickets and 20 movie tickets.

Probability of drawing a game ticket:

$$\frac{30}{50} = \frac{3}{5}$$

Probability of drawing a movie ticket:

$$\frac{20}{50} = \frac{2}{5}$$

1. You roll a 12-sided die that has the numbers 1 through 12 on its faces.

   a. What is the probability of getting an odd number?

   b. What is the probability of getting a number greater than 3?

   c. What is the probability of getting a 12?

   d. What is the probability of getting a number less than 6?

2. You are at a party with 24 other people. One door prize will be given.

   a. What is the probability that you will win?

   b. What is the probability that you will not win?

Application A26
(pages 216–217)

The school cafeteria chooses a sample of 100 students from the total student enrollment of 1200 to find out what lunch is the most popular.

Lunch	Favorable Votes
Soup	21
Salad	24
Sandwich	55

To predict how many students would prefer a sandwich, solve the proportion $\frac{55}{100} = \frac{p}{1200}$.

The prediction is 660 students.

3. A record company wants to find out what types of music people enjoy. The company chooses a sample of 500 people from a total market size of 800,000 people.

Music Choice	Favorable Votes
Classical	92
Jazz	125
Rock	283

How many people do you predict would prefer jazz music?

4. A company is developing a new magazine for high school students and wants to find out what student preferences are. The sample is 350 students from a market size of 700,000 students.
The number of favorable votes for sports is 125.

How many students do you predict want to read about sports?

Application A27
(page 224)

Find the mean. Round to the nearest tenth.

Find the mean:
21, 25, 30, 30, 26

Add: 21
      25
      30
      30
  + 26
   132

Divide by 5:

$$\frac{26.4}{5)132.0}$$

1. 47, 104, 60, 47, 70, 35

2. 8, 6, 3, 8, 15, 10

3. 1.2, 7.6, 12.6, 8.5, 7.6

4. 14.021, 17.3, 6.28, 11.005

5. $6,000, $15,000, $18,500, $21,000

6. On a four-day trip, Kevin drove 258 kilometers, 320 kilometers, 270 kilometers, and 384 kilometers. What was his mean daily distance?

7. A group of flashlight batteries were tested with a voltmeter. The following readings were found: 1.35, 1.55, 1.45, 1.60, 1.55. What is the mean?

Application A28
(page 225)

Find the median, mode, and range.

Find the median,
mode, and range:
7, 3, 20, 17, 7, 12

Arrange numbers in
order:
3, 7, 7, 12, 17, 20

To find the median,
average middle two
numbers:
$\frac{7 + 12}{2} = 9.5$

Mode: 7

Range: $20 - 3 = 17$

8. 11, 21, 18, 16, 18, 6

9. 150, 172, 112, 186, 98, 143, 215

10. 21, 25, 30, 30, 24, 22

11. 3.6, 4.2, 3.6, 2.7, 2.4

12. $25,000, $36,000, $25,000, $19,000

Answer each question.

13. Nine buildings have the following heights: 16 m, 18 m, 12 m, 25 m, 8 m, 14 m, 22 m, 11 m, 18 m. What is the median height?

14. Your grades on six tests are 84, 95, 62, 80, 75, 80. What is the mode? the range?

Application A29
(pages 238–239)

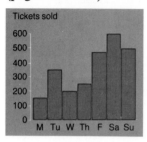

Saturday had the most sales.

The bar graph shows the estimated travel cost for a family of four traveling between two cities.

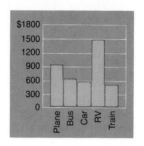

1. What is the least expensive method of travel?

2. Which methods of travel are close in price?

Application A30
(pages 248–249)

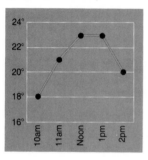

The lowest temperature was recorded at 10:00.

The line graph shows changes in population of two cities.

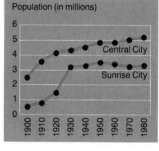

3. Which decade showed the greatest growth for Sunrise City?

4. Did Central City ever have a decrease in population?

5. What was the population of Central City in 1900?

Application A31
(pages 256–257)

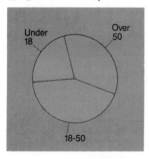

35% of the people were over 50 years old.

The circle graph shows workshop choices of 800 high school students.

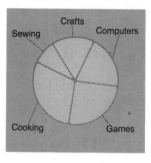

6. Which workshop had the lowest enrollment?

7. Which workshop had the highest enrollment?

8. Which workshop was more popular, computers or crafts?

9. Which two workshops together had 25% of the total enrollment?

Application A32
(pages 270–271)

Find the interest earned on $2000 at 5.5%.

$I = \$2000 \times 0.055$
$I = \$110$

Complete the table.

	1.	2.	3.	4.	5.
Interest Rate	8%	5.5%	4%	12%	10.25%
Principal	$900	$6750	$750	$4000	$500
Interest	?	?	?	?	?

6. Sally Roth has a credit card balance of $900. The monthly interest charge is 1.5%. How much interest does she pay?

Application A33
(pages 280–281)

Percent change $=\dfrac{\text{Amount of change}}{\text{Original amount}}$

Last year net sales were $30,000. This year net sales are $42,000.

Percent change $=\dfrac{\$12,000}{\$30,000}=40\%$

7. Sue Michaels bought an antique rocking chair for $120 and resold it ten years later for $180. What is the percent increase in the value of the chair?

8. Four years ago a calculator cost $80. Now it costs $36. What is the percent decrease in price?

9. A cab company reported that 640 more people used their cabs this year compared to last year. If 3200 people used their cabs last year, what is the percent increase?

Application A34
(pages 288–289)

Oak table regularly sells for $249.95. Sale price is $199.95.

D = R. P. − S. P.

D = $249.95 − $199.95
  = $50.00

D. Rate $=\dfrac{\text{D}}{\text{R. P.}}$

D. Rate $=\dfrac{\$50.00}{\$249.95}$

  = 20%

Complete the table.

	10.	11.	12.	13.
Regular Price	$23	$136	$27.88	$550
Sale Price	$21.85	$115.60	$20.91	?
Discount	$1.15	?	?	$165
Discount rate	?	?	?	30%

14. At a jewelry clearance sale, all items are marked 30% to 50% off. Jim bought a watch with a regular price of $175.

   a. What is the lowest sale price possible?

   b. What is the highest sale price possible?

Application Bank

Application A35
(pages 306–307)

In one day the temperature was ⁻6°C, rose 15°, and then dropped 7°.

What was the final temperature?

⁻6 + 15 + ⁻7 = 2

The temperature was 2°C.

Write a sum of positive or negative numbers to represent each problem. Then add to solve the problem.

1. The morning temperature was ⁻4°C. By afternoon the temperature had risen 6°, and then dropped 3°. What was the final temperature?

2. The afternoon temperature was 8°C. By evening the temperature had fallen 13°. What was the evening temperature?

3. A submarine which was 145 m below sea level rises 50 m. What is the depth of the submarine?

Application A36
(pages 316–317)

If the highest recorded temperature in Pennsylvania was 44°C and the lowest recorded temperature was ⁻41°C, find the difference.

44 − ⁻41 = 85

The difference is 85°.

Complete the table.

	4.	5.	6.	7.	8.
Place	Idaho	Maine	Ohio	Texas	Utah
Highest Temperature	48°C	41°C	45°C	49°C	47°C
Lowest Temperature	⁻51°C	⁻44°C	⁻39°C	⁻31°C	⁻46°C
Difference	?	?	?	?	?

Application A37 (pages 324–325)

Wind Speed (mph)	Thermometer Reading in Fahrenheit Degrees				
	30	20	10	0	⁻10
	Equivalent Calm-air Temperature				
10	16	4	⁻9	⁻24	⁻33
15	9	⁻5	⁻18	⁻32	⁻45
20	4	⁻10	⁻25	⁻39	⁻53
25	0	⁻15	⁻29	⁻44	⁻59

The temperature is 10°F. There is a wind of 15 miles per hour. The wind chill factor is ⁻18°F.

9. On a day when the wind is 25 miles per hour, the temperature is 20°F. What is the wind chill factor?

10. On Tuesday it was 10°F, with a 10-mile-per-hour wind. On Wednesday it was 20°F, with a 25-mile-per-hour wind.

   a. Which day felt colder?

   b. How much lower was the wind chill factor on the day that felt colder?

11. On a day when the temperature is 20°, the wind changes from 10 miles per hour to 20 miles per hour. How much colder does it seem?

Application A38
(pages 338–339)

Double v $2v$

x more than 5 $5 + x$

product of 7 and y $7y$

z minus 15 $z - 15$

Write as a variable expression.

1. 10 added to n

2. y decreased by z

3. the quotient of p divided by q

4. the product of c and d

5. x decreased by 19

6. the ratio of t to 12

Application A39
(pages 348–349)

Perimeter

$P = a + b + c + d + e$

Volume

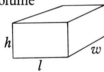

$V = lwh$

Write the formula for the perimeter.

7.

8.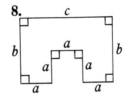

9.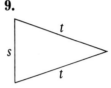

Write the formula for the volume.

10.

11.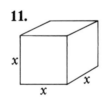

12.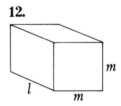

Application A40
(pages 356–357)

A number multiplied by eight is less than twenty-five.

$8n < 25$

Select the best answer.

13. The quotient of d divided by nine is greater than thirty.

   a. $d - 9 > 30$ **b.** $\frac{d}{9} > 30$ **c.** $\frac{9}{d} > 30$

14. If Joan were four years older, she would be twenty-eight years old.

   a. $28 - j = 24$ **b.** $j - 4 = 28$ **c.** $j + 4 = 28$

15. The price of an item decreased by six dollars is less than fifteen dollars.

   a. $15 < p - 6$ **b.** $6 + p < 15$ **c.** $p - 6 < 15$

Application Bank

Application A41
(page 370)

Graph the solution
of $x - 2 = {}^-1$

Solve.
$$x - 2 = {}^-1$$
$$x - 2 + 2 = {}^-1 + 2$$
$$x = 1$$

**Graph the solution of the given equation.
Use a number line.**

1. $y - 2 = 0$ 2. $n + 4 = {}^-8$

3. $x - 0 = 5$ 4. $p + 8 = 8$

5. $11 + t = 1$ 6. $x + 15 = 20$

7. ${}^-4 = 2 + n$ 8. $7 = x - 2$

9. $8 = t - 3$ 10. $x + 6 = {}^-2$

11. $4 = p + 9$ 12. $y - 4 = {}^-3$

13. ${}^-10 = {}^-30 + q$ 14. $x - 3 = 14$ 15. $b - 10 = {}^-25$

16. $m + {}^-14 = 2$ 17. $3 = {}^-9 + f$ 18. ${}^-16 = c - {}^-3$

19. ${}^-100 = {}^-98 + v$ 20. $45 + z = 26$ 21. ${}^-31 = x + {}^-28$

Application A42
(page 371)

Graph the solution
of $n \leq {}^-2$.

Solve.

${}^-2$ is a solution.
All numbers less
than ${}^-2$ are
solutions.

**Graph the solution of the given inequality.
Use a number line.**

22. $x < 7$ 23. $y \geq {}^-10$

24. $x \leq {}^-6$ 25. $t \geq 4$

26. $n < 0$ 27. $x \leq {}^-4$

28. $t \geq {}^-4$ 29. $q < 8$

30. $p > {}^-14$ 31. $y \geq 12$

32. $t < 20$ 33. $n > 25$

34. $p < {}^-9$ 35. $x \geq 1$

36. $m \geq {}^-11$ 37. $k \leq {}^-5$ 38. $b \geq {}^-1$

39. $t < {}^-9$ 40. $x > 7$ 41. $n \leq 0$

Application A43
(pages 380–381)

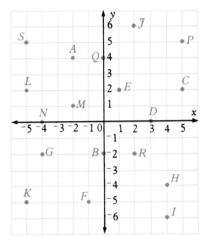

Point	A	B	C
Coordinates	$(^-2, 4)$	$(0, ^-2)$	$(5, 2)$

Give the coordinates of the point.

1. D
2. H
3. L
4. Q
5. C
6. F
7. R
8. S
9. J

Give the letter of the point described.

10. $(^-2, 4)$
11. $(4, ^-6)$
12. $(^-4, 0)$
13. $(1, 2)$
14. $(^-4, ^-2)$
15. $(0, ^-2)$
16. $(^-2, 1)$
17. $(5, 5)$
18. $(^-5, ^-5)$
19. $(5, 2)$
20. $(2, 6)$
21. $(0, 4)$
22. $(^-5, 2)$
23. $(3, 0)$
24. $(4, ^-4)$
25. $(^-1, ^-5)$
26. $(2, ^-2)$
27. $(^-5, 5)$

Application A44
(pages 388–389)

Complete the table. Then draw the graph using the ordered pairs.

$x = y + 2$

x	y	(x, y)
$^-2$	$^-4$	$(^-2, ^-4)$
0	$^-2$	$(0, ^-2)$
2	0	$(2, 0)$

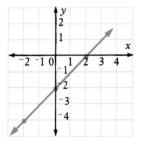

28. $x = y + 5$

x	y	(x, y)
$^-5$	$^-10$	?
0	$^-5$	?
5	0	?

29. $x - y = 3$

x	y	(x, y)
$^-2$	$^-5$	?
0	?	?
4	?	?

30. $x + 2y = 10$

x	y	(x, y)
$^-2$	?	?
0	?	?
2	?	?

31. $y = x - 4$

x	y	(x, y)
$^-4$	?	?
$^-2$	?	?
5	?	?

Table of Measures

METRIC

Length

10 millimeters (mm) = 1 centimeter (cm)

100 centimeters = 1 meter (m)

1000 meters = 1 kilometer (km)

Mass

1000 milligrams (mg) = 1 gram (g)

1000 grams = 1 kilogram (kg)

1000 kilograms = 1 tonne (t)

Capacity

1000 milliliters (mL) = 1 liter (L)

1000 cubic centimeters (cm³) = 1 liter

10 milliliters = 1 centiliter (cL)

10 deciliters (dL) = 1 liter

Temperature

0°C = freezing point of water

37°C = normal body temperature

100°C = boiling point of water

Area

100 square millimeters (mm²) = 1 square centimeter (cm²)

10,000 square centimeters = 1 square meter (m²)

10,000 square meters = 1 hectare (ha)

UNITED STATES CUSTOMARY

Length

12 inches (in.) = 1 foot (ft)

3 feet = 1 yard (yd)

5280 feet = 1 mile (mi)

Weight

16 ounces (oz) = 1 pound (lb)

2000 pounds = 1 ton (t)

Capacity

8 fluid ounces (fl oz) = 1 cup (c)

2 cups = 1 pint (pt)

2 pints = 1 quart (qt)

4 quarts = 1 gallon (gal)

Temperature

32°F = freezing point of water

98.6°F = normal body temperature

212°F = boiling point of water

Area

144 square inches (in.²) = 1 square foot (ft²)

9 square feet = 1 square yard (yd²)

43,560 square feet = 1 acre (A)

466

Key Ideas

acute angle (page 107)   An angle whose measure is less than 90°.

angle (page 107)   The figure formed by two rays that have the same endpoint.

annual percentage rate (APR) (page 285)   An index showing the relative cost of borrowing money.

area (page 170)   The measure of the space enclosed by a plane figure.

average (page 224)   The sum of the numbers in a set divided by the number of items in the set.

axes (page 380)   The two number lines that cross at a point labeled zero. The x–axis is horizontal, the y–axis vertical.

bar graph (page 238)   A presentation of data in which each quantity to be displayed is represented by the length of a bar.

base (page 171)   Any side of a triangle may be referred to as a base of the triangle. Also, the repeated factor in an exponential expression.

Celsius temperature (page 75)   A temperature scale based on the temperature range from the freezing point of water (0°C) to the boiling point of water (100°C).

center (page 125)   The point inside a circle which is the same distance from each point on the circle.

certain (page 206)   An event that is sure to occur.

chronological (page 84)   Referring to occurrence in time from the earliest to the most recent.

circle (page 125)   A closed curve in a plane whose points are all at equal distances from the center.

circle graph (page 256)   A graph which illustrates how a whole is divided. A circle, which represents the whole, is divided to show the parts of the whole.

circumference (page 156)   The distance around a circle.

common factor (page 103)   A number which is a factor of two or more numbers.

compound interest (page 275)   Interest earned not only on the original principal but also on the interest earned during previous interest periods.

Consumer Price Index (CPI) (page 220)   The current cost of goods and services that would have cost $100 in 1967. The CPI is an indicator of inflation and changes in the "cost of living."

coordinates (page 381)   The numbers in an ordered pair that locate a point.

customary units (page 38)   The traditional or common units of measurement, such as pound, foot, quart.

cylinder (page 189)   A space figure with a circular base. A cylinder is shaped like a can.

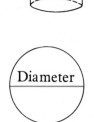

data processing (page 192)   Changing raw data into a more usable form.

decimal (page 3)   A numeral that has place value and a decimal point. For example: 35.2.

denominator (page 134)   In the fraction $\frac{3}{5}$, 5 is the denominator.

diameter (page 125)   A line segment connecting two points on a circle, through the center.

difference (page 16)   The result of subtracting one number from another. In the subtraction $37 - 12$, the difference is 25.

discount (page 288)   The difference between the regular selling price of an item and its sale price.

equation (page 354)   A mathematical sentence stating that two expressions are equal.

equilateral triangle (page 117)   A triangle in which all sides are equal.

equivalent measure (page 38)   The number of smaller units of measure whose sum equals a larger unit. For example, 12 inches = 1 foot.

estimate (page 16)   A reasonable guess.

exponent (page 346)   A number which shows how many times a base is to be used as a factor. In 2^5, 5 is the exponent and 2 is the base. $2^5 = 2 \cdot 2 \cdot 2 \cdot 2 \cdot 2$.

factors (page 102)   Two numbers that are multiplied to yield a product. In $4 \times 2 = 8$, 4 and 2 are the factors.

Fahrenheit temperature (page 75)   A temperature scale in which $32°F$ is the freezing point of water and $212°F$ is the boiling point of water.

finance charge (page 285)   The interest charge on money loaned, credit card accounts, and so on.

flow chart (page 328)   A diagram showing the order in which operations and decisions are to be carried out.

formula (page 348)   An equation that states a rule or relationship.

fraction (page 104)   A number written in the form $\frac{a}{b}$.

greatest common factor (GCF) (page 103)   The greatest whole number which is a factor of two or more given whole numbers.

higher terms (page 105)   To write a fraction in higher terms, multiply the numerator and denominator by the same number. In higher terms, $\frac{3}{4}$ is the same as $\frac{6}{8}$, $\frac{9}{12}$, $\frac{12}{16}$, and so on.

468

impossible (page 206)   An event that cannot occur at all.

inequalities (page 355)   Number statements that contain the symbols $>$, $<$, $\geq$, or $\leq$.

interest (page 270)   An amount earned on money in a savings account or a charge paid by a borrower for the use of money.

interest rate (page 270)   Interest expressed as a percent of the principal, based on one year.

intersecting lines (page 106)   Lines that meet at a common point.

inverse operation (page 378)   An arithmetic operation that "undoes" another operation. For example, subtraction is the inverse of addition; division is the inverse of multiplication.

isosceles triangle (page 117)   A triangle that has two equal sides and two equal angles.

kilowatt–hour (page 42)   The basic unit of measure of electrical power consumption.

least common denominator (LCD) (page 114)   The least common multiple of the denominators of two or more fractions. The LCD of $\frac{3}{8}$ and $\frac{5}{6}$ is 24.

least common multiple (LCM) (page 113)   The smallest positive number which is a multiple of two or more numbers.

line (page 106)   A straight mark that extends endlessly in opposite directions.

line graph (page 248)   A graph using lines to represent information.

lowest terms (page 104)   A fraction is in lowest terms if the greatest common factor of the numerator and the denominator is 1.

mean (page 224)   The average.

median (page 225)   When a set of numbers is arranged in order, the median is the middle number.

metric system (page 70)   A decimal system of measures based on the meter as the unit of length and the kilogram as the unit of mass.

mixed number (page 122)   A whole number plus a fraction. $1\frac{2}{3}$ is a mixed number.

mode (page 225)   The most frequently occurring number in a set of numbers.

multiple (page 36)   The product of a given number and a whole number. 6 is a multiple of 3.

negative number (page 302)   A number less than zero.

numerator (page 115)   In the fraction $\frac{2}{3}$, 2 is the numerator.

obtuse angle (page 107)   An angle whose measure is greater than $90°$ but less than $180°$.

parallel lines (page 106)   Lines in the same plane that do not intersect no matter how far they are extended.

parallelogram (page 124)   A quadrilateral that has two pairs of parallel sides.

percent (page 245)   Per hundred. $35\% = \frac{35}{100} = 0.35$.

percent change (page 280)   The change in a quantity expressed as a percent of the original amount.

perimeter (page 148)   The distance around a figure.

pi (π) (page 156)   The ratio of the circumference of a circle to the length of the diameter. π is approximately 3.14.

pictograph (page 252)   A graph using picture symbols to show amounts for comparison.

place value (page 2)   The value given to the place in which a digit appears. In 324, 3 is in the hundreds' place, 2 is the tens' place, and 4 is in the ones' place.

plane (page 106)   A flat surface extending endlessly.

point (page 106)   A position in space.

polygon (page 106)   A plane figure formed by line segments intersecting at their endpoints.

population (page 216)   An entire group of individuals, items, or scores from which a sample is taken.

positive number (page 302)   A number greater than zero.

principal (page 270)   The amount of money deposited or loaned.

probability (page 206)   A measure of chance.

product (page 60)   The result of a multiplication. In the multiplication 6×3, the product is 18.

program (page 192)   A set of instructions that a computer can recognize and follow. Programs are usually written in a computer language such as BASIC, FORTRAN, or COBOL.

proportion (page 204)   A statement that two ratios are equal.

Pythagorean theorem (page 139)   If a, b, and c are the lengths of the sides of a right triangle, with c the longest side, then
$a^2 + b^2 = c^2$.

quadrilateral (page 124)   A polygon that has four sides.

radius (page 125)   A line segment from any point on a circle to its center.

range (page 225)   The difference between the largest and smallest numbers in a group of numbers.

ratio (page 202)   Comparison of two numbers by division. The ratio $\frac{4}{5}$ can also be written 4 to 5 or 4:5.

ray (page 106)   A part of a line having one endpoint and extending endlessly in one direction.

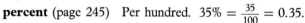

reciprocal (page 176) Two numbers are reciprocals if their product is 1. The reciprocal of $\frac{2}{5}$ is $\frac{5}{2}$.

rectangle (page 124) A quadrilateral that has four right angles.

rectangular prism (page 188) A box-shaped figure of which each face is a rectangle.

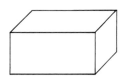

right angle (page 107) An angle whose measure is 90°.

right triangle (page 117) A triangle that has a right angle.

rounding (page 4) Rounding to the nearest 10 means approximating to the nearest multiple of 10. Rounding to the nearest 100 means approximating to the nearest multiple of 100, and so on.

sample (page 216) A selected group of individuals representative of a larger population.

scale drawing (page 211) A drawing that shows the correct shape of an object but differs in size.

semicircle (page 157) A half-circle as divided by a diameter.

square (page 124) A rectangle with 4 sides the same length.

square of a number (page 138) The product of a number multiplied by itself. The square of 5 is 25. This is written $5^2 = 25$.

square root (page 138) The factor of a number that when multiplied by itself gives the number. Because $3 \times 3 = 9$, the square root of 9 is 3.

sum (page 16) The result of addition. In the addition $12 + 29$, the sum is 41.

trapezoid (page 124) A quadrilateral that has one pair of parallel sides.

triangle (page 116) A polygon that has three sides.

unit price (page 228) The cost of an item expressed per unit of measure or count.

unit rate (page 222) Amount per unit.

Universal Product Code (UPC) (page 352) Code numbers, printed on retail merchandise, that are read by computerized check-out systems.

variable (page 334) A letter or symbol used to represent a number.

vertex (page 107) The common endpoint of the rays that form an angle.

vertical angles (page 107) Opposite angles, with equal measures, formed by intersecting lines.

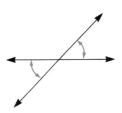

volume (page 188) The measure of the space a 3-dimensional figure contains.

whole number (page 2) The numbers in the set $\{0,1,2,3,\ldots\}$.

Index

PICTURE CREDITS

Book design: Don Leeds/Publishers' Graphics Inc.

Cover and title page photograph: © Robert J. Witkowski/The Image Bank

Technical Art: Feature page titles by Mary Jane Medved; Omnigraphics pp. 40, 42, 75, 81, 82, 108, 110, 118, 120, 126, 140, 149, 150, 159, 160, 172, 182, 185, 206, 207, 211, 228, 230, 233, 238, 239, 242, 248, 249, 252, 253, 256, 257, 260, 262, 263, 281, 298, 299, 302, 310, 328, 342, 460; all remaining technical art by Phil Carver & Friends.

Unit 1: pp. xii-1 © Lou Jones 1981; p. 8 © Alan Oransky 1981; p. 17 Peter Simon/Stock, Boston; p. 18 Mike Malyszko; p. 27 Owen Franken/Stock, Boston; p. 28 © Cradoc Bradshaw/Stock, Boston

Unit 2: pp. 32-33 © Gabe Palmer 1979/The Image Bank; p. 41 Owen Franken/Stock, Boston; p. 43 Frank Wing/Stock, Boston; p. 49 Mike Malyszko; p. 50 Koppers Co., Inc.; p. 51 Grant Heilman; pp. 52, 57 Mike Malyszko; p. 58 © Lou Jones 1980; pp. 59, 60 Mike Malyszko

Unit 3: pp. 64-65 © Guy Kern 1981; pp. 72, 73 © Alan Oransky 1981; p. 75 Grant Heilman; p. 83 Mike Malyszko; p. 85 James L. Ballard; p. 90 Mike Malyszko; p. 92 Dennis Brody/Stock, Boston

Unit 4: pp. 100-101 © Christopher Cunningham 1980; p. 109 Ellis Herwig/Stock, Boston; p. 111 NOAA map, NASA photo; p. 119 Owen Franken/Stock, Boston; p. 121 Edith G. Haun/Stock, Boston; p. 126 © John Lei/Stock, Boston

Unit 5: pp. 132-133 © David F. Hughes 1981/The Picture Cube; p. 141 Cary Wolinsky/Stock, Boston; p. 151 Mike Malyszko; p. 153 Edith G. Haun/Stock, Boston; p. 158 John Colwell/Grant Heilman; p. 159, 160 Owen Franken/Stock, Boston

Unit 6: pp. 164-165 Ira Kirschenbaum/Stock, Boston; p. 172 Mike Malyszko; p. 173 Marsha Fall Stuart/Global Focus; p. 182 Isaac Geib/Grant Heilman; p. 183 © 1979 Webbphotos; p. 190 © Bill Gillette 1979/Stock, Boston; p. 191 Burk Uzzle/Magnum. Courtesy Northwest Energy Co.; p. 192 Edith G. Haun/Stock, Boston

Unit 7: pp. 200-201 © Michael Melford 1980/The Image Bank; p. 208 Owen Franken/Stock, Boston; p. 209 John Coletti; p. 210 Donald Dietz/Stock, Boston; p. 211 Frank Siteman/Stock, Boston; pp. 218-219 John Coletti. Courtesy of Norma's Too, Boston, MA.; p. 221 Ellis Herwig/Stock, Boston; p. 226 James L. Ballard; p. 227 Joel Gordon(top), Elizabeth Crews/Stock, Boston(bottom)

Unit 8: pp. 232-233 Alan Oransky; p. 240 Mike Malyszko; p. 241 © Paul Johnson; p. 242 Jon L. Barkan/The Picture Cube; p. 250 © Frank Siteman 1979/The Picture Cube; p. 252 Owen Franken/Stock, Boston; p. 253 © Richard Wood 1980/The Picture Cube; p. 257 Donald Dietz/Stock, Boston; p. 258 © B. I. Ullman/Taurus(top), Marian Bernstein(bottom); p. 259 © Frank Siteman 1979/The Picture Cube

Unit 9: pp. 264-265 Alan Oransky. Courtesy Commonwealth Trading, Inc.; p. 270 © Frank Siteman 1979/Taurus; p. 272 Elizabeth Crews/Stock, Boston; p. 273 John Coletti; p. 282 © Craig Aurness/West Light; p. 283 © John Coletti; p. 284 Mike Malyszko; p. 290, 291 © John Coletti

Unit 10: pp. 300-301 R. Robert Abrams/Bruce Coleman Inc.; p. 307 © Paul Johnson; p. 308 © Paul Johnson(top), © E. Williamson/The Picture Cube(bottom); p. 317 Barry L. Runk/Grant Heilman; p. 318 © Joel Gordon 1978; p. 320 Fred Bavendam; p. 321 John Coletti. Courtesy of the Century House, Danvers, MA.; pp. 326, 327 John Coletti

Unit 11: pp. 332-333 Chuck O'Rear/West Light; p. 340 Grant Heilman; p 341 Ira Kirschenbaum/Stock, Boston; p. 350 © John Coletti; p. 351 © Lou Jones 1981(top), © B. I. Ullman/Taurus(bottom); p. 358 Mike Mazzaschi/Stock, Boston; p. 359 Eric A. Roth/The Picture Cube(top), James L. Ballard(bottom)

Unit 12: pp. 364-365 An Exxon photo; p. 372 J. L. Barkan/The Picture Cube; p. 373 Frank J. Staub/The Picture Cube; p. 375 © Paul Johnson; p. 382 P. Ellin/The Picture Cube; p. 383 © R. Terry Walker/The Picture Cube; p. 385 © John Coletti; p. 390 L. L. T. Rhodes/Taurus; p. 391 © Frank Siteman 1979/The Picture Cube(top left), © Mark Mittelman 1979/Taurus(top right), © Jean Wentworth 1980/The Picture Cube(bottom)

Many schools require students to pass a basic competency test in order to graduate from high school. The test below contains items similar to those on actual competency tests. Choose the correct answers.

1. $264 + 19 =$ __?__

   A. 283   **B.** 245

   C. 273   **D.** 383

2. $407 - 98 =$ __?__

   A. 409   **B.** 399

   C. 491   **D.** 309

3. $347 \times 28 =$ __?__

   A. 9716   **B.** 736

   C. 9266   **D.** 16,066

4. $26\overline{)3484}$

   A. 95 R14   **B.** 153

   C. 134   **D.** 109

5. $\frac{12}{54} =$ __?__

   A. $\frac{3}{5}$ **B.** $\frac{2}{9}$

   C. $\frac{1}{2}$ **D.** $\frac{5}{8}$

6. $\frac{3}{4} + \frac{1}{8} =$ __?__

   A. $\frac{4}{12}$ **B.** $\frac{7}{8}$

   C. $\frac{4}{8}$ **D.** $\frac{1}{3}$

7. $1\frac{1}{3} + 3\frac{1}{2} =$ __?__

   A. $4\frac{5}{6}$ **B.** $4\frac{2}{3}$

   C. $\frac{5}{6}$ **D.** $4\frac{1}{3}$

8. $6\frac{1}{8} - 2\frac{2}{8} =$ __?__

   A. $2\frac{7}{8}$ **B.** $4\frac{1}{8}$

   C. $3\frac{7}{8}$ **D.** $\frac{5}{8}$

9. $\frac{5}{8} \times \frac{2}{3} =$ __?__

   A. $\frac{10}{12}$ **B.** $\frac{5}{12}$

   C. $\frac{13}{5}$ **D.** $\frac{15}{16}$

10. $15 \div \frac{3}{5} =$ __?__

   A. $\frac{5}{5}$ **B.** $\frac{1}{9}$

   C. 9   **D.** 25

11. $\frac{2}{5} =$ __?__

   A. 0.25   **B.** 0.2

   C. 0.4   **D.** 0.04

12. $2.04 + 62.9 = \underline{\quad?\quad}$

  A. 64.94   **B.** 83.3

  C. 6.494   **D.** 8.33

---

13. $0.6 \times 4.23 = \underline{\quad?\quad}$

  A. 24.138   **B.** 253.8

  C. 2.538   **D.** 241.38

---

14. $6.8 \times 1{,}000 = \underline{\quad?\quad}$

  A. 6800   **B.** 0.6800

  C. 6.8000   **D.** 680,000

---

15. $4.2\overline{)10.92}$

  A. 0.26   **B.** 26

  C. 2.06   **D.** 2.6

---

16. Which is *least* expensive?

  A. 3 lb for $1.00

  B. 2 lb for 70¢

  C. 4 lb for $1.00

  D. 5 lb for $1.50

---

17. 20% of $84 = \underline{\quad?\quad}$

  A. 16.8   **B.** 1680

  C. 4.2   **D.** 168

---

18. 15 is 60% of what number?

  A. 9   **B.** 25

  C. 4   **D.** 0.25

19. Rob drove 115 mi at 55 miles per hour. About how long did he drive?

  A. 1.5 h   **B.** 2 h

  C. 2.5 h   **D.** 3 h

---

20. Dan bought 1.6 lb of meat at $2.50 per pound. What was the total cost?

  A. $40.00   **B.** $5.00

  C. $3.00   **D.** $4.00

Questions 21–22 refer to the rectangle below.

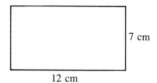

21. What is the area of the rectangle?

  A. 80 cm   **B.** 34 cm

  C. 84 cm   **D.** 84 cm²

22. What is the perimeter of the rectangle?

  A. 38 cm   **B.** 84 cm²

  C. 19 cm   **D.** 34 cm²

---

23. Mimi fell asleep at 1:15 A.M. and woke up at 9:30 A.M. How long did she sleep.?

  A. 8 h 45 min   **B.** 6 h 45 min

  C. 8 h 15 min   **D.** 7 h 30 min

24. Prize money of $3683.15 is to be divided equally among 19 people. Estimate how much each person will get.

A. $100   **B.** $150

C. $200   **D.** $500

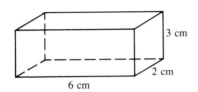

3 cm

2 cm

6 cm

25. What is the volume of the box shown?

A. 11 cm   **B.** 36 cm^2

C. 36 cm^3 **D.** 20 cm^3

26. Jim drove his car 85 mi, 92 mi, 67 mi, and 68 mi over the last four days. What was the average per day?

A. 68 mi   **B.** 78 mi

C. 70 mi   **D.** 90 mi

27. The sales tax in one state is 6%. How much tax will have to be paid on a purchase of $250?

A. $15   **B.** $25

C. $41.67   **D.** $12

28. Fred borrowed $500 at 10% interest per year. How much simple interest will be owed per year?

A. $5   **B.** $25

C. $10   **D.** $50

Questions 29–30 refer to the graph below.

Sources of Income
at Wholesome Farm

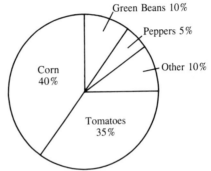

Green Beans 10%
Peppers 5%
Other 10%
Corn 40%
Tomatoes 35%

29. Which crop produces the greatest income?

A. Green beans

B. Corn

C. Tomatoes

D. Peppers

30. If the farm's income was $60,000 last year, how much income came from tomatoes?

A. $21,000   **B.** $24,000

C. $40,000   **D.** $35,000

479

Selected Answers

UNIT 1

Page 1 **1.** units, 8 **2.** hundredths, $\frac{6}{100}$ or 0.06 **3.** 180 **4.** 90.9 **5.** 95 **6.** 395 **7.** 232 **8.** 2419 **9.** 3128
10. 1056 **11.** 70 **12.** 6582 **13.** 24.2 **14.** 101.104 **15.** 64.2 **16.** 65.42 **17.** 110 **18.** 598.01 **19.** 890
20. 230 **21.** 300 **22.** 3200 **23.** 63.9 **24.** 93.6 **25.** 40 **26.** 25

Pages 2, 3 **1.** tens; 90 **3.** 4; 4000 **5.** tens; 70 **7.** hundreds, 600 **9.** units, 3 **11.** tens, 30 **13.** units, 5
15. hundreds, 300 **17.** millions; 3,000,000 **19.** units; 7 **21.** 5; 0.05 or $\frac{5}{100}$ **23.** 9; tens **25.** tenths; 0
27. thousandths; 0.005 or $\frac{5}{1000}$ **29.** hundredths, 0.06 **31.** thousandths, 0.005 **33.** hundredths, 0.09
35. hundreds, 600

Pages 4, 5 **1.** tens **3.** thousands **5.** 8 **7.** Yes **9.** Yes **11.** 490 **13.** 2110 **15.** 350 **17.** 100 **19.** 6100
21. 23,100 **23.** 3000 **25.** 62,800 **27.** 1,204,000 **29.** 6,437,000 **31.** 18,000 **33.** hundredths **35.** tenths **37.** 3
39. Yes **41.** Yes **43.** 0.4 **45.** 14.3 **47.** 0.6 **49.** 1.4 **51.** 6.08 **53.** 0.51 **55.** 0.90 **57.** 2.00 **59.** $12.40
61. $1695.90 **63.** $803.10 **65.** $9.40

Pages 6, 7 **1.** 563 **3.** 4110 **5.** 4338.2 **7.** 198.06 **9.** 586 **11.** 4221.2 **13.** 50.091 **15.** 92.36 **17.** 2984.20
19. 2.943 **21.** $40.61 **23.** 708.100, 710.52 **25.** Deluxe Wing Chair **27.** Yes

Pages 8, 9 **1.** 10 **3.** 6 **5.** Olvero; 520; 20.8 **7.** Olvero; 520; 20.8 **9.** Mann; 388; 15.5
11. Jones, Webster, Olvero, McKinney, Cohen, Sully **13.** 20.8 **15.** Sully

Page 10 *Extra!* **1.** merge **3.** tunnel **5.** bump **7.** children **9.** speed limit **11.** do not enter

Page 11 **1.** $1.34 **3.** $2.27 **5.** $1.87 **7.** Zone 2

Page 12 **1.** 8 **3.** 8 **5.** 8 **7.** 30 **9.** 70 **11.** 65 **13.** 72 **15.** 58 **17.** 89 **19.** 97 **21.** 436 **23.** 629 **25.** 898
27. 20 **29.** $697 **31.** $969 **33.** 9889 **35.** $9889 **37.** $9999

Page 13 **1.** 15 **3.** 18 **5.** 17 **7.** 945 **9.** 588 **11.** 787 **13.** $9092 **15.** $14,008 **17.** 112 **19.** 921
21. 56,833 **23.** 66,261 **25.** 949 **27.** 66 **29.** 264

Page 14 **1.** 6 **3.** 3 **5.** 2 **7.** 22 **9.** 72 **11.** 64 **13.** 822 **15.** 214 **17.** 221 **19.** 114 **21.** 551 **23.** 4141
25. 4311 **27.** $1511 **29.** $5102 **31.** 22,503 **33.** 56,183 **35.** 11,331 **37.** 41,312

Page 15 **1.** 395 **3.** 745 **5.** 383 **7.** 925 **9.** 304 **11.** 148 **13.** 459 **15.** 1888 **17.** 687 **19.** $304 **21.** $459
23. $45,734 **25.** $26,004 **27.** 31 **29.** 249 **31.** 159; 19

Pages 16, 17 **1.** 30 **3.** 70 **5.** 90 **7.** 680 **9.** 200 **11.** 200 **13.** 100 **15.** 800 **17.** 110 **19.** 490 **21.** 240
23. 21,000 **25.** 21,000 **27.** 13,000 **29.** 18,000 **31.** 30 **33.** 50 **35.** 1180 **37.** 100 **39.** 400 **41.** 800
43. 2400 **45.** 37,600 **47.** 35,000 **49.** 800

Pages 18, 19 **1.** 4 **3.** c-350 **5.** 274; 289; 334; 356 **7.** 1300 **9.** 1800 **11.** 800 **13.** C-100; C-300

Page 20 *Extra!* **1.** 13; 12; 10 **3.** Yes   *Page 21* **1.** $476.97; $572.37 **3.** $1785.84

Page 22 **1.** 69.8 **3.** 6.69 **5.** 9.98 **7.** 87.67 **9.** 6.061 **11.** 127.77 **13.** 82.580 **15.** 925.6 **17.** 110.482
19. 491.575 **21.** 4728.7083 **23.** $926.56 **25.** $3838.34 **27.** 16.03 **29.** $12.14 **31.** $10.87; $35.04

Page 23 **1.** 5.12 **3.** 21.2 **5.** 17.0 **7.** 33.3 **9.** 75.57 **11.** 2.79 **13.** 4.13 **15.** 21.56 **17.** 72.59 **19.** 18.163
21. 469.75 **23.** 1.799 **25.** $41.80 **27.** 77.191 **29.** 18.163 **31.** 23.282 **33.** 153.924 **35.** 120.06; 16.376

Pages 24, 25 **1.** 0.3 **3.** 3.1 **5.** 7.1 **7.** 7.09 **9.** 7.77 **11.** 2 **13.** 5 **15.** 8 **17.** 30 **19.** 100 **21.** 0.8
23. 14.8 **25.** 42.9 **27.** 74.7 **29.** 21.1 **31.** 124 **33.** $30 **35.** $10 **37.** $9653 **39.** 38.3 **41.** 445.2 **43.** 81.0
45. 39 **47.** 37 **49.** 340 **51.** $18 **53.** $415 **55.** 2.4 **57.** $32; no

Pages 26, 27 **1.** $5600 **3.** $820 **5.** $710 **7.** $1210 **9.** metallic blue, yellow, white/blue

Page 28 *Consumer Note* **1.** $4.50

1

Page 29 **1.** 1; 0.9 **3.** 10; 11; 10.6 **5.** 61,401 **7.** $925 **9.** 7891 **11.** 147,808 **13.** 342 **15.** 3383 **17.** 1643 **19.** 1617 **21.** 154.235 **23.** $1306.10 **25.** $13.98 **27.** 2.008 **29.** 190.03 **31.** 2.164

Page 30 **1.** 409 **3.** 17,954 **5.** 1.1119 **7.** 90 **9.** 470 **11.** 400 **13.** 900 **15.** 1.1 **17.** 8.3 **19.** 91 **21.** 147 **23.** 0.90 **25.** 63.14 **27.** $43 **29.** $137

UNIT 2

Page 33 **1.** 96 **2.** 86 **3.** 688 **4.** 5022 **5.** 48,120 **6.** 254,800 **7.** 24,836 **8.** 88,000 **9.** 2,933,041 **10.** 637,956 **11.** 32,970 **12.** 4,520,518 **13.** 96.138 **14.** 36.19 **15.** 0.0000938 **16.** 0.000091 **17.** 60 **18.** 12 **19.** 18 **20.** 52,800 **21.** 48 **22.** 6000 **23.** 28 **24.** 50 **25.** 8 **26.** 8 **27.** 16 **28.** 20

Page 34 **1.** 12 **3.** 0 **5.** 7 **7.** 30 **9.** 60 **11.** 0 **13.** 63 **15.** 84 **17.** 25 **19.** 93 **21.** 66 **23.** 404 **25.** 243 **27.** 222 **29.** 262 **31.** 369 **33.** 3101 **35.** 2824 **37.** 6208 **39.** 6004 **41.** 6090

Page 35 **1.** 96 **3.** 34 **5.** 84 **7.** 402 **9.** 963 **11.** 72 **13.** 50 **15.** 92 **17.** 381 **19.** 684 **21.** 723 **23.** 704 **25.** 546 **27.** 744 **29.** 528 **31.** 1708 **33.** 1356 **35.** 2106 **37.** 4278 **39.** 16,336

Pages 36, 37 **1.** 260 **3.** 780 **5.** 800 **7.** 49,800 **9.** 89,300 **11.** 1440 **13.** 5810 **15.** 1000 **17.** 7290 **19.** 4100 **21.** 148,800 **23.** 296,100 **25.** 144,900 **27.** 43,920 **29.** 183,380 **31.** 1,678,400 **33.** 1,188,200 **35.** 2,727,300 **37.** 17,307,200 **39.** 11,110,500 **41.** 409,600 **43.** 890,400 **45.** 4,388,000 **47.** 1,090,000 **49.** 7,908,000 **51.** 2,427,000 **53.** 1,230,000 **55.** 6,588,000 **57.** 45,630,000 **59.** 24,688,000 **61.** 82,460,000 **63.** 97,824,000 **65.** 128,572,000 **67.** 150,654,000 **69.** 262,435,000 **71.** 146,766,000 **73.** $506,933,000 **75.** $389,412,000

Pages 38, 39 **1.** 60 **3.** 15,840 **5.** 120 **7.** 1 **9.** 12 **11.** 3 **13.** in. **15.** ft **17.** in. **19.** mi **21.** 60 **23.** 3; 36 **25.** 30; 360 **27.** 3520; 10,560 **29.** 15,840; 47,520 **31.** 52,800; 158,400 **33.** 12; 16 **35.** 12; 14 **37.** 36; 53 **39.** 5 in.

Pages 40, 41 **1.** S. Hudson **3.** 6 mi **5.** 46 mi **7.** 69 **9.** 276

Page 42 *Consumer Note* 32,800 kW · h; 5390 kW · h

Page 44 **1.** 924 **3.** 4216 **5.** 4746 **7.** 4017 **9.** 23,692 **11.** 1922 **13.** 946 **15.** 840 **17.** 2580 **19.** 9920 **21.** 30,592 **23.** 40,376 **25.** 299,544 **27.** 617,348 **29.** 176,148 **31.** 190,850 **33.** 354,240 **35.** 989,901

Page 45 **1.** 25,628 **3.** 33,768 **5.** 40,474 **7.** 619,676 **9.** 994,640 **11.** 91,800 **13.** 92,448 **15.** 499,380 **17.** 545,836 **19.** 478,962 **21.** 1,040,608 **23.** 735,007 **25.** 19,611,438 **27.** 23,466,240 **29.** 55,381,173

Pages 46, 47 **1.** 960 **3.** 3000 **5.** 2800 **7.** 37,680 **9.** 36,400 **11.** 60,401 **13.** 583,035 **15.** 392,522 **17.** 170,962 **19.** 351,852 **21.** 7740 **23.** 3840 **25.** 8880 **27.** 68,250 **29.** 40,420 **31.** 827,400 **33.** 4,416,300 **35.** 1,071,600 **37.** 2,862,600 **39.** 1,706,800 **41.** 10,080 **43.** 8880 **45.** 1560 **47.** 432,322 **49.** 309,012 **51.** 748,752 **53.** 988,674 **55.** 1,779,975 **57.** 2,069,352 **59.** 894,600 **61.** 316,780 **63.** 69,422; 429,580 **65.** 3,586,297

Pages 48, 49 **1.** 112 **3.** 1600 **5.** 30,000 **7.** 2000 **9.** oz **11.** lb **13.** ounce **15.** ton **17.** ounce **19.** pound **21.** 160 **23.** 4000 **25.** 30,000 **27.** 21 **29.** 112; 122 **31.** 2 lb cake **33.** 1 lb 5 oz box **35.** 400 lb **37.** 2 oz

Pages 50, 51 **1.** Masonry sand; #1 gravel; #2 gravel **3.** 35 t **5.** 200 **7.** $7500 **9.** 70,000 **11.** 17,000 lb **13.** $8685 **15.** $10,919

Page 52 *Extra!* **1.** Yes **3.** No   *Page 53* **1.** $15 **3.** $215.60

Pages 54, 55 **1.** 17.283 **3.** 53.56 **5.** 3.392 **7.** 2769.48 **9.** 187.997 **11.** 66.65 **13.** 680.82 **15.** 1535.4 **17.** 396.657 **19.** 660.836 **21.** 14.7576 **23.** 31.6239 **25.** 19.3228 **27.** 1161.745 **29.** 4070.712 **31.** 22.69727 **33.** 35.74982 **35.** 40.36536 **37.** 0.07 **39.** 0.036 **41.** 0.084 **43.** 0.2418 **45.** 0.3021 **47.** 0.161 **49.** 0.204 **51.** 0.094 **53.** 0.009 **55.** 0.0435 **57.** 0.33332 **59.** 0.13392 **61.** 0.000482 **63.** 0.00356 **65.** 0.005238 **67.** 0.026145 **69.** 0.0000009

Pages 56, 57 **1.** 2 **3.** 2 **5.** qt **7.** 4 **9.** gal **11.** fl oz **13.** pt **15.** fl oz **17.** c **19.** fl oz **21.** 128; 16; 8 **23.** 32; 16; 8 **25.** 80; 40 **27.** $25.035 **29.** 6 c **31.** 700 mi **33.** 504

Pages 58, 59 **1.** 12 oz **3.** 1200; 425; 550; 360 **5.** 425 cases; 8000; 10,200 **7.** 37,872 **9.** 39,000 **11.** 10,920

Page 60 *Calculator Displays* Example: 9.5; 9.95; 9.995; 9.9995 **1.** 2.6; 2.66; 2.666; 2.66666666 **3.** 48; 48.8; 48.88; 48.88888888

Page 61 **1.** 648 **3.** 6776 **5.** 5304 **7.** 952 **9.** 574 **11.** 338 **13.** 7542 **15.** 3952 **17.** 2,682,616 **19.** 28,574 **21.** 7280 **23.** 28,380 **25.** 149,352 **27.** 5600 **29.** 4,398,636 **31.** 29.472 **33.** 18.886 **35.** 76.986 **37.** 0.00504 **39.** 4.9533

Page 62 **1.** 36 **3.** 7040 **5.** 720; 60 **7.** 68 **9.** 59 **11.** 112 **13.** 6000 **15.** 34,000 **17.** 91 **19.** 16 **21.** 16; 8 **23.** 40; 20

UNIT 3

Page 65 **1.** 52 **2.** 80 **3.** 77R1 **4.** 59R1 **5.** 6R27 **6.** 219R23 **7.** 42R56 **8.** 173R149 **9.** $113\frac{1}{8}$ **10.** $32\frac{3}{28}$ **11.** $18\frac{37}{49}$ **12.** $16\frac{317}{325}$ **13.** 9.5 **14.** 54.9 **15.** 71.3 **16.** 35.1 **17.** 70,123 m; 63 km; 2200 m; 0.112 km **18.** 7.8 cm, or 78 mm **19.** 0.105 **20.** 0.00078 **21.** 782 **22.** 23,500 **23.** 0.098 L; 98.3 mL; 0.731 L; 9813 mL

Page 66 **1.** 2 **3.** 7 **5.** 6 **7.** 20 **9.** 30 **11.** 31 **13.** 23 **15.** 42 **17.** 61 **19.** 91 **21.** 96 **23.** 64 **25.** 67 **27.** 144 **29.** 103 **31.** 870 **33.** 136 **35.** 781 **37.** 158

Page 67 **1.** 6R6 **3.** 4R4 **5.** 8R2 **7.** 4R1 **9.** 6R4 **11.** 21R1 **13.** 22R2 **15.** 10R4 **17.** 52R1 **19.** 71R3 **21.** 61R2 **23.** 23R5 **25.** 56R1 **27.** 91R2 **29.** 78R3 **31.** 617R2 **33.** 724R3 **35.** 547R5 **37.** 674R3

Pages 68, 69 **1.** 18 **3.** 63 **5.** 45 **7.** 23 **9.** 305 **11.** 66 **13.** 2187 **15.** 1813 **17.** 2751 **19.** 3R19 **21.** 2R4 **23.** 2R15 **25.** 26R6 **27.** 11R5 **29.** 4R14 **31.** 8R40 **33.** 8R2 **35.** 116R38 **37.** 173R37 **39.** 56R47 **41.** 80R9 **43.** 90R16 **45.** 66R18 **47.** 2 **49.** 1 **51.** 40 **53.** 20 **55.** 10 **57.** 3R37 **59.** 3R89 **61.** 1R8 **63.** 1R276 **65.** 22R54 **67.** 15R383 **69.** 6R7 **71.** 4R4 **73.** 440R1 **75.** 75R9 **77.** 63R10 **79.** 3561R21

Pages 70, 71 **1.** 300 **3.** 40 **5.** 12.476 **7.** meter **9.** kilometer **11.** meter **13.** 7560 **15.** 900 **17.** 150,000 **19.** 1.5 **21.** 14.896 **23.** 3.43 **25.** 4.413 **27.** 0.5 **29.** 12.74 **31.** 42 **33. a.** 12.8 km **b.** 7450 m **35.** 114,683 m; 84 km; 2.6 km; 1900 m **37.** 180 km **39.** $403.77

Pages 72, 73 **1.** 20 km **3. a.** 1000 m **b.** 250 m **c.** 200 m **d.** 500 m **5.** 100 **7.** 40 **9.** 100 **11.** 340

Page 74 *Consumer Note* $238.18

Page 75 **1.** 41 **3.** 100 **5.** 32 **7.** 70°C **9.** 100°C **11.** 20°C, 80°F, 100°F, 60°C

Pages 76, 77 **1.** $1\frac{1}{5}$ **3.** $2\frac{1}{4}$ **5.** $3\frac{1}{2}$ **7.** $8\frac{3}{9}$ **9.** $5\frac{1}{3}$ **11.** $68\frac{1}{5}$ **13.** $77\frac{1}{7}$ **15.** $95\frac{1}{4}$ **17.** $2\frac{21}{31}$ **19.** $1\frac{21}{69}$ **21.** $2\frac{16}{40}$ **23.** $5\frac{2}{19}$ **25.** $3\frac{2}{19}$ **27.** $22\frac{2}{34}$ **29.** $16\frac{45}{54}$ **31.** $3\frac{3}{73}$ **33.** $23\frac{4}{32}$ **35.** $13\frac{18}{30}$ **37.** $84\frac{26}{51}$ **39.** $58\frac{34}{84}$ **41.** $101\frac{4}{89}$ **43.** $306\frac{26}{30}$ **45.** $6\frac{31}{100}$ **47.** $3\frac{73}{300}$ **49.** $2\frac{1}{222}$ **51.** $2\frac{77}{250}$ **53.** $2\frac{67}{500}$ **55.** $20\frac{6}{423}$ **57.** $5\frac{65}{107}$ **59.** $1\frac{150}{744}$ **61.** $8\frac{6}{803}$ **63.** $9\frac{2}{395}$ **65.** $7\frac{39}{391}$ **67.** $3\frac{8}{764}$ **69.** $11\frac{8}{832}$ **71.** $8\frac{3}{847}$ **73.** $32\frac{15}{541}$ **75.** $68\frac{114}{231}$ **77.** $215\frac{27}{323}$

Pages 78, 79 **1.** 4.1 **3.** 2.6 **5.** 20.9 **7.** 9.7 **9.** 2.2 **11.** 7.3 **13.** 13.5 **15.** 8.2 **17.** 86.6 **19.** 49.8 **21.** 16.5 **23.** 11.8 **25.** 19.2 **27.** 8.2 **29.** 4.6 **31.** 5.3 **33.** 9.5 **35.** 8.4 **37.** 25.1 **39.** 50.5 **41.** 95.8 **43.** 63.0 **45.** 6.53 **47.** 1.80 **49.** 1.29 **51.** 1.13 **53.** 1.33 **55.** 11.86 **57.** 7.17 **59.** 2.65 **61.** 0.44 **63.** 1.03 **65.** 32.27 **67.** 6.17 **69.** 4.31 **71.** 6.33 **73.** 1.92 **75.** 1.14 **77.** 3.01 **79.** 8.56 **81.** 5.18 **83.** 6.76

Pages 80, 81 **1.** 47.8 **3.** 7.748 **5.** 871.2 **7.** 400 **9.** 25 **11.** 2.3 **13.** 1.8 **15.** 0.43 **17.** 6.03 **19.** 0.42 **21.** 25 **23.** 192 **25.** 760 **27.** 7500; 7.5 **29.** 43,890; 43.89 **31.** 36.7; 3.67 **33. a.** 3.6 **b.** 1 **c.** 0.2 **d.** 10.7 **e.** 6 **35. a.** 12 **b.** 10 **c.** 9 **37.** 106.4 mm, or 10.64 cm

Pages 82, 83 **1. a.** 2500 **b.** 1500 **c.** 1000 **3.** 2500; 1500; 1000 **5.** 45,000 cm **7.** 40,000 cm **9.** Yes

Page 84 *Extra!* **1.** 13 **3.** 156 **5.** 36 **7.** 324   *Page 85* **1.** $33\frac{5}{9}$ **3.** $68\frac{5}{6}$ **5.** $30\frac{19}{27}$ **7.** $230\frac{2}{38}$

Pages 86, 87 **1.** 74 **3.** 1.8 **5.** 0.4 **7.** 940 **9.** 91 **11.** 4970 **13.** 7.1 **15.** 41.5 **17.** 0.8 **19.** 1.4 **21.** 2.5 **23.** 5.4 **25.** 2.4 **27.** 14.6 **29.** 2.0 **31.** 9.3 **33.** 0.2 **35.** 11.5 **37.** 80.2 **39.** 74.3 **41.** 18 **43.** 3210 **45.** 3.0 **47.** 7.2 **49.** 1300.0 **51.** 1200.0 **53.** 21.4 **55.** 249.4 **57.** 0.3 **59.** 7.33 **61.** 21.43 **63.** 2.41 **65.** 6.76

Page 88 **1.** 0.4 kg **3.** 550 g **5.** 75,000 **7.** 6398 **9.** 55,600 **11.** 490 **13.** 18.374

Page 89 **1.** 5.6 **3.** 0.5 **5.** 0.075 **7.** 7030 **9.** 18,300 **11.** 15 L; 14,360 mL; 0.483 L; 67.8 mL **13.** $270.00

Pages 90, 91 **1.** 4 **3. a.** 70 g **b.** 1.5 kg **c.** 125 mL **d.** 40 g **5.** 125 mL; 1125 mL **7.** 10 mL; 90 mL **9.** 20 mL; 180 mL **11.** 7 **13.** 1 **15.** $1.78 **17.** $2.37 **19.** $1.78 **21.** $39.11

Page 92 *Consumer Note* **1.** 23.5 MPG **3.** 27.1 MPG

Page 93 **1.** 16 **3.** 64 **5.** 864 **7.** 10R32 **9.** 114R55 **11.** 14R534 **13.** 9R36 **15.** 8R235 **17.** $11\frac{6}{82}$ **19.** $2\frac{2}{492}$ **21.** $3\frac{51}{270}$ **23.** $12\frac{160}{395}$ **25.** 8.4 **27.** 136.4 **29.** 4.83 **31.** 15.85 **33.** 10.1 **35.** 1.4 **37.** 600.0 **39.** 600.0 **41.** 83.04 **43.** 0.02 **45.** 1137.00 **47.** 729.00

Page 94 **1.** 951 **3.** 75,000 **5.** 20 **7.** 398 **9.** 0.325 **11.** 0.9 **13.** 6.2 **15.** 23,323 m; 19 km; 9.6 km; 3572 m **17.** 60; 0.06 **19.** 73,520; 73.52 **21.** 58.1; 0.0581 **23.** Wen-Ying **25.** 7 **27.** 65.727 **29.** 1.9 **31.** 1.3 **33.** 700 **35.** 680 **37.** 375.1 **39.** 10

Cumulative Review: Units 1–3

Pages 96, 97, 98 *Skills* **1.** units, 3 **3.** tenths, $\frac{3}{10}$ or 0.3 **5.** 150 **7.** 0.5 **9.** 79 **11.** 893 **13.** 102 **15.** 1450 **17.** 51 **19.** 413 **21.** 181 **23.** 7670 **25.** 36.7 **27.** 765.776 **29.** 24.9 **31.** 216.754 **33.** 86 **35.** 669 **37.** 2049 **39.** 1704 **41.** 31,860 **43.** 287,200 **45.** 31,716 **47.** 19,300 **49.** 237,475 **51.** 5,415,084 **53.** 98,022 **55.** 4,303,424 **57.** 447.2 **59.** 0.0336 **61.** 113 **63.** 356 **65.** 224R1 **67.** 16R8 **69.** 10R46 **71.** 33R144 **73.** $2\frac{23}{36}$ **75.** 495 **77.** 4.3 **79.** 1.5 **81.** 2.2 **83.** 564.2

Pages 98, 99 *Applications* **1.** 472 **3.** 8.6036 **5.** 340 **7.** 2000 **9.** 15.4 **11.** $4 **13.** 12 **15.** 123 **17.** 48 **19.** 22 **21.** 24 **23.** 28 **25.** 6000 m **27.** 360 km **29.** 6 **31.** 0.35 kg **33.** 0.015 kg

UNIT 4

Page 101 **1.** 1, 2, 4, 7, 14, 28 **2.** 6 **3.** $\frac{3}{8}$ **4.** 15 **5.** 8, 16, 24, 32 **6.** 30 **7.** 15 **8.** $\frac{3}{24}, \frac{20}{24}$ **9.** $6\frac{9}{10}$ **10.** $\frac{19}{3}$ **11.** $\overline{FG}, \overline{GH}, \overline{HF}$ **12.** obtuse **13.** 45° **14.** radius

Page 102 **1.** 1, 2, 3, 4, 6, 12 **3.** 1, 2, 3, 4, 6, 8, 12, 24 **5.** 1, 3 **7.** 1, 5 **9.** 1, 2 **11.** 1, 7 **13.** 1, 3, 5, 15 **15.** 1, 2, 5, 10 **17.** 1, 2, 3, 4, 6, 12 **19.** 1, 2, 3, 4, 6, 8, 12, 24 **21.** 1, 23 **23.** 1, 37 **25.** 1, 3, 9, 27 **27.** 1, 2, 5, 10, 25, 50 **29.** 1, 59 **31.** 1, 2, 4, 5, 8, 10, 16, 20, 40, 80 **33.** 1, 2, 7, 14, 49, 98

Page 103 **1.** 1, 3; 3 **3.** 1, 2, 4; 4 **5.** 1, 5; 5 **7.** 2 **9.** 1 **11.** 4 **13.** 3 **15.** 2 **17.** 9 **19.** 3 **21.** 2 **23.** 15 **25.** 13 **27.** 6 **29.** 5 **31.** 15 **33.** 1 **35.** 8

Page 104 **1.** 3 **3.** 2 **5.** 4 **7.** $\frac{1}{3}$ **9.** $\frac{1}{2}$ **11.** $\frac{1}{2}$ **13.** $\frac{1}{2}$ **15.** $\frac{2}{5}$ **17.** $\frac{3}{7}$ **19.** $\frac{5}{6}$ **21.** $\frac{9}{16}$ **23.** $\frac{7}{10}$ **25.** $\frac{1}{7}$ **27.** $\frac{2}{5}$ **29.** $\frac{4}{7}$ **31.** $\frac{4}{11}$ **33.** $\frac{12}{13}$ **35.** $7\frac{1}{3}$ **37.** $7\frac{2}{3}$ **39.** $4\frac{2}{5}$ **41.** $6\frac{1}{2}$ **43.** $8\frac{1}{2}$ **45.** $14\frac{4}{5}$ **47.** $24\frac{2}{3}$

Page 105 **1.** 3 **3.** 3 **5.** 3 **7.** 3 **9.** $\frac{18}{30}$ **11.** $\frac{36}{44}$ **13.** 2 **15.** 21 **17.** 72 **19.** 32 **21.** 48 **23.** 60 **25.** 190 **27.** 480 **29.** 9 **31.** 75

Page 106 **9.** $\overleftrightarrow{AC}$, $\overleftrightarrow{AB}$; $\overleftrightarrow{AB}$, $\overleftrightarrow{BD}$; $\overleftrightarrow{BD}$, $\overleftrightarrow{DC}$; $\overleftrightarrow{DC}$, $\overleftrightarrow{AC}$ **11.** A

Page 107 **1.** acute **3.** right **5.** acute **7.** right

Pages 108, 109 **1.** North-south **3.** 1st Ave., 3rd Blvd. **5.** 3.8 mi **7.** 2.4 mi **9.** 3.8 mi **11.** 4.0 mi **13.** 2.4 mi; 14 min

Page 110 Consumer Note **1.** $220.00 **3.** $148.75 *Page 111* **1.** 70°F **3.** 50°F **5.** 2

Page 112 **1.** 14, 21, 28 **3.** 16, 20, 24 **5.** 48, 60, 72 **7.** 1, 2, 3, 4 **9.** 5, 10, 15 **11.** 9, 18, 27, 36 **13.** 12, 24, 36, 48 **15.** 16, 32, 48, 64 **17.** 20, 40, 60, 80 **19.** 22, 44, 66, 88 **21.** 26, 52, 78, 104 **23.** 40, 80, 120, 160 **25.** 50, 100, 150, 200 **27.** 32, 64, 96, 128 **29.** 36, 72, 108, 144 **31.** 51, 102, 153, 204 **33.** 300, 600, 900, 1200 **35.** 203, 406, 609, 812 **37.** 310, 620, 930, 1240 **39.** 299, 598, 897, 1196 **41.** 250, 500, 750, 1000

Page 113 **1.** 6 **3.** 12 **5.** 6 **7.** 45 **9.** 36 **11.** 16 **13.** 60 **15.** 50 **17.** 60 **19.** 72 **21.** 144 **23.** 84 **25.** 36 **27.** 45 **29.** 60 **31.** 84 **33.** 96

Page 114 **1.** 6, 12, 18, 24 **3.** 14, 28, 42, 56 **5.** 17, 34, 51, 68 **7.** 23, 46, 69, 92 **9.** 36, 72, 108, 144 **11.** 24 **13.** 84 **15.** 80 **17.** 96 **19.** 20 **21.** 12 **23.** 18 **25.** 28 **27.** 36 **29.** 48 **31.** 126 **33.** 36 **35.** 112 **37.** 72 **39.** 12 **41.** 36 **43.** 28 **45.** 100 **47.** 84

Page 115 **1.** 4; 3 **3.** 8; 10 **5.** 30; 12 **7.** 28; 15 **9.** $\frac{5}{10}$, $\frac{6}{10}$ **11.** $\frac{15}{18}$, $\frac{14}{18}$ **13.** $\frac{21}{28}$, $\frac{18}{28}$ **15.** $\frac{15}{24}$, $\frac{22}{24}$ **17.** $\frac{8}{14}$, $\frac{9}{14}$ **19.** $\frac{77}{112}$, $\frac{40}{112}$ **21.** $\frac{10}{48}$, $\frac{9}{48}$ **23.** $\frac{15}{36}$, $\frac{14}{36}$ **25.** $\frac{21}{48}$, $\frac{20}{48}$ **27.** $\frac{15}{72}$, $\frac{52}{72}$ **29.** $\frac{10}{15}$, $\frac{9}{15}$, $\frac{8}{15}$ **31.** $\frac{35}{56}$, $\frac{32}{56}$, $\frac{12}{56}$

Pages 116, 117 **1.** 50° **3.** 30° **5.** 40° **7.** 40° **9.** 65° **11.** 79° **13.** 100° **15.** Right, isosceles **17.** Right **19.** 70.5° **21.** No **23.** 90°, Right

Pages 118, 119 **1.** 160′ **3.** 80′ **5.** 80′ **7.** 380′ **9.** 740′ **11.** $6290 **13.** 20′ **15.** 240′ **17.** $2040 **19.** log fence **21.** $1500 **23.** $330 **25.** $10,800

Page 120 Extra! **1.** 30; $\frac{15}{30}$, $\frac{18}{30}$, $\frac{25}{30}$, $\frac{20}{30}$ **3.** 60; $\frac{25}{60}$, $\frac{44}{60}$, $\frac{26}{60}$, $\frac{42}{60}$

Page 122 **1.** $\frac{2}{3}$ **3.** $\frac{1}{3}$ **5.** $\frac{1}{3}$ **7.** $4\frac{4}{7}$ **9.** $13\frac{1}{6}$ **11.** $18\frac{4}{7}$ **13.** 2 **15.** $1\frac{1}{3}$ **17.** 3 **19.** 1 **21.** $3\frac{2}{5}$ **23.** 3 **25.** 1 **27.** 1 **29.** $4\frac{1}{2}$ **31.** 1 **33.** $1\frac{1}{3}$ **35.** $2\frac{1}{8}$ **37.** 5 **39.** 1 **41.** $8\frac{4}{9}$ **43.** $11\frac{1}{4}$ **45.** 15 **47.** 1 **49.** 21 **51.** 1 **53.** $2\frac{8}{47}$

Page 123 **1.** 20 **3.** 54 **5.** 70 **7.** $\frac{41}{5}$ **9.** $\frac{57}{8}$ **11.** $\frac{13}{4}$ **13.** $\frac{14}{3}$ **15.** $\frac{14}{5}$ **17.** $\frac{14}{1}$ **19.** $\frac{13}{5}$ **21.** $\frac{75}{8}$ **23.** $\frac{40}{7}$ **25.** $\frac{38}{7}$ **27.** $\frac{53}{6}$ **29.** $\frac{42}{11}$ **31.** $\frac{23}{1}$ **33.** $\frac{194}{21}$ **35.** $\frac{69}{17}$ **37.** $\frac{438}{27}$ **39.** $\frac{109}{1}$ **41.** $\frac{506}{25}$ **43.** $\frac{417}{1}$ **45.** $\frac{241}{41}$ **47.** $\frac{38}{1}$ **49.** $\frac{532}{43}$ **51.** $\frac{1231}{50}$ **53.** $\frac{1163}{40}$ **55.** $\frac{2085}{52}$ **57.** $\frac{948}{43}$ **59.** $\frac{963}{20}$

Pages 124, 125 **1.** parallelogram **3.** trapezoid **5.** parallelogram **7.** square **9.** radius **11.** radius **13.** F **15.** A, B, C, E, G **17.** E, G **19.** 5 cm

Pages 126, 127 **1.** 3 **3.** 10 **5.** 27 **7.** 12 **9.** $117.00 **11.** 12; 27.00 **13.** 424.80 **15.** 800.80

Page 129 **1.** $\frac{1}{2}$ **3.** $\frac{3}{4}$ **5.** $\frac{5}{6}$ **7.** $5\frac{1}{3}$ **9.** $4\frac{5}{6}$ **11.** $9\frac{9}{16}$ **13.** 3 **15.** 36 **17.** 40 **19.** 24 **21.** 16 **23.** 8 **25.** 35 **27.** 36 **29.** 45 **31.** 84 **33.** $\frac{2}{4}$, $\frac{1}{4}$ **35.** $\frac{9}{24}$, $\frac{16}{24}$ **37.** $\frac{8}{36}$, $\frac{9}{36}$ **39.** $\frac{21}{42}$, $\frac{14}{42}$, $\frac{6}{42}$ **41.** $2\frac{2}{5}$ **43.** 1 **45.** $7\frac{1}{2}$ **47.** $6\frac{1}{3}$ **49.** 52 **51.** 17 **53.** $\frac{22}{7}$ **55.** $\frac{35}{8}$ **57.** $\frac{20}{3}$ **59.** $\frac{135}{4}$ **61.** $\frac{102}{13}$ **63.** $\frac{172}{21}$

Page 130 **1.** $\overleftrightarrow{BA}$, $\overleftrightarrow{DC}$ **3.** $\overline{BC}$ or $\overline{AB}$ or $\overline{CD}$ **5.** right **7.** 35° **9.** 73.5° **11.** 48° **13.** trapezoid **15.** 3 cm, 6 cm

UNIT 5

Page 133 **1.** $1\frac{2}{9}$ **2.** $\frac{1}{2}$ **3.** $11\frac{8}{9}$ **4.** $14\frac{4}{7}$ **5.** $1\frac{7}{60}$ **6.** $1\frac{7}{30}$ **7.** $11\frac{5}{18}$ **8.** $11\frac{13}{40}$ **9.** $\frac{3}{5}$ **10.** $1\frac{2}{7}$ **11.** $7\frac{4}{7}$ **12.** $1\frac{7}{9}$ **13.** $\frac{5}{36}$ **14.** $\frac{13}{72}$ **15.** $\frac{27}{35}$ **16.** $6\frac{29}{30}$ **17.** 49 **18.** 15 **19.** Yes **20.** 20 ft **21.** 15.7 in. **22.** 69.08 yd

Page 134 **1.** $\frac{3}{4}$ **3.** $\frac{2}{3}$ **5.** $\frac{4}{5}$ **7.** 6 **9.** $1\frac{1}{3}$ **11.** $1\frac{2}{3}$ **13.** $1\frac{1}{2}$ **15.** $4\frac{2}{3}$ **17.** $5\frac{1}{2}$ **19.** $\frac{2}{5}$ **21.** $\frac{1}{2}$ **23.** $\frac{2}{3}$ **25.** 1 **27.** $1\frac{2}{3}$ **29.** $1\frac{2}{5}$ **31.** $1\frac{3}{5}$ **33.** $\frac{2}{3}$ **35.** $\frac{3}{4}$

Page 135 **1.** $\frac{2}{3}$ **3.** $\frac{1}{4}$ **5.** $\frac{1}{5}$ **7.** $2\frac{1}{2}$ **9.** $6\frac{1}{2}$ **11.** $7\frac{1}{4}$ **13.** $1\frac{1}{3}$ **15.** $1\frac{1}{3}$ **17.** $2\frac{1}{3}$ **19.** $1\frac{1}{3}$ **21.** $2\frac{1}{7}$ **23.** $1\frac{1}{3}$ **25.** 6 **27.** $5\frac{4}{9}$ **29.** $9\frac{1}{3}$ **31.** 18 **33.** $14\frac{3}{7}$ **35.** $8\frac{4}{5}$

Page 136 **1.** 4 **3.** 15 **5.** $\frac{3}{6}$, $\frac{5}{6}$ **7.** $\frac{12}{16}$, $\frac{11}{16}$ **9.** $1\frac{1}{6}$ **11.** $1\frac{11}{18}$ **13.** $1\frac{5}{9}$ **15.** $1\frac{14}{45}$ **17.** $1\frac{11}{84}$ **19.** $1\frac{13}{33}$ **21.** $1\frac{7}{48}$ **23.** $1\frac{13}{60}$ **25.** 2

Page 137 **1.** $\frac{10}{14}$, $\frac{7}{14}$ **3.** $\frac{9}{15}$, $\frac{8}{15}$ **5.** $4\frac{2}{3}$ **7.** $1\frac{3}{4}$ **9.** $7\frac{1}{12}$ **11.** $6\frac{33}{35}$ **13.** $9\frac{38}{63}$ **15.** $11\frac{49}{52}$ **17.** $12\frac{11}{14}$ **19.** $9\frac{16}{63}$ **21.** $15\frac{67}{84}$ **23.** $23\frac{41}{60}$

Page 138 **1.** 16 **3.** 49 **5.** 100 **7.** 36 **9.** 289 **11.** 20.25 **13.** 5 **15.** 6 **17.** 9 **19.** 11 **21.** 20

Page 139 **1.** 5 **3.** 164 **5.** yes **7.** no **9.** yes

Pages 140, 141 **1.** 10 ft **3.** 4 **5.** yes **7.** 12 **9.** 10 **11.** 16 **13.** 10, $135.00 **15.** $4.70

Page 142 *Career Clippings* **1.** 4:10 **3.** 4:45 **5.** 4:05 *Page 143* **1.** yes **3.** yes **5.** yes **7.** no **9.** yes

Page 144 **1.** $\frac{1}{4}$ **3.** $\frac{1}{4}$ **5.** $\frac{1}{2}$ **7.** $\frac{6}{7}$ **9.** $\frac{3}{5}$ **11.** $\frac{12}{25}$ **13.** $\frac{1}{3}$ **15.** $\frac{2}{3}$ **17.** $\frac{1}{5}$ **19.** $\frac{2}{9}$ **21.** $\frac{1}{3}$ **23.** $\frac{1}{2}$ **25.** $\frac{3}{7}$ **27.** $\frac{1}{3}$ **29.** $\frac{2}{5}$ **31.** $\frac{4}{7}$ **33.** $\frac{4}{9}$ **35.** $\frac{2}{3}$

Page 145 **1.** $\frac{1}{2}$ **3.** $\frac{3}{5}$ **5.** $\frac{2}{5}$ **7.** $3\frac{1}{3}$ **9.** $10\frac{7}{10}$ **11.** $20\frac{3}{7}$ **13.** $1\frac{1}{2}$ **15.** $3\frac{1}{4}$ **17.** $5\frac{2}{3}$ **19.** $8\frac{3}{4}$ **21.** $15\frac{1}{2}$ **23.** $2\frac{1}{8}$ **25.** $25\frac{3}{5}$ **27.** $13\frac{16}{25}$ **29.** $6\frac{1}{3}$

Page 146 **1.** 4 **3.** 5 **5.** 15 **7.** $5\frac{2}{3}$ **9.** $19\frac{3}{5}$ **11.** $6\frac{3}{5}$ **13.** $3\frac{1}{3}$ **15.** $17\frac{1}{10}$ **17.** $3\frac{8}{15}$ **19.** $7\frac{1}{20}$ **21.** $1\frac{1}{5}$ **23.** $1\frac{2}{9}$ **25.** $2\frac{3}{8}$ **27.** $8\frac{4}{9}$

Page 147 **1.** $7\frac{5}{4}$ **3.** $9\frac{11}{8}$ **5.** $17\frac{17}{12}$ **7.** $9\frac{3}{4}$ **9.** $6\frac{4}{5}$ **11.** $2\frac{1}{2}$ **13.** $2\frac{3}{4}$ **15.** $1\frac{6}{7}$ **17.** $11\frac{4}{5}$ **19.** $11\frac{5}{7}$ **21.** $15\frac{5}{7}$ **23.** $12\frac{36}{37}$ **25.** $13\frac{94}{99}$

Pages 148, 149 **1.** 3.9 **3.** 9.14 **5.** $10\frac{1}{8}$ **7.** 100 ft **9.** $13\frac{1}{2}$ mi **11.** 66 km **13.** 80 ft **15.** 32.2 km **17.** 5.5 in. **19.** 42 in. **21.** 4605 ft

Pages 150, 151 **1. a.** 10 ft **b.** 10 ft **c.** 12 ft **d.** 8 ft **3.** 48 ft **5.** 160 ft **7.** 18 ft **9.** $1120 **11.** $40 **13.** $1300 **15.** 44, $52.80 **17.** 18, $41.40

Page 152 *Consumer Note* **1.** 2 **3.** 2 **5.** 0   *Page 153* 8:43, 9:50, 8:10, 8:50; 43:18

Page 154 **1.** 10 **3.** 36 **5.** $\frac{9}{12}$, $\frac{8}{12}$ **7.** $\frac{9}{24}$, $\frac{20}{24}$ **9.** $\frac{1}{4}$ **11.** $\frac{1}{18}$ **13.** $\frac{7}{12}$ **15.** $\frac{17}{24}$ **17.** $\frac{7}{24}$ **19.** $\frac{23}{60}$ **21.** $\frac{1}{48}$

23. $\frac{17}{42}$

Page 155 **1.** $7\frac{6}{8}$, $6\frac{1}{8}$ **3.** $4\frac{5}{12}$, $1\frac{3}{12}$ **5.** $3\frac{2}{35}$ **7.** $5\frac{1}{24}$ **9.** $4\frac{11}{21}$ **11.** $17\frac{13}{14}$ **13.** $7\frac{17}{48}$ **15.** $8\frac{9}{10}$ **17.** $19\frac{7}{48}$

19. $29\frac{3}{56}$ **21.** $5\frac{17}{48}$

Pages 156, 157 **1.** 6.3 **3.** 22.0 **5.** 14 cm **7.** 30 m **9.** 1.5 yd **11.** 3000 ft **13.** 18.8 ft **15.** 40.8 m
17. 77.2 cm **19.** 162.0 yd **21.** 27,318 mi **23.** 35.7 ft **25.** 59.7 m

Pages 158, 159 **1.** 2 ft **3.** 10 **5.** 6.28 ft **7.** no **9.** 162.72 in. **11.** yes   *Page 160* *Extra!* **1.** $625 **3.** $2053.69

Page 161 **1.** 5 **3.** $9\frac{2}{3}$ **5.** $14\frac{1}{2}$ **7.** $4\frac{13}{14}$ **9.** $14\frac{1}{40}$ **11.** $43\frac{1}{9}$ **13.** $4\frac{1}{2}$ **15.** $2\frac{2}{9}$ **17.** $12\frac{2}{25}$ **19.** $3\frac{2}{3}$ **21.** $4\frac{5}{6}$

23. $23\frac{1}{3}$ **25.** $\frac{22}{45}$ **27.** $\frac{19}{56}$ **29.** $\frac{7}{45}$ **31.** $1\frac{3}{4}$ **33.** $6\frac{23}{40}$ **35.** $10\frac{11}{16}$

Page 162 **1.** 121 **3.** 12 **5.** 16 **7.** yes **9.** no **11.** 22 in. **13.** 360 ft **15.** 6.5 in., 40.8 in. **17.** 0.35 cm, 2.2 cm

UNIT 6

Page 165 **1.** $\frac{1}{6}$ **2.** $10\frac{4}{5}$ **3.** $1\frac{3}{7}$ **4.** $6\frac{3}{4}$ **5.** $\frac{8}{5}$ **6.** $\frac{2}{5}$ **7.** $\frac{9}{10}$ **8.** $\frac{1}{6}$ **9.** $\frac{1}{18}$ **10.** 6 **11.** $\frac{2}{5}$ **12.** $\frac{4}{9}$
13. 36 in.^{2} **14.** 84 cm^2 **15.** 78.5 ft^2 **16.** 30.28 ft^2 **17.** 30 ft^3 **18.** 12.56 in.^{3}

Page 166 **1.** 2 **3.** 9 **5.** 7 **7.** $\frac{1}{6}$ **9.** $\frac{8}{15}$ **11.** $\frac{20}{63}$ **13.** $\frac{4}{15}$ **15.** $\frac{24}{55}$ **17.** $\frac{5}{16}$ **19.** $\frac{3}{40}$ **21.** $\frac{11}{72}$ **23.** $\frac{9}{50}$

25. $\frac{7}{15}$ **27.** $\frac{3}{5}$ **29.** $\frac{7}{15}$ **31.** $\frac{1}{4}$ **33.** $\frac{5}{7}$

Page 167 **1.** 2 **3.** 1 **5.** 12 **7.** $1\frac{1}{3}$ **9.** $1\frac{1}{8}$ **11.** $1\frac{1}{2}$ **13.** $\frac{5}{7}$ **15.** $1\frac{3}{5}$ **17.** 2 **19.** 10 **21.** $4\frac{4}{9}$ **23.** $7\frac{1}{5}$

25. $4\frac{1}{6}$ **27.** $\frac{1}{2}$ **29.** $4\frac{1}{2}$ **31.** 20 **33.** $3\frac{1}{2}$

Page 168 **1.** $\frac{5}{4}$ **3.** $\frac{20}{3}$ **5.** $\frac{55}{6}$ **7.** $\frac{17}{16}$ **9.** $\frac{34}{15}$ **11.** $1\frac{3}{5}$ **13.** $2\frac{2}{3}$ **15.** $2\frac{2}{5}$ **17.** $1\frac{7}{18}$ **19.** $1\frac{11}{32}$ **21.** $2\frac{14}{15}$

23. $1\frac{1}{12}$ **25.** $2\frac{2}{5}$ **27.** $1\frac{7}{9}$ **29.** $\frac{23}{110}$ **31.** $\frac{4}{5}$ **33.** $1\frac{1}{9}$ **35.** $\frac{2}{5}$ **37.** $4\frac{1}{5}$ **39.** $\frac{23}{72}$

Page 169 **1.** $\frac{11}{5}$ **3.** $\frac{23}{8}$ **5.** $\frac{19}{8}$ **7.** $\frac{137}{16}$ **9.** $\frac{7}{1}$ **11.** $5\frac{2}{15}$ **13.** $2\frac{11}{12}$ **15.** $6\frac{2}{15}$ **17.** $8\frac{2}{5}$ **19.** $4\frac{1}{5}$ **21.** $7\frac{5}{7}$

23. $32\frac{4}{7}$ **25.** $14\frac{2}{7}$ **27.** $8\frac{1}{2}$ **29.** $7\frac{1}{4}$ **31.** $9\frac{1}{2}$ **33.** $8\frac{1}{5}$ **35.** 35 **37.** $25\frac{19}{60}$

Page 170 **1.** 504 **3.** $118\frac{1}{8}$ **5.** 36.757 **7.** 140 cm^2 **9.** 10.5 mi^2 **11.** 19,360 ft^2 **13.** square **15.** 96,725 mi^2

Page 171 **1.** 48 m^2 **3.** 7 ft^2 **5.** 3000 yd^2 **7.** 20 m^2 **9.** 125 cm^2

Pages 172, 173 **1.** 2 **3.** 2 **5.** 4 ft by 8 ft **7. a.** 16 **b.** 20 **9.** 936 ft^2 **11.** 904 ft^2 **13.** 11 **15.** 34

Page 174 *Extra!* **1.** $\frac{5}{8}$ **3.** $1\frac{13}{15}$ **5.** $\frac{3}{10}$ **7.** $\frac{4}{15}$

Page 175 Total assets: $12,661,000; total liabilities: $11,461,000; owner's equity: $1,200,000; total liabilities and owner's equity: $12,661,000

Page 176 **1.** 1 **3.** 1 **5.** $\frac{4}{3}$ **7.** $\frac{8}{7}$ **9.** $\frac{10}{9}$ **11.** $\frac{7}{18}$ **13.** 3 **15.** $\frac{5}{3}$ **17.** $\frac{5}{8}$ **19.** $\frac{16}{15}$ **21.** $\frac{19}{16}$ **23.** $\frac{1}{6}$ **25.** $\frac{1}{4}$

27. $\frac{1}{8}$ **29.** $\frac{1}{203}$ **31.** $\frac{1}{34}$ **33.** $\frac{2}{3}$ **35.** $\frac{5}{19}$ **37.** $\frac{9}{25}$

Page 177 **1.** $\frac{7}{4}$ **3.** $\frac{11}{7}$ **5.** $\frac{31}{8}$ **7.** $\frac{4}{3}$ **9.** $\frac{6}{1}$, or 6 **11.** $\frac{12}{7}$ **13.** $\frac{2}{3}$ **15.** $5\frac{1}{4}$ **17.** $3\frac{1}{3}$ **19.** $\frac{24}{35}$ **21.** 3 **23.** $1\frac{22}{27}$
25. 5 **27.** $3\frac{6}{7}$ **29.** $\frac{5}{28}$ **31.** $\frac{9}{80}$ **33.** $2\frac{6}{7}$ **35.** $1\frac{3}{22}$ **37.** $\frac{5}{6}$ **39.** $\frac{1}{3}$ **41.** $1\frac{1}{5}$

Page 178 **1.** $\frac{1}{4}$ **3.** $\frac{1}{11}$ **5.** $\frac{1}{16}$ **7.** $\frac{5}{4}$ **9.** $\frac{11}{10}$ **11.** $\frac{11}{47}$ **13.** 8 **15.** $4\frac{1}{2}$ **17.** $4\frac{4}{5}$ **19.** 5 **21.** 30 **23.** $8\frac{8}{9}$
25. $\frac{1}{16}$ **27.** $\frac{3}{8}$ **29.** $\frac{5}{32}$ **31.** $\frac{1}{6}$ **33.** $\frac{1}{20}$ **35.** $\frac{1}{15}$ **37.** $7\frac{1}{2}$ **39.** 6 **41.** 28 **43.** 810 **45.** $\frac{9}{20}$ **47.** $1\frac{7}{8}$

Page 179 **1.** $\frac{5}{4}$ **3.** $\frac{20}{7}$ **5.** $\frac{8}{3}$ **7.** $\frac{29}{10}$ **9.** $\frac{37}{16}$ **11.** 18 **13.** $1\frac{3}{4}$ **15.** $3\frac{1}{5}$ **17.** 36 **19.** $5\frac{1}{7}$ **21.** $\frac{4}{51}$ **23.** $\frac{1}{6}$
25. $\frac{1}{9}$ **27.** $4\frac{7}{32}$ **29.** $1\frac{1}{5}$ **31.** $13\frac{1}{6}$ **33.** $3\frac{5}{6}$

Page 180 **1.** 36 **3.** $20\frac{1}{4}$ **5.** 314 cm² **7.** 38.5 m² **9.** 120.7 cm² **11.** 12.56 m² **13.** circle

Page 181 **1.** 54 in.² **3.** 84 ft² **5.** 56 cm² **7.** 68 yd² **9.** 114.24 mi²

Pages 182, 183 **1.** Square **3.** 4 **5.** 640,000 ft² **7.** 14.7 acres **9.** 12.5 acres **11.** 470.4 **13.** $3386.88
15. $753.12

Page 184 *Consumer Note* **1.** $519.50 **3.** $337.00

Page 185 **1.** Central **3.** Eastern **5.** Mt., Pacific **7.** Central and Mountain **9.** 8:00 A.M. **11.** 7:00 A.M.
13. 9:00 A.M. **15.** 8:00 A.M.

Page 186 **1.** 11 **3.** 8 **5.** 2 **7.** 1 **9.** 2 **11.** 5 **13.** $1\frac{1}{3}$ **15.** $3\frac{2}{3}$ **17.** $5\frac{3}{4}$ **19.** $4\frac{1}{3}$ **21.** $1\frac{1}{8}$ **23.** $\frac{3}{5}$ **25.** $\frac{7}{54}$
27. $\frac{5}{36}$ **29.** 4 **31.** $1\frac{13}{35}$ **33.** $1\frac{11}{64}$ **35.** $3\frac{6}{7}$ **37.** $3\frac{32}{45}$ **39.** $1\frac{1}{7}$ **41.** $1\frac{1}{5}$

Page 187 **1.** 14 **3.** 31 **5.** 31 **7.** 24 **9.** 6 **11.** $\frac{4}{35}$ **13.** $\frac{3}{11}$ **15.** $\frac{7}{46}$ **17.** $\frac{14}{19}$ **19.** 1 **21.** 3 **23.** 2 **25.** $1\frac{1}{14}$
27. $2\frac{2}{3}$ **29.** $2\frac{1}{2}$ **31.** $1\frac{10}{11}$ **33.** 10 **35.** $5\frac{1}{4}$ **37.** $2\frac{13}{46}$ **39.** $\frac{74}{75}$ **41.** $3\frac{57}{91}$ **43.** $9\frac{23}{90}$

Page 188 **1.** 96 **3.** 2860 **5.** 73 **7.** 162 cm³ **9.** 36 ft³ **11.** 2880 cm³ **13.** 1st carton **15.** 13,500 gallons

Page 189 **1.** 942 **3.** 502.4 **5.** 169.56 cm³ **7.** 1946.8 m³ **9.** 2nd can

Pages 190, 191 **1.** 6 **3.** 50 ft **5.** 392,500 ft³ **7.** 17,662,500 gallons **9.** 18 million gallons
11. 4 million gallons

Page 193 **1.** $\frac{1}{6}$ **3.** $\frac{1}{14}$ **5.** $\frac{15}{64}$ **7.** $\frac{9}{22}$ **9.** $\frac{11}{18}$ **11.** $2\frac{5}{8}$ **13.** $13\frac{1}{3}$ **15.** $2\frac{5}{8}$ **17.** $9\frac{1}{15}$ **19.** $\frac{1}{5}$ **21.** 6 **23.** $\frac{4}{7}$
25. $13\frac{1}{2}$ **27.** 30 **29.** $\frac{11}{48}$ **31.** 15 **33.** $\frac{1}{18}$ **35.** $13\frac{3}{5}$ **37.** $\frac{4}{65}$ **39.** $\frac{1}{4}$ **41.** $\frac{5}{77}$ **43.** $3\frac{1}{3}$ **45.** $1\frac{13}{56}$ **47.** $5\frac{3}{5}$
49. $3\frac{5}{9}$

Page 194 **1.** $30\frac{7}{8}$ in.² **3.** 168 yd² **5.** 56 m² **7.** 76,610 cm² **9.** 7.065 mi² **11.** 3629.84 m² **13.** $43\frac{1}{2}$ in.²
15. 3420 in.³ **17.** 15,498 in.³ **19.** b

Page 195 **1.** $\frac{5}{21}$ **3.** $4\frac{2}{3}$ **5.** $2\frac{5}{14}$ **7.** $6\frac{7}{8}$ **9.** $\frac{4}{7}$ **11.** $\frac{14}{15}$ **13.** $\frac{1}{20}$ **15.** 16 **17.** $\frac{3}{8}$ **19.** $\frac{2}{3}$ **21.** 81 in.²
23. 78.5 cm² **25.** 52.26 m³

8

Pages 196, 197 Skills **1.** $\frac{1}{2}$ **3.** $2\frac{3}{4}$ **5.** 6 **7.** 21 **9.** $\frac{2}{6}, \frac{3}{6}$ **11.** $\frac{13}{20}, \frac{15}{20}, \frac{8}{20}$ **13.** $1\frac{4}{5}$ **15.** $9\frac{3}{4}$ **17.** 27

19. $\frac{31}{8}$ **21.** $\frac{138}{5}$ **23.** $6\frac{1}{5}$ **25.** $15\frac{1}{11}$ **27.** $10\frac{11}{35}$ **29.** $101\frac{7}{9}$ **31.** $6\frac{5}{8}$ **33.** $11\frac{1}{10}$ **35.** $5\frac{5}{6}$ **37.** $8\frac{1}{3}$ **39.** $\frac{26}{45}$

41. $\frac{8}{15}$ **43.** $1\frac{13}{24}$ **45.** $13\frac{35}{36}$ **47.** $\frac{1}{9}$ **49.** $\frac{4}{7}$ **51.** $9\frac{1}{2}$ **53.** $18\frac{6}{7}$ **55.** $1\frac{1}{3}$ **57.** $3\frac{3}{4}$ **59.** $\frac{11}{16}$ **61.** 2

Pages 198, 199 Applications **1.** $\overleftrightarrow{AC}, \overleftrightarrow{BD}$ **3.** A, B, D, C **5.** obtuse **7.** 30° **9.** 70° **11.** parallelogram **13.** 7
15. 12 **17.** 17 cm **19.** 40 cm **21.** 28.26 cm **23.** 18.84 ft **25.** 90 ft² **27.** 115 cm² **29.** 615.44 in.²
31. 216 m³ **33.** 673.92 cm³ **35.** 28.26 in.³ **37.** 150.72 ft³

UNIT 7

Page 201 **1.** $\frac{3}{7}$ **2.** $\frac{17}{30}$ **3.** yes **4.** no **5.** $a = 16$ **6.** $d = 16$ **7.** $n = 12\frac{1}{2}$ **8.** $\frac{12}{2} = \frac{t}{3}$ **9.** $\frac{310}{4} = \frac{c}{3}$

10. $\frac{150}{2} = \frac{x}{3}, x = 225$ **11.** 2 errors/page **12.** 1.2 touchdowns/game **13.** $\frac{3}{4}$ **14.** 600 **15.** 24 **16.** 25

Page 202 **1.** $\frac{1}{9}$ **3.** $\frac{1}{2}$ **5.** $\frac{6}{13}$ **7.** 46:190 **9.** 28:73 **11.** $\frac{2}{3}$ **13.** $\frac{7}{12}$ **15.** $\frac{1}{12}$ **17.** $\frac{19}{20}$ **19.** $\frac{2}{7}$ **21.** $\frac{1}{3}$

23. a. 9:6 **b.** 6:4 **c.** 6:37 **d.** 9:37 **e.** 4:37

Page 203 **1.** 9 **3.** 3 **5.** 50 **7.** yes **9.** yes **11.** yes **13.** no **15.** yes **17.** no **19.** no **21.** yes **23.** yes
25. no

Pages 204, 205 **1.** 6 **3.** 2 **5.** × **7.** 1 **9.** no **11.** no **13.** yes **15.** yes **17.** yes **19.** no **21.** no **23.** yes
25.–35. Answers will vary. Examples are given. **25.** $\frac{6}{10}, \frac{9}{15}$ **27.** $\frac{2}{8}, \frac{3}{12}$ **29.** $\frac{8}{10}, \frac{12}{15}$ **31.** $\frac{10}{16}, \frac{15}{24}$ **33.** $\frac{6}{14}, \frac{9}{21}$

35. $\frac{12}{14}, \frac{18}{21}$ **37.** 5 **39.** 40 **41.** 81 **43.** $n = 8$ **45.** $b = 6$ **47.** $c = 49$ **49.** $m = 6\frac{2}{3}$ **51.** $a = 3$

53. $d = 26\frac{1}{4}$ **55.** $y = 2\frac{22}{25}$ **57.** $n = 1\frac{5}{9}$ **59.** $z = 5\frac{1}{4}$ **61.** $b = 52$ **63.** $z = 17\frac{1}{5}$ **65.** $c = 5$ **67.** $d = 15$

69. $c = 12$ **71.** $x = 4$ **73.** $m = 2$ **75.** $b = 10\frac{1}{2}$ **77.** $m = 21\frac{3}{5}$ **79.** $x = 51$ **81.** $u = 108$ **83.** $u = 115$

85. $p = 18$ **87.** $c = 12$ **89.** $v = 143$

Pages 206, 207 **1.** $\frac{1}{2}$ **3.** 1 **5.** $\frac{1}{4}$ **7.** 0 **9.** $\frac{1}{4}$ **11. a.** $\frac{3}{5}$ **b.** $\frac{2}{5}$ **13. a.** $\frac{1}{6}$ **b.** $\frac{1}{2}$ **c.** $\frac{1}{2}$ **d.** $\frac{2}{3}$ **e.** 1 **f.** 0

g. $\frac{1}{3}$ **h.** $\frac{2}{3}$ **15.** 25 **17.** Earl

Pages 208, 209 **1.** 0.08 **3.** 0.05 **5.** 0.03 **7.** 16 **9.** 13 **11.** Precision Computer **13.** National Oil
15. Answers will vary. For example, you might recommend American Metals because it has a very low P/E and a
medium dividend yield ratio.

Page 210 Calculator Displays **1.** $x = 27$ **3.** $h = 4.4$ **5.** $a = 15.75$

Page 211 kitchen: 4.5 m; 3.75 m study: 2.5 cm, 3.75 m; 2.5 cm, 3.75 m bathroom: 2 cm, 3 m; 1.5 cm, 2.25 m
living room: 3.5 cm, 5.25 m; 3 cm, 4.5 m entry: 2 cm, 3 m; 1.5 cm, 2.25 m

Pages 212, 213 **1.** $\frac{100}{20}$ **3.** $\frac{3}{120}$ **5.** $\frac{36}{2}$ **7.** $\frac{630}{3}$ **9.** $\frac{120}{10}$ **11.** $\frac{39}{4}$ **13.** $\frac{70}{3}$ **15.** $\frac{110}{7}$ **17.** $\frac{4}{262}$ **19.** $\frac{380}{40}$

21. $\frac{42}{7}$ **23.** $\frac{36}{8}$ **25.** $\frac{7}{430}$ **27.** $\frac{5}{3}$ **29.** $\frac{4}{2}$ **31.** $\frac{12}{5}$ **33.** $\frac{4}{15}$ **35.** $\frac{93}{17}$ **37.** 6 **39.** t **41.** $\frac{825}{30} = \frac{n}{6}$

43. $\frac{46}{3} = \frac{138}{m}$ **45.** $\frac{3}{2} = \frac{9}{w}$ **47.** $\frac{36}{2} = \frac{p}{10}$ **49.** $\frac{8}{4} = \frac{z}{15}$ **51.** $\frac{243}{7} = \frac{e}{10}$ **53.** $\frac{768}{2} = \frac{m}{11}$

Pages 214, 215 **1.** s **3.** 50 **5.** n **7.** 21 **9.** b **11.** 8 **13.** $x = 24$ **15.** $a = 2$ **17.** $s = 18\frac{1}{4}$ **19.** $n = 1\frac{1}{2}$

21. $t = 10\frac{1}{2}$ **23.** $u = 13\frac{1}{8}$ **25.** $\frac{350}{25} = \frac{x}{4}$, $x = 56$ **27.** $\frac{30}{14} = \frac{75}{t}$, $t = 35$ **29.** $\frac{38.40}{8} = \frac{t}{40}$, $t = 192.00$

31. $\frac{36}{1} = \frac{180}{n}$, $n = 5$ **33.** $\frac{475}{1} = \frac{b}{12}$, $b = 5700$ **35.** $\frac{4}{8} = \frac{k}{10}$, $k = 5$ **37.** $\frac{1}{15} = \frac{p}{45}$, $p = 3$

39. $\frac{8}{4} = \frac{r}{10}$, $r = 20$ **41.** $\frac{560}{2} = \frac{t}{24}$, $t = 6720$ **43.** $\frac{18}{3} = \frac{138}{s}$, $s = 23$ **45.** $\frac{246}{82} = \frac{q}{1}$, $q = 3$

Pages 216, 217 **1.** 5922 **3.** 42,230 **5.** 78,309 **7.** $n = 7$ **9.** $n = 640$ **11.** 81,086; 66,343; 110,571

Pages 218, 219 **1.** $25,000 **3.** $100,000 **5.** $51,500

Page 220   Consumer Note **1.** $22,643.07 **3.** $29.75 **5.** $5.25

Pages 222, 223 **1.** $s = 5$ **3.** $x = 10$ **5.** $n = 9$ **7.** $m = 6.8$ **9.** 35 words/min **11.** 68 km/h
13. 312.5 mi/min **15.** 273 jelly beans/jar **17.** 2 windows/h **19.** 8.6 repairs/d **21.** 60 people/room
23. 460 mi/d **25.** A **27.** Train **29.** Joseph **31.** Pecans **33.** Anna **35.** Same cost/oz **37.** Copper

Page 224 **1.** 233 **3.** 276 **5.** 10 **7.** 20.8 **9.** 82 **11.** 324 **13.** 131.8 **15.** 31 miles per gallon

Page 225 **1.** 18, 24, 28, 36, 40 **3.** 4666, 4674, 4682 **5.** 51; 17 **7.** 409; 104 **9.** 385; 456 **11.** 7000; 7000
13. $75; $28 **15.** $420; $340 **17.** 255 lb; 255 lb

Pages 226, 227 **1.** $760 **3.** $550 **5.** $4200 **7.** $760 **9.** $132 **11.** $1200 **13.** 20¢/mi

Page 228   Consumer Note **1.** 5.5¢ per oz, 5.3¢ per oz; 53¢ for 10 oz **3.** 5.4¢ per oz, 5.2¢ per oz; $1.82 for 35 oz

Page 229 **1.** yes **3.** no **5.** no **7.** no **9.** yes **11.** $b = 3$ **13.** $m = 20$ **15.** $m = 83\frac{3}{4}$ **17.** $t = 2\frac{2}{5}$ **19.** $\frac{6}{40}$

21. $\frac{240}{4}$ **23.** $\frac{64}{40}$ **25.** $\frac{1470}{3} = \frac{t}{40}$, $t = 19{,}600$ **27.** $\frac{w}{5} = \frac{696}{6}$, $w = 580$ **29.** 240 mi/min
31. 65 windows/building **33.** A

Page 230 **1.** $\frac{1}{2}$ **3.** $\frac{1}{2}$ **5.** $\frac{1}{4}$ **7.** $\frac{1}{8}$ **9.** 5.5 **11.** 312.5 **13.** 45; 40; 65

UNIT 8

Page 233 **1.** 0.83 **2.** 0.95 **3.** $\frac{3}{4}$ **4.** $\frac{21}{25}$ **5.** 3.83 **6.** 7.38 **7.** $2\frac{3}{5}$ **8.** $7\frac{11}{20}$ **9.** 60 **10.** 35 **11.** 8.6 **12.** 37.5
13. 0.42 **14.** 0.038 **15.** $\frac{3}{5}$ **16.** $\frac{18}{25}$ **17.** 1984 **18.** 5000 **19.** salary **20.** research

Page 234 **1.** 0.58 **3.** 0.53 **5.** 0.01 **7.** 0.84 **9.** 0.5 **11.** 0.75 **13.** 0.4 **15.** 0.8 **17.** 0.83 **19.** 0.3 **21.** 0.67
23. 0.15 **25.** 0.65 **27.** 0.08 **29.** 0.28 **31.** 0.01 **33.** 0.14 **35.** 0.44 **37.** 0.19 **39.** 0.125 **41.** 0.625
43. 0.233 **45.** 0.175 **47.** 0.063 **49.** 0.611

Page 235 **1.** $\frac{1}{5}$ **3.** $\frac{1}{2}$ **5.** 1 **7.** $\frac{1}{5}$ **9.** $\frac{9}{20}$ **11.** $\frac{3}{4}$ **13.** $\frac{21}{250}$ **15.** $\frac{12}{125}$ **17.** $\frac{13}{40}$ **19.** $\frac{1}{10}$ **21.** $\frac{1}{2}$ **23.** $\frac{9}{10}$
25. $\frac{2}{25}$ **27.** $\frac{1}{20}$ **29.** $\frac{1}{4}$ **31.** $\frac{12}{25}$ **33.** $\frac{3}{4}$ **35.** $\frac{43}{50}$ **37.** $\frac{6}{25}$ **39.** $\frac{1}{8}$ **41.** $\frac{11}{40}$ **43.** $\frac{7}{8}$ **45.** $\frac{9}{40}$ **47.** $\frac{3}{8}$ **49.** $\frac{1}{40}$
51. $\frac{29}{500}$ **53.** $\frac{13}{125}$ **55.** $\frac{81}{125}$ **57.** $\frac{18}{125}$

Page 236 **1.** $\frac{9}{4}$ **3.** $\frac{51}{8}$ **5.** $\frac{63}{5}$ **7.** 3.25 **9.** 7.13 **11.** 1.22 **13.** 5.8 **15.** 3.6 **17.** 12.75 **19.** 16.63 **21.** 13.38
23. 24.83 **25.** 28.78 **27.** 3.44 **29.** 8.73 **31.** 7.556 **33.** 9.462 **35.** 6.583 **37.** 5.444 **39.** 7.225

Page 237 **1.** $8\frac{1}{4}$ **3.** $4\frac{2}{3}$ **5.** $9\frac{1}{3}$ **7.** $16\frac{1}{3}$ **9.** $20\frac{2}{3}$ **11.** $5\frac{1}{4}$ **13.** $8\frac{11}{25}$ **15.** $15\frac{1}{4}$ **17.** $\frac{9}{20}$ **19.** $\frac{3}{5}$ **21.** $6\frac{1}{5}$
23. $7\frac{2}{5}$ **25.** $7\frac{1}{5}$ **27.** $1\frac{31}{50}$ **29.** $4\frac{1}{50}$ **31.** $7\frac{1}{25}$ **33.** $40\frac{1}{4}$ **35.** $26\frac{4}{25}$ **37.** $4\frac{3}{5}$ **39.** $8\frac{2}{5}$ **41.** $5\frac{1}{25}$ **43.** $9\frac{5}{8}$
45. $8\frac{7}{40}$ **47.** $42\frac{107}{250}$ **49.** $67\frac{41}{125}$ **51.** $174\frac{32}{125}$ **53.** $456\frac{33}{40}$

Pages 238, 239 **1.** college **3.** 45 **5.** about 220 **7.** 5'3"–5'5", 5'6"–5'8", 5'9"–5'11" **9.** December
11. 660 kW·h **13.** April, May, September, November, December **15.** Alaska **17.** 14–17

Pages 240, 241 **1.** 183 **3.** $.27 **5.** 2424 **7.** 156 **9.** $30.39 **11.** $77.03

Page 242 *Calculator Displays* **1.** 0.777... **3.** 0.232323...

Page 243 **1.** 305.31, 299.99, 349.99 **3.** $291.84, $257.67, $204.72, $304.72, $346.47

Page 244 **1.** 20 **3.** 4 **5.** 10 **7.** 21 **9.** 30 **11.** 20 **13.** $\frac{25}{100}$ **15.** $\frac{20}{100}$ **17.** $\frac{5}{100}$ **19.** $\frac{80}{100}$ **21.** $\frac{30}{100}$
23. $\frac{15}{100}$ **25.** $\frac{6}{100}$ **27.** $\frac{18}{100}$ **29.** $\frac{48}{100}$ **31.** $\frac{68}{100}$ **33.** $\frac{64}{100}$ **35.** $\frac{58}{100}$ **37.** $\frac{52}{100}$ **39.** $\frac{192}{100}$ **41.** $\frac{158}{100}$ **43.** $\frac{300}{100}$
45. $\frac{128}{100}$ **47.** $\frac{130}{100}$

Page 245 **1.** $\frac{50}{100}$ **3.** $\frac{20}{100}$ **5.** $\frac{10}{100}$ **7.** $\frac{55}{100}$ **9.** $\frac{120}{100}$ **11.** $\frac{300}{100}$ **13.** $\frac{74}{100}$ **15.** $\frac{84}{100}$ **17.** 8% **19.** 3%
21. 10% **23.** 12% **25.** 19% **27.** 45% **29.** 76% **31.** 40% **33.** 80% **35.** 36% **37.** 70% **39.** 85% **41.** 150%
43. 375% **45.** 22% **47.** 120% **49.** 160% **51.** 75% **53.** 175% **55.** 98%

Page 246 **1.** $\frac{23}{100}$ **3.** $\frac{237}{100}$ **5.** $\frac{600}{100}$ **7.** 3% **9.** 9% **11.** 17% **13.** 33% **15.** 83% **17.** 49% **19.** 64%
21. 90% **23.** 470% **25.** 790% **27.** 508% **29.** 909% **31.** 849% **33.** 700% **35.** 4.5% **37.** 2.3% **39.** 42.5%
41. 66.3% **43.** 0.8% **45.** 0.9% **47.** 629.4% **49.** 20.9% **51.** 0.29% **53.** 235.4%

Page 247 **1.** 36% **3.** 93% **5.** 0.8 **7.** 0.583 **9.** 12.5% **11.** 20% **13.** 80% **15.** 87.5% **17.** 30% **19.** 48%
21. 70% **23.** 55% **25.** 325% **27.** 612.5% **29.** 262.5% **31.** 890% **33.** 1040% **35.** 675%

Pages 248, 249 **1.** $125,000 **3.** $75,000 **5.** 6th **7.** 8th **9.** 17% **11.** Dec-Jan **13.** Dec, Apr, May **15.** 1960
17. 1910 and 1930, 1940 and 1950 **19.** 158 cm **21.** Jane **23.** between 15 and 16

Pages 250, 251 **1.** haddock, chicken, tuna, egg, hamburger **3.** 215 **5.** 1161 **7.** 3.32 **9. a.** 62 **b.** 59 **c.** 121
d. 530

Page 252 *Extra!* **1.** News Review; Teen Times   *Page 253* **1.** 35 units/mL **3.** 15 units/mL **5.** 3 hours

Page 254 **1.** 5% **3.** 9% **5.** 17% **7.** 65% **9.** 25.8% **11.** 238% **13.** 0.04 **15.** 0.03 **17.** 0.02 **19.** 0.09
21. 0.11 **23.** 0.16 **25.** 0.23 **27.** 0.56 **29.** 1.00 **31.** 7.00 **33.** 1.80 **35.** 1.95 **37.** 4.32 **39.** 3.06 **41.** 0.005
43. 0.002 **45.** 0.062 **47.** 0.076 **49.** 0.023 **51.** 0.0478 **53.** 0.789 **55.** 0.2356 **57.** 0.313 **59.** 0.502
61. 0.00403 **63.** 0.00374 **65.** 0.00145 **67.** 0.02672

Page 255 **1.** 0.12 **3.** 0.372 **5.** 0.0015 **7.** 0.049 **9.** 2.5 **11.** $\frac{4}{25}$ **13.** $2\frac{7}{50}$ **15.** $\frac{1}{8}$ **17.** $\frac{103}{250}$ **19.** $1\frac{3}{4}$
21. $\frac{3}{50}$ **23.** $\frac{3}{100}$ **25.** $\frac{9}{100}$ **27.** $\frac{3}{20}$ **29.** $\frac{6}{25}$ **31.** $\frac{1}{4}$ **33.** $\frac{14}{25}$ **35.** $\frac{11}{20}$ **37.** $\frac{1}{2}$ **39.** $\frac{4}{5}$ **41.** $\frac{16}{25}$ **43.** $\frac{21}{25}$
45. $\frac{9}{25}$ **47.** $\frac{1}{8}$ **49.** $1\frac{4}{5}$ **51.** $\frac{3}{200}$ **53.** $2\frac{9}{10}$ **55.** $8\frac{3}{20}$ **57.** $5\frac{27}{50}$ **59.** 8

Pages 256, 257 **1.** clothing **3.** car, other **5.** 63% **7.** inventory **9.** 33¢ **11.** meat **13.** meat, vegetables
15. 67% **17.** sports **19.** meals, television

Pages 258, 259 **1.** $32.94 **3.** $69.55 **5.** $2.98 **7.** 2 **9.** $22\frac{1}{2}$ **11.** $370.95

Page 260 *Consumer Note* **1.** housing **3.** transportation **5.** food, housing, transportation **7.** $16,500

Page 261 **1.** 0.44 **3.** 0.78 **5.** 0.71 **7.** 0.625 **9.** 0.667 **11.** $\frac{3}{4}$ **13.** $\frac{1}{5}$ **15.** $\frac{1}{8}$ **17.** 2.125 **19.** 6.889 **21.** 9.4
23. 12.75 **25.** $3\frac{1}{25}$ **27.** $11\frac{7}{40}$ **29.** 26% **31.** 437% **33.** 83% **35.** 61% **37.** 45% **39.** 32.5% **41.** 775%
43. 437.5% **45.** 0.059 **47.** 1.65 **49.** 0.075 **51.** 4.08 **53.** $\frac{1}{20}$ **55.** $\frac{3}{4}$ **57.** $2\frac{2}{5}$ **59.** $6\frac{7}{50}$

Page 262 **1.** 200 **3.** June, November; April, December **5.** 3% **7.** 4:00 **9.** 10%

Page 265 **1.** $16\% \times 75 = 12$ **2.** $35\% \times 92 = 32.2$ **3.** 6%, 82, 4.92 **4.** 21 **5.** 70 **6.** 25% **7.** 85%
8. $\frac{75}{100} = \frac{30}{40}$ **9.** $\frac{N}{100} = \frac{22}{83}$ **10.** \$3.58 **11.** 150% **12.** \$6268.20

Page 266 **1.** 24 **3.** 30 **5.** $46\% \times 85 = 39.1$ **7.** $13\% \times 25 = 3.25$ **9.** $30 = 50\% \times 60$ **11.** $24 = 30\% \times 80$
13. $36 = N\% \times 67$ **15.** $N = 10\% \times 99$ **17.** $N\% \times 69 = 31$ **19.** $N = 55\% \times 148$ **21.** $16.5\% \times 18 = 2.97$
23. $48\% \times 79 = N$ **25.** $N = 150\% \times 88$ **27.** $N = 15\% \times 90$

Page 267 **1.** $10\% \times 73 = 7.3$ **3.** $N = 25\% \times 152$ **5.** 60% **7.** N% **9.** 55 **11.** 304 **13.** 32.8 **15.** N
17. 3%, 94, 2.82

Pages 268, 269 **1.** $\frac{75}{100}$ **3.** $\frac{86}{100}$ **5.** $\frac{99}{100}$ **7.** 9 **9.** $4\frac{19}{20}$ **11.** 1 **13.** 12 **15.** 42 **17.** 2 **19.** 123 **21.** 72
23. 15 **25.** 39 **27.** 0.23 **29.** 0.07 **31.** 3.5 **33.** 8.75 **35.** 208.5 **37.** 22.2 **39.** 7.5 **41.** 44.1 **43.** 7.83
45. 12.18 **47.** 30 **49.** 112 **51.** 187 **53.** 57 **55.** 39 **57.** 208 **59.** 90.5 **61.** 3920 **63.** 1005

Pages 270, 271 **1.** 0.12 **3.** 0.075 **5.** 0.75 **7.** 0.006 **9.** \$10.56 **11.** \$53.37 **13.** \$18.90 **15.** \$170.37
17. \$43.75, \$668.75 **19.** \$13.32, \$203.57 **21.** \$575.81, \$8801.65 **23. a.** \$1400 **b.** \$11,400 **25. a.** \$27.21
b. \$1387.91

Pages 272, 273 **1.** \$6169 **3.** \$377.64 **5.** \$74.028 **7.** \$22.20 **9.** \$79 **11.** \$413.32 **13.** \$22.20 **15.** \$892.16

Page 274 Consumer Note **1.** \$5925

Pages 276, 277 **1.** $\frac{12}{25}$ **3.** $\frac{1}{2}$ **5.** $\frac{19}{20}$ **7.** $\frac{1}{2}$ **9.** 80 **11.** 3000 **13.** 240 **15.** 80 **17.** 1400 **19.** 5400 **21.** $33\frac{1}{3}$
23. $\frac{1}{4}$ **25.** 45 **27.** 0.65 **29.** 0.6 **31.** 65.6 **33.** 80 **35.** 600 **37.** 4260 **39.** 692 **41.** 1636 **43.** 385 **45.** 620
47. 33.3 **49.** 3520 **51.** 1500 **53.** 15.2 **55.** 138 **57.** 70

Pages 278, 279 **1.** 0.27 **3.** 0.83 **5.** 1.41 **7.** 0.80 **9.** 40% **11.** 6% **13.** 107% **15.** 55% **17.** 9.6% **19.** 50%
21. 25% **23.** 75% **25.** 25% **27.** 50% **29.** 37.5% **31.** 62.5% **33.** 12.5% **35.** 37.5% **37.** 62.5% **39.** 0.5%
41. 0.25% **43.** 0.8% **45.** 0.4% **47.** 500% **49.** 300% **51.** 500% **53.** 250% **55.** 37.5% **57.** 18% **59.** 75%

Pages 280, 281 **1.** 0.4 **3.** 0.8 **5.** 56% **7.** 27% **9.** 120% **11.** 21%

Pages 282, 283 **1.** \$27,000 **3.** \$1998 **5.** 90 **7.** \$195 **9.** \$28.50

Page 284 Extra! **1.** 920 **3.** 13.5% *Page 285* **1.** 14.5% **3.** 13.5%

Pages 286, 287 **1.** $\frac{N}{100}$; 4; 9 **3.** $\frac{N}{100} = \frac{9}{72}$ **5.** $\frac{N}{100} = \frac{51}{93}$ **7.** $\frac{N}{100} = \frac{13}{65}$ **9.** $\frac{N}{100} = \frac{17}{221}$ **11.** $\frac{N}{100} = \frac{180}{276}$
13. 25% **15.** 40% **17.** 75% **19.** 10% **21.** 38% **23.** $\frac{42}{100}$ **25.** $\frac{163}{100}$ **27.** $\frac{17}{100}$ **29.** $\frac{32}{100}$; N; 90
31. $\frac{43}{100}$; 77; N **33.** $\frac{24}{100} = \frac{80}{N}$ **35.** $\frac{42}{100} = \frac{N}{730}$ **37.** $\frac{60}{100} = \frac{N}{200}$ **39.** 54 **41.** 350 **43.** 70 **45.** 132

Pages 288, 289 **1. a.** \$13.25 **b.** \$39.75 **3. a.** \$61.50 **b.** \$92.25 **c.** \$30.75 **5.** \$28.25 **7. a.** \$17 **b.** 10%

Pages 290, 291 **1.** \$8.00 **3.** \$198.00 **5.** \$73.00 **7.** \$585

Page 292 Calculator Displays **1.** soil **3.** boggle **5.** bog **7.** sizzles

Page 293 **1.** 7%; 80; 5.6 **3.** 50%; 148; 74 **5.** 12 **7.** $25\frac{9}{10}$ **9.** 47.5 **11.** 88.9 **13.** 50 **15.** 160 **17.** 20%
19. 375% **21.** $\frac{N}{100} = \frac{11}{55}$ **23.** $\frac{62}{100} = \frac{60}{N}$ **25.** 28 **27.** 72

Page 294 **1.** \$5.82 **3.** \$451.50 **5. a.** 233% **b.** 338% **c.** 228% **7.** 30%

Cumulative Review: Units 7–9

Pages 296, 297 Skills **1.** $\frac{5}{8}$ **3.** $\frac{50}{365}$ **5.** no **7.** yes **9.** a = 15 **11.** f = 6 **13.** $\frac{10}{2} = \frac{n}{4}$, n = 20

15. 52 mi/h **17.** 0.08 **19.** 0.42 **21.** 0.075 **23.** 0.889 **25.** $\frac{9}{25}$ **27.** $\frac{3}{40}$ **29.** 2.56 **31.** 2.833 **33.** $9\frac{1}{50}$
35. $10\frac{1}{200}$ **37.** 24 **39.** 17 **41.** 210 **43.** 51% **45.** 0.2% **47.** 85% **49.** 0.13 **51.** $\frac{9}{20}$ **53.** $8\frac{1}{2}$
55. N = 17% × 73 **57.** 60%; 105; 63 **59.** 46 **61.** 750 **63.** 48% **65.** $\frac{N}{100} = \frac{17}{55}$

Pages 298, 299 Applications **1. a.** $\frac{2}{3}$ **b.** $\frac{1}{3}$ **3.** $175.34 **5.** 51, 22 **7.** Puma **9.** about 18
11. 1980–81, 1983–84 **13.** 1980, 1982, 1983 **15.** lifeguards **17.** maintenance **19.** $6.57 **21.** 75%
23. a. $31.99 **b.** $127.96

UNIT 10

Page 301 **1.** ⁻235 **2.** 14 **3.** > **4.** < **5.** ⁻14 **6.** ⁻25 **7.** 11 **8.** ⁻4 **9.** ⁻4 **10.** ⁻13 **11.** 11 **12.** ⁻3
13. 21 **14.** 12 **15.** ⁻108 **16.** ⁻8 **17.** 34 **18.** 7 **19.** 20 **20.** ⁻30 + 14; 16 ft deep **21.** ⁻25 − ⁻42; $17
22. 16

Page 302 **1.** ⁺10 **3.** ⁻60 **5.** ⁺30 **7.** ⁺20 **9.** ⁺50 **11.** ⁻15 **13.** ⁻150 **15.** ⁻500

Page 303 **1.** > **3.** < **5.** ⁻4 **7.** ⁻12 **9.** < **11.** < **13.** < **15.** > **17.** 1 **19.** 2 **21.** ⁻11

Page 304 **1.** 5 **3.** 5 **5.** 2 **7.** ⁻4 **9.** ⁻5 **11.** ⁻3 **13.** 10 **15.** ⁻17 **17.** 16 **19.** 13 **21.** 16 **23.** ⁻18
25. ⁻18 **27.** ⁻14

Page 305 **1.** 1 **3.** 1 **5.** ⁻4 **7.** 4 **9.** 4 **11.** 0 **13.** ⁻4 **15.** ⁻5 **17.** 5 **19.** 8 **21.** ⁻10 **23.** 7 **25.** ⁻2 **27.** 6

Pages 306, 307 **1.** ⁻4 **3.** ⁻8 **5.** ⁻25 **7.** ⁻14 **9.** 65 **11.** 8000 **13.** ⁻6
15. ⁻1300 + 1500; 200 ft above sea level **17.** 238 + ⁻25 + ⁻42 + 36; $207 **19.** 7400 + 2200 + ⁻3600; 6000 ft
21. 50,000 + ⁻125,000 + ⁻35,000 + 85,000; loss of $25,000

Pages 308, 309 **1.** 12 ft **3.** 2:15 **5.** 9:02 A.M. **7.** 18 ft **9.** $1\frac{1}{2}$, 15 **11.** 1, $12\frac{1}{2}$ **13.** 12:02 P.M.

Page 310 Calculator Displays **1.** ⁻37 **3.** ⁻1030 **5.** ⁻434

Page 311 **1.** $13,480; $508.80 **3.** $18,715; $822.90 **5. a.** $17,495 **b.** $749.70 **c.** $524.85

Page 312 **1.** ⁻4 **3.** ⁻6 **5.** ⁻13 **7.** ⁻17 **9.** ⁻6 **11.** 5 **13.** ⁻11 **15.** 18 **17.** ⁻21 **19.** ⁻6 **21.** ⁻2 **23.** ⁻16
25. ⁻3 **27.** ⁻20 **29.** ⁻13 **31.** ⁻8 **33.** ⁻9 **35.** ⁻10 **37.** ⁻4 **39.** ⁻8 **41.** ⁻30

Page 313 **1.** 2 **3.** 5 **5.** 15 **7.** 36 **9.** 3 **11.** 11 **13.** 11 **15.** ⁻3 **17.** 14 **19.** 15 **21.** 19 **23.** 10 **25.** 8
27. ⁻6 **29.** 17 **31.** ⁻8 **33.** ⁻7 **35.** ⁻15 **37.** 7 **39.** ⁻7 **41.** ⁻7 **43.** 4

Page 314 **1.** 63 **3.** 60 **5.** 7 **7.** 40 **9.** 40 **11.** 42 **13.** 56 **15.** 20 **17.** 91 **19.** 8 **21.** 4 **23.** 7 **25.** 2
27. 8 **29.** 16

Page 315 **1.** ⁻12 **3.** ⁻12 **5.** ⁻24 **7.** ⁻21 **9.** ⁻36 **11.** ⁻56 **13.** ⁻48 **15.** ⁻63 **17.** ⁻105 **19.** ⁻161 **21.** ⁻280
23. ⁻8 **25.** ⁻7 **27.** ⁻10 **29.** ⁻9 **31.** ⁻15 **33.** ⁻26 **35.** ⁻25

Pages 316, 317 **1.** 7 **3.** ⁻2 **5.** 29 **7.** 328 **9.** ⁺5 **11.** ⁻5,000,000 **13.** ⁺5 **15.** 12 − ⁻3; $15 million increase
17. 32 − ⁻175; 207° increase **19.** ⁻73 − ⁻30; 43 m lower **21.** ⁻1 − 4; dropped 5°

Pages 318, 319 **1.** 15.6 yr **3.** 26.5 yr **5.** 8.8 **7.** 4.1 **9.** 0.98 **11.** 0.95 **13.** 0.68 **15.** 1920–1940 **17.** 56%
19. 70

Page 320 Calculator Displays **1.** $2.00 **3.** $14.40   *Page 321* **1.** Yes

Page 322 **1.** 13 **3.** 2 **5.** 9 **7.** 12 **9.** 9 **11.** 2 **13.** 8 **15.** 17

Page 323 **1.** 4 **3.** 5 **5.** 4 **7.** 10 **9.** 0 **11.** 3 **13.** ⁻16 **15.** ⁻23 **17.** 80 **19.** 7 **21.** ⁻17 **23.** 0 **25.** 5
27. 12

Page 324 **1.** 10:00 A.M. **3.** 6 h   *Page 325* **1.** 40°F **3.** 29°

Pages 326, 327 **1.** 520 mi **3.** $317 **5.** $5 **7.** 20¢ **9.** $208 **11.** $504 **13.** one-way

Page 328 Computer Bits $14.80; $18.20; $27.37; $7.86; $26.20

Page 329 **1.** ⁻12 **3.** ⁺10 **5.** ⁺109 **7.** < **9.** < **11.** ⁻4 **13.** 26 **15.** 14 **17.** ⁻17 **19.** ⁻19 **21.** 3 **23.** 2 **25.** 10 **27.** 7 **29.** ⁻3 **31.** ⁻16 **33.** 3 **35.** ⁻4 **37.** 2 **39.** 48 **41.** 2 **43.** ⁻60 **45.** ⁻7 **47.** 60 **49.** 12 **51.** 2

Page 330 **1.** 30 + 15 + ⁻10; 35 ft **3.** ⁻100 − 300; 400 ft down **5.** ⁻1000 − 700; $1700 down **7.** ⁻10°

UNIT 11

Page 333 **1.** add 3 to c **2.** subtract 7 from x **3.** multiply b by 5 **4.** multiply 11 by 4 **5.** 0 **6.** 36 **7.** $9n$ **8.** 3 **9.** ⁻$2c$ **10.** $5n − 4$ **11.** 3^3 **12.** r^3s **13.** 5 **14.** 20 **15.** 2 **16.** ⁻5 **17.** 1, 2, 3 **18.** ⁻1, 0 **19.** 3 **20.** $3x$ **21.** $16 + n$ **22.** $P = 2a + b + c$ **23.** b

Page 334 **1.** 1, y, 3 **3.** subtract a from 16 **5.** add w to 2 **7.** subtract 4 from z **9.** divide 12 by s **11.** add 4 to d, multiply by 5 **13.** add f to m, divide by 4 **15.** add 6 to w, subtract from 18 **17.** add 9 to k, divide by 7, multiply by 2

Page 335 **1.** y **3.** d; m **5.** multiply n by 9 **7.** multiply n by m **9.** add 1 to x, multiply by 4 **11.** add b to a, multiply by 3 **13.** multiply m by 2, subtract 3, multiply by 16

Pages 336, 337 **1.** ⁻2 **3.** 3 **5.** ⁻49 **7.** 4 **9.** 13 **11.** 13 **13.** 10 **15.** 6 **17.** 1 **19.** 9 **21.** 5 **23.** 6 **25.** 1 **27.** 27 **29.** 3 **31.** 2 **33.** 0 **35.** 6 **37.** 4 **39.** 30 **41.** 8 **43.** ⁻2 **45.** 27 **47.** 17 **49.** 11 **51.** 3 **53.** 13 **55.** ⁻4 **57.** 5 **59.** ⁻4 **61.** ⁻5 **63.** 24 **65.** ⁻8 **67.** 7 **69.** 6 **71.** 28 **73.** ⁻18 **75.** 17

Pages 338, 339 **1.** add x and y **3.** twice y **5.** twice x, plus y **7.** twice a, plus 1 **9.** $x + 3$ **11.** $a − 4$ **13.** $2c$ **15.** $c + 2a$ **17.** $t − 9$ **19.** $w + 7$ **21.** $10 + x$ **23.** $4 + y$ **25.** $n + 5$ **27.** $x − 10$ **29.** $n − 9$ **31.** $10x$ **33.** $2n$ **35.** $x + 2y$ **37.** $\frac{m}{b}$ **39.** $2x + 2$ **41.** $2xy$ **43.** $3y$ **45.** $15 − b$ **47.** $3 + y$

Pages 340, 341 **1.** 400 cal **3.** 20 cal **5.** 400 cal **7.** 540,000 cal **9.** 500,000 cal **11.** 3,600,000 cal

Page 342 Extra! **1.** 1 cube 3 sides painted, 4 cubes 4 sides painted, 1 cube 5 sides painted **3.** 2 cubes 3 sides painted, 4 cubes 4 sides painted

Page 343 **1.** $750, dep.; 20% **3.** $14,040, app.; 30%

Page 344 **1.** ⁻6 **3.** 0 **5.** 0 **7.** 4 **9.** x **11.** $12y$ **13.** 0 **15.** $3x$ **17.** 1 **19.** 0 **21.** $30n$ **23.** 0 **25.** 0 **27.** n

Page 345 **1.** Yes **3.** No **5.** No **7.** No **9.** $6x$ **11.** a **13.** $2x + 2y$ **15.** $12a$ **17.** ⁻$2x$ **19.** m **21.** $5b + 2$ **23.** $15s$ **25.** $8x$ **27.** $21n$ **29.** $6x$ **31.** $5x + y$ **33.** $n + 1$ **35.** 0 **37.** 0 **39.** 0

Page 346 **1.** 2 **3.** 2 **5.** 4 **7.** xy, xy **9.** 5^3 **11.** 2^4 **13.** a^2 **15.** $(ab)^2$ **17.** $(y + 2)^2$ **19.** $3x^2$ **21.** $(x − 5)^2$ **23.** $(rs)^2$ **25.** x^3y^2 **27.** $4n^2$ **29.** $(3xy)^2$ **31.** $5x^2y$ **33.** $7x^2y^2$ **35.** $3x^2y^3$ **37.** $4(a − b)^2$ **39.** $6x^3y$

Page 347 **1.** 8 **3.** 72 **5.** 3 **7.** 52 **9.** 1 **11.** 4 **13.** 36 **15.** ⁻6 **17.** 4 **19.** ⁻7 **21.** 7 **23.** $\frac{1}{2}$ **25.** 1 **27.** 8 **29.** 4 **31.** 100 **33.** 50 **35.** 36 **37.** 16 **39.** 60 **41.** 100 **43.** 100 **45.** 81 **47.** 49

Pages 348, 349 **1.** a **3.** b **5.** $x + y + z$ **7.** $3a$ **9.** $4s$ **11.** $m + n + 2p$ **13.** $2x + 4y$ **15.** $6d$ **17.** a^3 **19.** a^2b **21.** 78 cm²

Page 351 **1.** $4\frac{1}{2}$ **3.** 6 **5.** $2\frac{1}{2}$ **7.** $36 **9.** $36 **11.** $17.50 **13.** $60 **15.** $107.75

Page 353 **7.** $14,300.00 **9c.** $134.65 **11.** $14,673.49 **13.** $2,360.15 **15.** $2,261.00

Page 354 **1.** 5 **3.** 1 **5.** 2 **7.** 40 **9.** Yes **11.** No **13.** Yes **15.** 2 **17.** 3 **19.** 7 **21.** 4 **23.** 2 **25.** 2 **27.** 4 **29.** 15 **31.** 2 **33.** 21 **35.** ⁻2 **37.** ⁻2 **39.** 6 **41.** 5 **43.** 6 **45.** 9

Page 355 **1.** > **3.** < **5.** < **7.** > **9.** 1, 2, 4 **11.** ⁻1 **13.** ⁻1, 0 **15.** ⁻1, 0, 1 **17.** 0, 1, 2, 4 **19.** ⁻1, 0, 1, 2, 4 **21.** 4 **23.** 2, 4 **25.** ⁻1, 0, 1 **27.** 2, 4 **29.** ⁻1, 0, 1, 2, 4 **31.** ⁻1, 0 **33.** none **35.** 1, 2, 4 **37.** ⁻1, 0, 1, 2, 4 **39.** 1, 2, 4

Pages 356, 357 **1.** $x + 3$ **3.** $3x$ **5.** a **7.** b **9.** b **11.** c **13.** d **15.** a **17.** d

14

Pages 358, 359 **1.** $18,000 **3.** $276.70 **5.** $6640.80 **7.** $840 **9.** $1680 **11.** 34%

Page 360 Consumer Note $1125

Page 361 **1.** subtract *b* from 20 **3.** add *m* to 6 **5.** subtract 2 from *f*, divide by 3 **7.** 15 **9.** 17 **11.** $^-$3 **13.** 2 **15.** 9 **17.** 1 **19.** 13 **21.** *n* **23.** 11*x* **25.** $^-$4*y* **27.** 8*x* + 6 **29.** 6^3 **31.** a^2b **33.** $7a^2b$ **35.** 8 **37.** 16 **39.** $^-$32 **41.** 108 **43.** 3 **45.** 5 **47.** 1 **49.** 26 **51.** 3, 4 **53.** 3, 4

Page 362 **1.** 9 + *t* **3.** 3 + 2*n* **5.** *m* × 16 **7.** 2*a* + *b* + *c* **9.** 2*m* + 2*n* **11.** e^2f **13.** d **15.** b

UNIT 12

Page 365 **1.** 19 **2.** 38 **3.** 9 **4.** 10 **5.** 7 **6.** 8 **7.** 40 **8.** $^-$48 **9.** 35 **10.** 10 **11.** yes **12.** no **13.** 3 **14.** 4 **15.** 5 **16.** 8 **18.** (1,1) **19.** ($^-$1, $^-$2) **20.** ($^-$2, 0) **21.** ($^-$2, 0); $^-$2, (0, $^-$2); $^-$4, (2, $^-$4)

Pages 366, 367 **1.** 1 **3.** 15 **5.** 16 **7.** $^-$14 **9.** 12 **11.** 18 **13.** 45 **15.** 1 **17.** 41 **19.** 37 **21.** 40 **23.** 10 **25.** 18 **27.** 1 **29.** 4 **31.** 9 **33.** 7 **35.** 23 **37.** 27 **39.** 32 **41.** 30 **43.** 12 **45.** 7 **47.** 8 **49.** 6 **51.** 10 **53.** 7 **55.** 13 **57.** 16 **59.** 18 **61.** 8 **63.** 15

Pages 368, 369 **1.** 3 **3.** 2 **5.** 5 **7.** 6 **9.** 4 **11.** 5 **13.** 5 **15.** 8 **17.** 2 **19.** 0 **21.** 8 **23.** $^-$1 **25.** $^-$1 **27.** $^-$7 **29.** 43 **31.** 10 **33.** 0 **35.** 0 **37.** $^-$21 **39.** $^-$2 **41.** $^-$5 **43.** 0 **45.** 4 **47.** 5 **49.** $^-$18 **51.** 8 **53.** $^-$3 **55.** 0 **57.** 1 **59.** 4 **61.** 7 **63.** 8 **65.** $^-$6 **67.** 6 **69.** 8 **71.** $^-$9 **73.** 0 **75.** 9 **77.** 0 **79.** $^-$6 **81.** 15 **83.** 12 **85.** 13 **87.** 4 **89.** $^-$12

Page 370 **9.** 1 **11.** 5 **13.** 6 **15.** $^-$1 **17.** $^-$2 **19.** 4 **21.** 1 **23.** 0 **25.** $^-$6 **27.** $^-$1 **29.** $^-$4 **31.** $^-$2 **33.** 6 **35.** 7 **37.** 9 **39.** 3 **41.** 6 **43.** 10

Page 373 **1.** 0.6 mi **3.** 5 mi **5.** 2250 ft **7.** Eagle Lakes Trail, 5.6 **9.** Sawmill Meadows Trail, 5.5 **11.** Canby Meadow, 5 **13.** 43.5 mi

Page 374 Extra! **1.** $\left[\dfrac{(n + 10)2}{4} - 5\right]2$; The answer is always the number picked.

Page 375 **1.** $38,800

Page 376 **1.** 1 **3.** *x* **5.** *x* **7.** *y* **9.** 3 **11.** 4 **13.** 9 **15.** 11 **17.** 13 **19.** 25 **21.** $^-$2 **23.** 5 **25.** $^-$7 **27.** $^-$5 **29.** 8 **31.** $^-$4 **33.** 4 **35.** 20

Page 377 **1.** 1 **3.** 1 **5.** 2 **7.** *x* **9.** 9 **11.** $^-$2 **13.** 12 **15.** 64 **17.** $^-$30 **19.** 56 **21.** 0 **23.** 12 **25.** 100 **27.** 36 **29.** 121 **31.** 1000 **33.** 80 **35.** 108

Page 378 **1.** subt. **3.** subt. **5.** div. **7.** 12 **9.** 10 **11.** 8 **13.** 28 **15.** 245 **17.** 450 **19.** 0 **21.** $^-$4 **23.** 49 **25.** $^-$2 **27.** 112 **29.** 99 **31.** $^-$4 **33.** 15 **35.** $10\frac{1}{2}$

Page 379 **1.** 22 **3.** 10 **5.** 3 **7.** $^-$3 **9.** $^-$4 **11.** 6 **13.** $^-$2 **15.** 1 **17.** 2 **19.** 5 **21.** 2 **23.** 3 **25.** 5 **27.** 2 **29.** 1 **31.** $^-$3 **33.** 4 **35.** $^-$4 **37.** 7 **39.** 12

Pages 380, 381 **1.** 3 **3.** 4; 4; 4, 4 **5.** 3, down 3; $^-$3 **7.** 2, 2 down; 2, $^-$2 **9.** 4 down; 0, $^-$4 **11.** 2 **13.** 2, $^-$3 **15.** F **17.** M **19.** J **21.** H **23.** K **25.** R **35.** a rectangle

Pages 382, 383 **1.** 2.9, 3.7 **3.** 2.0, 2.6 **5.** 1.9 **7.** 7 **9.** 14 **11.** 20.3 **13.** 15.4 **15.** 21.7 **17.** 35,280 **19.** 6.7

Page 384 Extra! **1.** Across the rose bush. **3.** Twenty paces north of the apple tree.

Page 385 **1.** $3055

Page 386 **1.** True **3.** False **5.** yes **7.** yes **9.** yes **11.** yes **13.** yes **15.** yes **17.** yes **19.** yes

Page 387 **1.** 4 **3.** 4 **5.** 5 **7.** $^-$1 **9.** 4 **11.** 3 **13.** 0 **15.** 10 **17.** 0 **19.** 0 **21.** 7 **23.** 7 **25.** $^-$2 **27.** 5 **29.** 8 **31.** 9 **33.** 4 **35.** 3 **37.** $^-$2 **39.** 21 **41.** $^-$8

Pages 388, 389 **1.** 2 **3.** 8 **5.** 1 **7.** 5 **9.** 4 **11.** 2 **13.** No **15.** No **17.** Yes **19.** ($^-$1, 3); 4, (0, 4); 5, (1, 5); 6, (2, 6) **21.** ($^-$2, $^-$4); $^-$2, ($^-$1, $^-$2); 0, (0, 0); 2, (1, 2); 4, (2, 4) **23.** ($^-$1, 6); 4, (0, 4); 2, (1, 2); 0, (2, 0); $^-$2, (3, $^-$2)

Pages 390, 391 **1.** 1976, 1981–1984 **3.** $24,589 **5.** 24 **7.** 40% **9.** $8351

Page 392 *Consumer Note* **1.** $1419 **3.** $2600

Page 393 **1.** 14 **3.** 19 **5.** 17 **7.** 9 **9.** ⁻9 **11.** ⁻9 **13.** 3 **15.** ⁻7 **17.** 0 **19.** ⁻18 **21.** ⁻20 **23.** 0 **25.** 14 **27.** ⁻6 **29.** ⁻5 **31.** 3 **33.** 8 **35.** 1 **37.** yes **39.** no **41.** no **43.** 7 **45.** ⁻1 **47.** 0

Page 394 **1.** 7 **3.** ⁻5 **9.** 0 **11.** ⁻2, ⁻4 **17.** (⁻2, 9); 5, (0, 5); ⁻1, (3, ⁻1)

Pages 396, 397, 398 *Skills* **1.** ⁺325 **3.** > **5.** > **7.** ⁻5 **9.** ⁻27 **11.** 8 **13.** 8 **15.** ⁻1 **17.** ⁻2 **19.** 14 **21.** 12 **23.** 12 **25.** 27 **27.** ⁻32 **29.** ⁻6 **31.** 33 **33.** ⁻9 **35.** 24 **37.** subt. *b* from 22 **39.** add *d* to *c*, mult. by 2 **41.** 9 times 6 **43.** 0 **45.** ⁻2 **47.** *x* **49.** *c* **51.** 11*n* **53.** 5*y* − 5*x* **55.** a^4 **57.** $(a − b)^2$ **59.** ⁻9 **61.** ⁻24 **63.** 9 **65.** 10 **67.** 1, 2, 4 **69.** 4 **71.** 1 **73.** 6 **75.** ⁻22 **77.** ⁻3 **79.** 35 **81.** ⁻26 **83.** 2 **85.** no **87.** ⁻3 **89.** ⁻2

Pages 398, 399 *Applications* **1.** 355 + ⁻205 + ⁻125 + 98; $123 **3.** ⁻20 − ⁻5; dropped 15° **5. a.** 1:00 P.M. **b.** 12 noon **c.** 3:00 P.M. **7.** *x* − 8 **9.** $\frac{m}{n}$ **11.** *x* + 2 **13.** 18 ft² **15.** b **17.** ⁻6 **25.** 6, (0, 6); 7, (1, 7); 8, (2, 8)

SKILL BANK

Page 405 **1.** tens, 40 **3.** units, 0 **5.** hundreds, 400 **7.** hundreds, 0 **9.** hundredths, 0.07 **11.** tenths, 0.9 **13.** thousandths, 0.009 **15.** tens, 70 **17.** hundredths, 0.05 **19.** tenths, 0.5 **21.** thousandths, 0.008 **23.** tenths, 0 **25.** hundreds, 400 **27.** 1500 **29.** 5000 **31.** 6900 **33.** 4400 **35.** 60 **37.** 90 **39.** 380 **41.** 160 **43.** 18,000 **45.** 11,000 **47.** 4000 **49.** 10.4 **51.** 40.0 **53.** 0.5 **55.** 0.1 **57.** 3.5 **59.** 27.35 **61.** 5640.06 **63.** 0.53 **65.** 11.00 **67.** 82.48

Page 406 **1.** 95 **3.** 39 **5.** 88 **7.** 79 **9.** 69 **11.** 79 **13.** 83 **15.** 179 **17.** 288 **19.** 684 **21.** 758 **23.** 889 **25.** 6879 **27.** 4875 **29.** 73 **31.** 110 **33.** 141 **35.** 131 **37.** 717 **39.** 354 **41.** 424 **43.** 701 **45.** 990 **47.** 1265 **49.** 1000 **51.** 1553 **53.** 12,144 **55.** 9381

Page 407 **1.** 82 **3.** 53 **5.** 22 **7.** 60 **9.** 300 **11.** 532 **13.** 605 **15.** 661 **17.** 110 **19.** 5321 **21.** 1002 **23.** 1352 **25.** 5141 **27.** 5312 **29.** 65 **31.** 15 **33.** 246 **35.** 476 **37.** 488 **39.** 215 **41.** 37 **43.** 2491 **45.** 1986 **47.** 3990 **49.** 15,628 **51.** 36,794 **53.** 67,566

Page 408 **1.** 99.32 **3.** 1001.1 **5.** 442.585 **7.** 1037.07 **9.** 8077.353 **11.** 125.161 **13.** 138.826 **15.** $304.25 **17.** $403.70 **19.** 23.09 **21.** 332.2 **23.** 14.08 **25.** 30.476 **27.** 46.264 **29.** 4908.819 **31.** 51.493 **33.** $13.86 **35.** $980.35 **37.** $1877.18

Page 409 **1.** 46 **3.** 93 **5.** 406 **7.** 336 **9.** 2084 **11.** 0 **13.** 3009 **15.** 126 **17.** 78 **19.** 1173 **21.** 386 **23.** 47,034 **25.** 55,496 **27.** 810 **29.** 2880 **31.** 37,520 **33.** 10,800 **35.** 212,400 **37.** 339,500 **39.** 11,690,000 **41.** 73,188,000 **43.** 42,035,000 **45.** 21,476,000 **47.** 39,785,000

Page 410 **1.** 2128 **3.** 2784 **5.** 49,176 **7.** 52,260 **9.** 21,280 **11.** 69,839 **13.** 63,525 **15.** 171,192 **17.** 106,200 **19.** 174,466 **21.** 378,738 **23.** 647,775 **25.** 192,528 **27.** 623,832 **29.** 190,872 **31.** 150,528 **33.** 82,565 **35.** 628,340 **37.** 86,240 **39.** 67,896 **41.** 4,473,656 **43.** 2,317,020 **45.** 5,782,384 **47.** 3,630,060 **49.** 30,425,715 **51.** 20,092,728 **53.** 40,235,256 **55.** 40,681,216

Page 411 **1.** 12,420 **3.** 32,160 **5.** 282,100 **7.** 292,400 **9.** 306,030 **11.** 43,500 **13.** 1,601,340 **15.** 7,734,040 **17.** 852,600 **19.** 3,481,000 **21.** 4,848,606 **23.** 4,100,576 **25.** 13,269,200 **27.** 22,411,000 **29.** 745.86 **31.** 429.562 **33.** 78.4707 **35.** 249.106 **37.** 1592.94 **39.** 117.719 **41.** 48.9676 **43.** 0.0468 **45.** 0.0832 **47.** 0.04008 **49.** 0.18903 **51.** 0.30553321 **53.** 0.000502 **55.** 0.002626

Page 412 **1.** 15 **3.** 15 **5.** 57 **7.** 29 **9.** 152 **11.** 185 **13.** 755 **15.** 698 **17.** 804 **19.** 847 **21.** 628 **23.** 733 **25.** 133 **27.** 545 **29.** 203 **31.** 181 **33.** 9 R3 **35.** 18 R2 **37.** 84 R6 **39.** 94 R4 **41.** 85 R3 **43.** 47 R4 **45.** 468 R4 **47.** 564 R1 **49.** 654 R3 **51.** 847 R4 **53.** 236 R4 **55.** 841 R2 **57.** 343 R1 **59.** 203 R1 **61.** 459 R1

Page 413 **1.** 1 R10 **3.** 1 R6 **5.** 24 R6 **7.** 6 R30 **9.** 70 R3 **11.** 129 R19 **13.** 112 R64 **15.** 73 R41 **17.** 2 R150 **19.** 1 R322 **21.** 20 R26 **23.** 15 R225 **25.** 68 R91 **27.** 115 R13 **29.** 362 R629 **31.** 3246 R222

33. $1\frac{27}{49}$ **35.** $1\frac{3}{91}$ **37.** $8\frac{52}{82}$ **39.** $38\frac{4}{22}$ **41.** $100\frac{21}{47}$ **43.** $142\frac{50}{61}$ **45.** $161\frac{8}{26}$ **47.** $117\frac{41}{52}$ **49.** $2\frac{118}{426}$
51. $2\frac{108}{290}$ **53.** $19\frac{163}{416}$ **55.** $16\frac{178}{319}$ **57.** $16\frac{255}{532}$ **59.** $113\frac{420}{620}$ **61.** $172\frac{51}{381}$

Page 414 **1.** 3.4 **3.** 8.8 **5.** 260.7 **7.** 26.9 **9.** 11.8 **11.** 11.5 **13.** 112.8 **15.** 162.0 **17.** 2.86 **19.** 14.80 **21.** 21.02 **23.** 12.83 **25.** 3.66 **27.** 2.32 **29.** 27.45 **31.** 7.93 **33.** 1.8 **35.** 1.3 **37.** 1.0 **39.** 1.0 **41.** 11.9 **43.** 1.8 **45.** 2.82 **47.** 10.92 **49.** 13.54 **51.** 3.15 **53.** 9.93 **55.** 24.66 **57.** 21.48 **59.** 801.23

Page 415 **1.** 5, no, no, 2; 1, 2, 5, 10 **3.** 1, 5 **5.** 1, 2, 4, 8 **7.** 1, 2, 5, 10 **9.** 1, 2, 7, 14 **11.** 1, 2, 3, 6, 9, 18 **13.** 1, 2, 4, 7, 14, 28 **15.** 1, 2, 3, 4, 6, 9, 12, 18, 36 **17.** 1, 3, 5, 9, 15, 45 **19.** 1, 2, 29, 58 **21.** 1, 2, 4, 8, 16, 32, 64 **23.** 1, 73 **25.** 1, 2, 4, 19, 38, 76 **27.** 1 **29.** 3 **31.** 2 **33.** 5 **35.** 4 **37.** 3 **39.** 8 **41.** 3 **43.** 8 **45.** 8 **47.** 14 **49.** 36 **51.** 2 **53.** 5

Page 416 **1.** $\frac{1}{4}$ **3.** $\frac{1}{2}$ **5.** $\frac{1}{3}$ **7.** $\frac{1}{3}$ **9.** $\frac{1}{2}$ **11.** $\frac{1}{2}$ **13.** $\frac{2}{7}$ **15.** $\frac{3}{4}$ **17.** $\frac{1}{4}$ **19.** $\frac{3}{4}$ **21.** $8\frac{3}{4}$ **23.** $8\frac{2}{3}$ **25.** $2\frac{1}{9}$
27. $12\frac{1}{3}$ **29.** $20\frac{5}{6}$ **31.** $\frac{4}{6}$ **33.** $\frac{4}{12}$ **35.** 5 **37.** 3 **39.** 24 **41.** 9 **43.** 45 **45.** 28 **47.** 30 **49.** 48 **51.** 4 **53.** 21 **55.** 24

Page 417 **1.** 16, 20, 24 **3.** 40, 50, 60 **5.** 2, 4, 6, 8 **7.** 8, 16, 24, 32 **9.** 11, 22, 33, 44 **11.** 14, 28, 42, 56 **13.** 18, 36, 54, 72 **15.** 21, 42, 63, 84 **17.** 26, 52, 78, 104 **19.** 40, 80, 120, 160 **21.** 45, 90, 135, 180 **23.** 56, 112, 168, 224 **25.** 6 **27.** 12 **29.** 30 **31.** 60 **33.** 30 **35.** 90 **37.** 120 **39.** 16 **41.** 60 **43.** 48 **45.** 6 **47.** 18 **49.** 36 **51.** 30 **53.** 24

Page 418 **1.** 6 **3.** 42 **5.** 6 **7.** 8 **9.** 12 **11.** 12 **13.** 30 **15.** 24 **17.** 30 **19.** 36 **21.** 15 **23.** 24 **25.** 30 **27.** 5, 4 **29.** $\frac{3}{8}, \frac{4}{8}$ **31.** $\frac{4}{6}, \frac{5}{6}$ **33.** $\frac{3}{6}, \frac{5}{6}$ **35.** $\frac{14}{21}, \frac{6}{21}$ **37.** $\frac{16}{36}, \frac{27}{36}$ **39.** $\frac{5}{18}, \frac{4}{18}$ **41.** $\frac{7}{32}, \frac{20}{32}$ **43.** $\frac{5}{15}, \frac{9}{15}$
45. $\frac{30}{36}, \frac{27}{36}, \frac{10}{36}$ **47.** $\frac{3}{20}, \frac{8}{20}, \frac{10}{20}$

Page 419 **1.** 3 **3.** $2\frac{1}{4}$ **5.** $1\frac{1}{3}$ **7.** $3\frac{1}{3}$ **9.** $3\frac{1}{5}$ **11.** $2\frac{2}{9}$ **13.** 4 **15.** $1\frac{7}{9}$ **17.** $2\frac{2}{5}$ **19.** $1\frac{5}{9}$ **21.** $2\frac{1}{2}$ **23.** $10\frac{1}{2}$
25. $1\frac{3}{5}$ **27.** $3\frac{1}{3}$ **29.** 2 **31.** 1 **33.** $1\frac{23}{40}$ **35.** $20\frac{1}{10}$ **37.** $\frac{10}{3}$ **39.** $\frac{41}{8}$ **41.** $\frac{57}{7}$ **43.** $\frac{20}{3}$ **45.** $\frac{43}{6}$ **47.** $\frac{73}{8}$
49. $\frac{64}{5}$ **51.** $\frac{63}{4}$ **53.** $\frac{199}{8}$ **55.** $\frac{51}{10}$ **57.** $\frac{49}{16}$ **59.** $\frac{81}{20}$ **61.** $\frac{125}{12}$ **63.** $\frac{221}{18}$ **39.** $\frac{205}{8}$

Page 420 **1.** $\frac{4}{5}$ **3.** $\frac{6}{7}$ **5.** $1\frac{1}{5}$ **7.** $\frac{5}{7}$ **9.** $1\frac{3}{4}$ **11.** $1\frac{2}{3}$ **13.** $\frac{2}{3}$ **15.** $\frac{3}{4}$ **17.** $1\frac{4}{15}$ **19.** 9 **21.** $10\frac{1}{2}$ **23.** $6\frac{2}{9}$
25. 17 **27.** $49\frac{1}{4}$ **29.** $9\frac{1}{4}$ **31.** $16\frac{1}{4}$ **33.** $8\frac{1}{6}$

Page 421 **1.** $\frac{5}{6}$ **3.** $1\frac{4}{15}$ **5.** $1\frac{1}{2}$ **7.** $\frac{7}{20}$ **9.** $1\frac{1}{60}$ **11.** $1\frac{4}{39}$ **13.** $1\frac{11}{18}$ **15.** $5\frac{3}{4}$ **17.** $7\frac{7}{8}$ **19.** $5\frac{19}{20}$ **21.** $39\frac{11}{24}$
23. $89\frac{49}{72}$ **25.** $76\frac{19}{22}$ **27.** $31\frac{1}{15}$

Page 422 **1.** $\frac{1}{3}$ **3.** $\frac{2}{3}$ **5.** $\frac{2}{5}$ **7.** $\frac{1}{4}$ **9.** $\frac{3}{5}$ **11.** $\frac{3}{10}$ **13.** $\frac{1}{4}$ **15.** $\frac{1}{4}$ **17.** $2\frac{2}{3}$ **19.** $4\frac{1}{3}$ **21.** $3\frac{2}{3}$ **23.** $14\frac{1}{3}$
25. $3\frac{3}{7}$ **27.** $2\frac{1}{4}$ **29.** $1\frac{1}{5}$

Page 423 **1.** $5\frac{1}{4}$ **3.** $7\frac{5}{6}$ **5.** $10\frac{2}{3}$ **7.** $2\frac{1}{4}$ **9.** $2\frac{5}{8}$ **11.** $2\frac{2}{9}$ **13.** $10\frac{4}{9}$ **15.** $3\frac{5}{12}$ **17.** $1\frac{3}{5}$ **19.** $1\frac{2}{3}$ **21.** $2\frac{4}{5}$
23. $11\frac{5}{7}$ **25.** $7\frac{3}{4}$ **27.** $1\frac{2}{3}$ **29.** $5\frac{1}{2}$

Page 424 **1.** $\frac{1}{4}$ **3.** $\frac{1}{9}$ **5.** $\frac{13}{18}$ **7.** $\frac{1}{3}$ **9.** $\frac{11}{18}$ **11.** $\frac{11}{15}$ **13.** $\frac{13}{33}$ **15.** $\frac{9}{40}$ **17.** $\frac{7}{8}$ **19.** $2\frac{5}{9}$ **21.** $3\frac{3}{4}$ **23.** $10\frac{17}{24}$ **25.** $8\frac{29}{36}$ **27.** $1\frac{1}{14}$ **29.** $3\frac{7}{8}$

Page 425 **1.** $\frac{1}{9}$ **3.** $\frac{1}{30}$ **5.** $\frac{4}{63}$ **7.** $\frac{7}{10}$ **9.** $\frac{35}{48}$ **11.** $\frac{5}{12}$ **13.** $\frac{2}{3}$ **15.** $\frac{3}{14}$ **17.** $\frac{28}{45}$ **19.** $\frac{1}{28}$ **21.** $\frac{5}{24}$ **23.** $\frac{11}{20}$ **25.** $\frac{15}{22}$ **27.** $1\frac{2}{5}$ **29.** $3\frac{1}{3}$ **31.** $4\frac{1}{2}$ **33.** 10 **35.** $9\frac{3}{8}$ **37.** $6\frac{2}{5}$ **39.** $1\frac{3}{4}$ **41.** $1\frac{3}{4}$ **43.** $2\frac{1}{2}$ **45.** $1\frac{5}{6}$ **47.** $2\frac{7}{16}$

Page 426 **1.** $1\frac{5}{8}$ **3.** $\frac{37}{56}$ **5.** $\frac{7}{24}$ **7.** $1\frac{1}{6}$ **9.** $1\frac{1}{3}$ **11.** $2\frac{11}{20}$ **13.** $1\frac{27}{35}$ **15.** 1 **17.** $\frac{1}{2}$ **19.** $\frac{17}{20}$ **21.** $\frac{17}{66}$ **23.** $3\frac{1}{2}$ **25.** $7\frac{25}{42}$ **27.** $2\frac{4}{7}$ **29.** 8 **31.** 4 **33.** $2\frac{17}{24}$ **35.** $2\frac{19}{25}$ **37.** $9\frac{3}{5}$ **39.** 13 **41.** $5\frac{1}{2}$ **43.** $4\frac{1}{8}$

Page 427 **1.** $\frac{1}{3}$ **3.** $\frac{1}{15}$ **5.** 2 **7.** $\frac{5}{4}$ **9.** $\frac{3}{7}$ **11.** $\frac{5}{6}$ **13.** $\frac{3}{4}$ **15.** $\frac{6}{23}$ **17.** $\frac{16}{43}$ **19.** $\frac{1}{2}$ **21.** $\frac{4}{5}$ **23.** $2\frac{2}{5}$ **25.** 3 **27.** $\frac{25}{48}$ **29.** $\frac{9}{70}$ **31.** $\frac{5}{11}$ **33.** 50 **35.** 36 **37.** $4\frac{1}{2}$ **39.** $10\frac{1}{2}$ **41.** $1\frac{5}{7}$ **43.** 3 **45.** $\frac{1}{6}$ **47.** $\frac{1}{3}$

Page 428 **1.** $12\frac{4}{7}$ **3.** $21\frac{3}{5}$ **5.** $2\frac{14}{27}$ **7.** $\frac{4}{57}$ **9.** $\frac{10}{91}$ **11.** $2\frac{2}{3}$ **13.** $\frac{41}{90}$ **15.** $1\frac{2}{5}$ **17.** $1\frac{5}{22}$ **19.** $\frac{7}{16}$ **21.** 1 **23.** $\frac{5}{12}$ **25.** $\frac{9}{14}$ **27.** $1\frac{17}{55}$ **29.** $\frac{85}{98}$

Page 429 **1.** $12{:}5, \frac{12}{5}$ **3.** $30{:}33, \frac{30}{33}$ **5.** $\frac{1}{2}$ **7.** $\frac{1}{4}$ **9.** $\frac{5}{6}$ **11.** no **13.** no **15.** no **17.** yes **19.** no **21.** yes **23.** yes **25.** $a = \frac{2}{5}$ **27.** $y = 4$ **29.** $c = 36$ **31.** $s = 12$ **33.** $x = 1$ **35.** $s = 50$ **37.** $n = 45$ **39.** $x = 3$

Page 430 **1.** $\frac{72}{16}$ **3.** $\frac{3}{2}$ **5.** $\frac{3}{5} = \frac{x}{10}$ **7.** $\frac{10}{4} = \frac{40}{m}$ **9.** $\frac{3}{8} = \frac{b}{32} > b = 12$ **11.** $\frac{120}{4} = \frac{m}{10} > m = 300$ **13.** $\frac{15}{1} = \frac{d}{9} > d = 135$ **15.** 6 repairs/day **17.** $1\frac{1}{4}$ hits/game **19.** B

Page 431 **1.** 0.33 **3.** 0.17 **5.** 0.2 **7.** 0.11 **9.** 0.1 **11.** 0.32 **13.** 0.7 **15.** 0.28 **17.** $\frac{1}{5}$ **19.** $\frac{1}{2}$ **21.** $\frac{7}{100}$ **23.** $\frac{3}{50}$ **25.** $\frac{1}{4}$ **27.** $\frac{9}{20}$ **29.** $\frac{3}{8}$ **31.** $\frac{31}{40}$ **33.** 1.5 **35.** 5.17 **37.** 4.2 **39.** 8.11 **41.** 12.25 **43.** 17.11 **45.** 2.08 **47.** 6.05 **49.** $9\frac{1}{10}$ **51.** $6\frac{3}{10}$ **53.** $6\frac{8}{25}$ **55.** $8\frac{14}{25}$ **57.** $30\frac{7}{20}$ **59.** $9\frac{3}{10}$

Page 432 **1.** $\frac{75}{100}$ **3.** $\frac{60}{100}$ **5.** $\frac{20}{100}$ **7.** $\frac{25}{100}$ **9.** $\frac{30}{100}$ **11.** $\frac{35}{100}$ **13.** $\frac{56}{100}$ **15.** $\frac{26}{100}$ **17.** $\frac{410}{100}$ **19.** $\frac{287}{100}$ **21.** $\frac{252}{100}$ **23.** 6% **25.** 5% **27.** 9% **29.** 8% **31.** 15% **33.** 18% **35.** 75% **37.** 40% **39.** 96% **41.** 56% **43.** 100% **45.** 4% **47.** 5% **49.** 57% **51.** 48% **53.** 68% **55.** 10% **57.** 503% **59.** 425% **61.** 542% **63.** 804% **65.** 386%

Page 433 **1.** 25% **3.** 50% **5.** 80% **7.** 35% **9.** 32% **11.** 425% **13.** 820% **15.** 912.5% **17.** 0.05 **19.** 0.10 **21.** 0.17 **23.** 0.45 **25.** 1.10 **27.** 1.30 **29.** 1.80 **31.** 0.034 **33.** 0.018 **35.** $\frac{2}{25}$ **37.** $\frac{1}{20}$ **39.** $\frac{7}{100}$ **41.** $\frac{19}{500}$ **43.** $\frac{69}{1000}$ **45.** $\frac{3}{8}$ **47.** $\frac{3}{25}$ **49.** $\frac{9}{50}$ **51.** $\frac{1}{4}$ **53.** $1\frac{1}{10}$ **55.** $1\frac{1}{20}$ **57.** $1\frac{3}{10}$ **59.** $1\frac{37}{100}$ **61.** $1\frac{3}{4}$

Page 434 **1.** $10\% \times 40 = 4$ **3.** $20\% \times 60 = 12$ **5.** $4\% \times 40 = 1.6$ **7.** $5\% \times 400 = 20$ **9.** $70\% \times 80 = 56$ **11.** $85\% \times 40 = 34$ **13.** $75\% \times 400 = 300$ **15.** $N = 5\% \times 600$ **17.** $N = 20\% \times 100$ **19.** $N = 65\% \times 600$ **21.** $N = 55\% \times 180$ **23.** 15% **25.** 50% **27.** $N\%$ **29.** $N\%$ **31.** 300 **33.** 50 **35.** N **37.** N **39.** 25%; 200; 50

18

Page 435 **1.** 2 **3.** 13 **5.** 3 **7.** 6 **9.** $36\frac{2}{5}$ **11.** 8 **13.** 27 **15.** 400 **17.** 800 **19.** 400 **21.** 700 **23.** $33\frac{1}{3}$
25. 96 **27.** 77 **29.** 32 **31.** 31.8

Page 436 **1.** 40% **3.** 75% **5.** 80% **7.** 100% **9.** 50% **11.** 75% **13.** 43.5% **15.** 65.2% **17.** $\frac{N}{100} = \frac{20}{40}$
19. $\frac{N}{100} = \frac{60}{80}$ **21.** 50% **23.** 50% **25.** $\frac{80}{100} = \frac{N}{20}$ **27.** $\frac{60}{100} = \frac{N}{35}$ **29.** 40 **31.** 80

Page 437 **1.** $^+30$ **3.** $^-15$ **5.** $^+10$ **7.** $^-20$ **9.** $^+10$ **11.** $^+800$ **13.** $^+80$ **15.** $^-15$ **17.** $^+20$ **19.** $^-80$ **21.** $^+17$
23. $^+15$ **25.** $^+2$ **27.** $>$ **29.** $<$ **31.** $<$ **33.** $>$ **35.** $>$ **37.** $^-1$ **39.** 4 **41.** 56 **43.** $^-25$ **45.** $^-96$ **47.** $^-16$

Page 438 **1.** 10 **3.** 11 **5.** $^-7$ **7.** 9 **9.** 15 **11.** $^-11$ **13.** 16 **15.** 17 **17.** $^-15$ **19.** 11 **21.** $^-20$ **23.** 19
25. 38 **27.** 42 **29.** $^-44$ **31.** $^-50$ **33.** $^-4$ **35.** 6 **37.** $^-2$ **39.** 8 **41.** 8 **43.** 0 **45.** 0 **47.** 10 **49.** 6 **51.** $^-6$
53. $^-5$ **55.** $^-50$ **57.** 1 **59.** 0 **61.** 1 **63.** 3

Page 439 **1.** 1 **3.** 5 **5.** 8 **7.** $^-13$ **9.** $^-20$ **11.** $^-9$ **13.** $^-9$ **15.** $^-7$ **17.** $^-3$ **19.** $^-15$ **21.** $^-15$ **23.** $^-5$
25. $^-47$ **27.** 5 **29.** $^-84$ **31.** 12 **33.** 8 **35.** 5 **37.** $^-3$ **39.** $^-3$ **41.** $^-2$ **43.** $^-1$ **45.** 4 **47.** 7 **49.** 1 **51.** 2
53. 8 **55.** 40 **57.** 16 **59.** 174

Page 440 **1.** 12 **3.** 7 **5.** 24 **7.** 12 **9.** 24 **11.** 21 **13.** 9 **15.** 3 **17.** 14 **19.** 2 **21.** 5 **23.** 4 **25.** 8
27. 8 **29.** 31 **31.** $^-10$ **33.** $^-28$ **35.** $^-30$ **37.** $^-200$ **39.** $^-99$ **41.** $^-63$ **43.** $^-140$ **45.** $^-160$ **47.** $^-168$
49. $^-7$ **51.** $^-1$ **53.** $^-21$ **55.** $^-2$ **57.** $^-25$ **59.** $^-33$

Page 441 **1.** 19 **3.** $^-12$ **5.** $^-3$ **7.** 4 **9.** 9 **11.** 16 **13.** $^-1$ **15.** 13 **17.** 6 **19.** 10 **21.** 3 **23.** 2 **25.** 9
27. 42 **29.** 20 **31.** 12 **33.** 100 **35.** $^-25$ **37.** 28 **39.** 18

Page 442 **1.** Add n to 7. **3.** Multiply 5 by m. **5.** Divide 12 by w. **7.** Subtract 1 from b. Multiply by 3.
9. Multiply 16 by y. **11.** Subtract r from n. Add 7. **13.** Multiply 17 by d. Subtract c.
15. Add 6 to w. Multiply by 8. **17.** Multiply b by c. Add a. **19.** Multiply a by 6. **21.** Multiply 4 and m.
23. Multiply 3 and d. Subtract from s. **25.** Add m to 1. Multiply by 2. **27.** Multiply 10 and m. Multiply by c.
29. Multiply 7 and n. Subtract 8. Multiply by a. **31.** Divide c by d. Multiply by 4.
23. Add f and i. Divide by 2. Multiply by a. **35.** 27 **37.** $^-6$ **39.** 10 **41.** 3 **43.** 70 **45.** 1 **47.** 7 **49.** $^-2$
51. 1

Page 443 **1.** x **3.** 0 **5.** $3n$ **7.** 1 **9.** 8 **11.** 6 **13.** b **15.** 0 **17.** 0 **19.** 1 **21.** 0 **23.** $15n$ **25.** $10y$
27. $16x$ **29.** $13m$ **31.** $7x + y$ **33.** $3x$ **35.** $6 - z$ **37.** 0 **39.** $4 + 3z$ **41.** $8n$ **43.** $13m - 3$ **45.** $19x - 3$
47. 6^3 **49.** 5^2a **51.** x^2y **53.** $7x^2y$ **55.** $9x^2y$ **57.** $9x^2y^2$ **59.** $5a^2b^3$ **61.** $6(a + x)^2$

Page 444 **1.** 9 **3.** 4 **5.** 24 **7.** 25 **9.** 36 **11.** $^-54$ **13.** 16 **15.** 1 **17.** 0 **19.** 125 **21.** 9 **23.** 14 **25.** 2
27. 5 **29.** 3 **31.** $^-4$ **33.** 0 **35.** $^-1$ **37.** 1, 2 **39.** 2 **41.** $^-3$ **43.** 1, 0, $^-1$, $^-2$, $^-3$ **45.** $^-2$, $^-1$, 0, 1, 2
47. 0, $^-1$, $^-2$, $^-3$ **49.** 2, 1, 0, $^-1$, $^-2$, $^-3$ **51.** 1, 2 **53.** $^-1$, $^-2$, $^-3$

Page 445 **1.** 19 **3.** 38 **5.** 18 **7.** 47 **9.** 39 **11.** 18 **13.** 24 **15.** 24 **17.** 45 **19.** 11 **21.** 42 **23.** 17
25. 80 **27.** 144 **29.** 6 **31.** 2 **33.** 21 **35.** 3 **37.** 36 **39.** 5 **41.** $^-5$ **43.** 35 **45.** $^-18$ **47.** $^-36$ **49.** 90
51. 84

Page 446 **1.** 8 **3.** 12 **5.** 4 **7.** 3 **9.** 4 **11.** $^-7$ **13.** $^-9$ **15.** $^-5$ **17.** $^-3$ **19.** $^-1$ **21.** 0 **23.** $^-16$ **25.** $\frac{1}{4}$
27. $\frac{1}{18}$ **29.** $\frac{1}{15}$ **31.** $\frac{1}{5}$ **33.** 27 **35.** $^-4$ **37.** 42 **39.** 36 **41.** $^-28$ **43.** 300 **45.** 0 **47.** $^-63$ **49.** 100 **51.** 60
53. $^-54$ **55.** $^-70$ **57.** 120 **59.** 100 **61.** 124 **63.** 0

Page 447 **1.** 14 **3.** 10 **5.** 3 **7.** 22 **9.** 144 **11.** $^-6$ **13.** $^-6$ **15.** $^-2$ **17.** $^-49$ **19.** $^-20$ **21.** $^-539$ **23.** 2
25. 6 **27.** 3 **29.** 2 **31.** 3 **33.** 5 **35.** 5 **37.** 2 **39.** 3 **41.** $^-3$ **43.** 9 **45.** $^-3$

Page 448 **1.** yes **3.** no **5.** yes **7.** no **9.** yes **11.** yes **13.** no **15.** no **17.** no **19.** yes **21.** yes **23.** 10
25. $^-3$ **27.** $^-6$ **29.** 16 **31.** 11 **33.** $^-10$ **35.** 1 **37.** 18 **39.** $^-1$ **41.** 24 **43.** $^-7$

APPLICATION BANK

Page 449 **1.** 52.036 **3.** 9.875 **5.** $250 **7.** $1000 **9.** $8050 **11.** 50,680 **13.** 15,000 **15.** 64,000 **17.** 100 **19.** 4700 **21.** 157 **23.** $1380 **25.** 22.2 **27.** 17.7

Page 450 **1.** 36; 41 **3.** 324; 330 **5.** 4 ft **7.** 192; 197 **9.** 130 oz **11.** 12; 13 **13.** 250 mi **15.** $2.37 **17.** $26.88

Page 451 **1.** 5800 **3.** 1.6 **5.** 10,400 **7.** 2000 m, 6.7 km, 31,000 m, 72 km **9.** 0.12 m, 120 mm **11.** 8.5 m, 850 cm **13.** 14 cm **15.** 94,000 **17.** 10.32 **19.** 0.6 kg **21.** 6.5 **23.** 46,000 **25.** 62.1 mL, 0.075 L, 0.9 L, 962 mL

Page 452 **1.** $\overleftrightarrow{AC}$, $\overleftrightarrow{FD}$ **3.** $\overleftrightarrow{FD}$, $\overleftrightarrow{BE}$ **5.** E **13.** obtuse **15.** acute

Page 453 **1.** 100° **3.** 25° **5.** right **7.** yes **9.** 60° **11.** square **13.** parallelogram **15.** 8.4 cm

Page 454 **1.** 49 **3.** 625 **5.** 0.0016 **7.** 2500 **9.** 10,000 **11.** 1 **13.** 13 **15.** 25 **17.** 21 **19.** no **21.** yes **23.** no **25.** no

Page 455 **1.** 12 cm **3.** 39 cm **5.** 34 m **7.** 48 cm **9.** 28.8 ft **11.** 131.4 m **13.** 10 ft **15.** 4.2 km; 26.376 km **17.** 226.08 cm

Page 456 **1.** 180 cm² **3.** 250.24 m² **5.** $76\frac{1}{8}$ in.² **7.** 108 ft² **9.** 5.76 m² **11.** 26 ft² **13.** 72 ft² **15.** 254.3 cm² **17.** 38.5 m² **19.** 706.5 m² **21.** 31,400 cm²

Page 457 **1.** 51 in.² **3.** 64.26 ft² **5.** 560 cm³ **7.** 510 ft³ **9.** 720 in.³ **11.** 401.92 in.³ **13.** 1073.88 cm³ **15.** 549.5 cm³

Page 458 **1. a.** $\frac{1}{2}$ **b.** $\frac{3}{4}$ **c.** $\frac{1}{12}$ **d.** $\frac{5}{12}$ **3.** 200,000 **5.** 250,000

Page 459 **1.** 60.5 **3.** 7.5 **5.** $15,125 **7.** 1.5 **9.** 150, no mode, 117 **11.** 3.6, 3.6, 1.8 **13.** 16 m

Page 460 **1.** $30,000,000 **3.** train **5.** 1975–1976 **7.** 1920–1930 **9.** over 50 **11.** sewing **13.** computers

Page 461 **1.** $72 **3.** $30 **5.** $51.25 **7.** $72 **9.** 55% **11.** 5% **13.** $20.40; 150% **15.** $385

Page 462 **1.** $^-4 + 6 + \,^-3$; $^-1°$C **3.** $75 + \,^-20 + \,^-12$; $43 **5.** $^-145 + 50$; 95 m **7.** 85° **9.** 80° **11.** $^-15°$F **13.** 14°

Page 463 **1.** $10 + n$ **3.** $\frac{p}{q}$ **5.** $x - 19$ **7.** $17 + r$ **9.** $P = x + 2y + z$ **11.** $P = s + 2t$ **13.** $V = x^3$ **15.** b **17.** c

Page 464 **1.** 2 **3.** 5 **5.** $^-10$ **7.** $^-6$ **9.** 11 **11.** $^-5$ **13.** 20 **15.** $^-15$

Page 465 **1.** (3, 0) **3.** $(^-5, 2)$ **5.** (5, 2) **7.** $(2, \,^-2)$ **9.** (2, 6) **11.** I **13.** E **15.** B **17.** P **19.** $(^-5, \,^-10)$, $(0, \,^-5)$, (5, 0) **21.** 6, $(^-2, 6)$; 5, (0, 5); 4, (2, 4) **23.** $^-4$, $(^-5, \,^-4)$; $^-1$, $(^-2, \,^-1)$; 4, (3, 4)